Science of
Tea Technology

About the Editors

Dr. P.S. Ahuja, FNA
Director
CSIR-Institute of Himalayan Bioresource Technology
(Council of Scientific & Industrial Research)
Post Box No. 6 , Palampur (H.P.)- 176061, India.
E-mail: director@ihbt.res.in; psahuja@ihbt.res.in

Dr. Arvind Gulati
Chief Scientist
CSIR-Institute of Himalayan Bioresource Technology
(Council of Scientific & Industrial Research)
Post Box No. 6, Palampur (H.P.)- 176061, India.
E-mail: arvindgulati@ihbt.res.in

Dr. R.D. Singh
Chief Scientist
CSIR-Institute of Himalayan Bioresource Technology
(Council of Scientific & Industrial Research)
Post Box No.0000 6, Palampur (H.P.)- 176061, India.
E-mail: rdsingh@ihbt.res.in

Dr. R.K. Sud
Principal Scientist
CSIR-Institute of Himalayan Bioresource Technology
(Council of Scientific & Industrial Research)
Post Box No. 6, Palampur (H.P.)- 176061, India.
E-mail: rksud@ihbt.res.in

Dr. R.C. Boruah, FNA Sc
Outstanding Scientist
CSIR-North East Institute of Science & Technology
(Council of Scientific & Industrial Research)
Jorhat- 785006, Assam.
E-mail: boruahrc@rrljorhat.res.in

Science of Tea Technology

Editors

Dr. P.S. Ahuja, Dr. Arvind Gulati,
Dr. R.D. Singh, Dr. R.K. Sud
CSIR-Institute of Himalayan Bioresource Technology (IHBT),
Palampur (H.P.), India

Dr. R.C. Boruah
CSIR-NEIST, Jorhat (Assam)

SCIENTIFIC PUBLISHERS

popular. Advances in the field of biotechnology and molecular biology has revolutionised the tea improvement programmes. Under a joint project, different Indian premier institutes have finger-printed the entire tea germplasm and shortlisted the core collections for adding efficiency to breeding programmes in the future. Advances in elucidating tissue culture technology, catechins biosynthetic pathway, development of molecular markers to identify adulterants in tea and major plant responses during winter dormancy and drought response, efficient protocols for somatic embryogenesis for the production of synthetic seeds and protocols for development of transgenic tea plants will have their impact in near future. CSIR-IHBT has been engaged in applied, fundamental and high science cutting edge technology ever since its establishment in the year 1983-84 and has contributed immensely for the global tea industry. This book provides information on some of the latest aspects of tea husbandry, physiology, biochemistry, manufacture and biotechnology in its chapters. Chapter-wise highlights are highlighted hereunder:

1. History of Tea in India (Brajinder Singh and R.K. Sud)

Tea was discovered by the Chinese Emperor Shen Nung in 2737 BC and thereafter became a popular beverage and health drink in China. It was introduced to Japan in the early part of eighth century but the rest of the world was ignorant about tea until about middle of seventeenth century. Interesting stories of origin of the tea plant and how the British introduced the Chinese tea in India and subsequently discovered an India tea variety in Assam jungle have been described. Tea was introduced to India in 1774 by a Botanist, Mr. Robert Fortune. Later on, in 1823 a native species of tea plant growing wild in Assam was discovered by Major Robert Bruce. Commercial cultivation began on 12 February, 1839 when the world's first privately owned tea company – the Assam Tea Company – was established at Jorhat, Assam with the directives from British Parliament. Today majority of tea plantations around the world are based upon Assam type tea. A detailed account of setting up and development of tea plantations in different parts of India with specific account of Kangra valley in Himachal Pradesh, which is lesser known distinctive flavoured tea, has been portrayed.

2. Genetic Improvement of Tea (S. Singh *et al.*)

Like any other agricultural crop, availability of high yielding quality clones with greater tolerance to pest, diseases and environmental stresses is a basic need for sustaining production levels. With the introduction of improved clonal cultivars, the tea industry has shown significant and sustained progress over the years. However, at present tea production has stagnated and is beset with new problems which are widespread or region specific. Improvements in terms of quality and productivity can only be achieved through development of new varieties. Therefore, tea breeding programs in the world need to be given impetus through germplasm appraisal, use of molecular markers associated with the traits of interest for early selection in segregating progenies, redefining parameters constituting plant ideotype and utilizing tools of phenomics for evaluating plant responses to environmental variations and stresses. Descriptive details of all these parameters are highlighted.

3. Tea Applied Genomics: Utilizing Genetic Variability and Current Genomic Tools to Enhance Productivity (R.K. Sharma *et al.*)

Genetic improvement of tea involves identification, characterization, evaluation, domestication, maintenance and utilization of germplasm for development of superior plant material. Conventional breeding programme in tea, however, limited by long-gestation period, out-breed nature and self-incompatibility. This chapter links status of emerging molecular genomic information that can expedite the genetic improvement in tea and hence increase the productivity. This will provide a background of possibilities for modern tea breeding together with some current efforts for development of sequence based markers and links genetic diversity of existing gene pools, identification of diverse parental groups, efficient phenotyping to support operational breeding. Preliminary attempts on quantitative trait locus (QTL) mapping in tea were also reviewed and prospective is also provided to exploit the power of association genetics to dissect quantitative traits. Challenges and opportunities owing to advent of next generation sequencing (NGS) technologies to generate genome wide makers and to integrate genomic information into directional selective breeding are also discussed.

4. Tea Propagation (R.K. Sud)

One of the reasons for the stagnation of the tea production in India is that majority of the plantations have crossed their peak production phase and therefore call for replanting and new planting in a phased manner. Overall requirement of nursery tea plants for new planting, replanting and infilling and interplanting is enormous as presently, nearly 12,000 plants are planted in one hectare area. Both clonal and seed-raised plants are needed to cope up the climatic, environmental and biotic stresses. Nursery techniques involve several steps starting from appropriate soil to erecting nursery structure, suitable seed/cutting to hardening of the raised plants. Adoption of recent advances in each step is necessary to grow plants in the shortest possible time with a minimum of losses. Growing composite plants with scions-cuttings of high yielding clones grafted on hardy rootstock-cuttings is gaining momentum. The Chapter provides latest information on all aspects of plant propagation.

5. Tea Seed and its Biology (Amita Bhattacharya *et al.*)

In tea, biclonal and polyclonal seeds are produced through hybridization between two or more superior clones. These play an important role in the selection of new varieties. The complete reproductive cycle from flower initiation to anthesis and seed maturity as well as the physiology of seed development involving endogenous changes in hormone and subsequent germination pattern have been discussed. Tea seeds are recalcitrant and cannot withstand desiccation and dehydration. Thus, the storage of seeds is unconventional and requires the maintenance of appropriate moisture levels. Seeds are also the source of saponins, edible oil, cakes, cosmetics, insecticides, molluscicides etc.

6. Young Tea Management (R.D. Singh)

This chapter discusses raising of young tea plantation for early canopy cover to facilitate manual or mechanical harvesting of tea shoots. It deals with time and method of pruning, bending and pegging, tipping, and use of plant growth regulators as pruning agents in young tea. Method and importance of mulching as well as intercropping in young tea has been also discussed.

7. Tea Pruning (R.K. Sud)

Pruning is referred to as a "necessary evil" in tea parlance as it causes severe crop loss on the year and involves extra expenditure. Yet it is a fundamental need to maintain vegetative regeneration and sustained productivity. Appropriate pruning cycle i.e. pruning with different skiffing-types in different years, is a key to strike a balance with yield and quality as per the market needs. Understanding of physiological basis of pruning is imperative to rationalise source-sink relationship and formulate the pruning cycle. Time of pruning, linked with maximum root starch reserve period, is the most important factor for minimising mortality and achieving better post-prune recovery.

Interaction of pruning times with insect-pests and diseases is another significant factor. Rejuvenation pruning may be adopted as a short measure to revive the aging plantation but it cannot substitute replanting especially in Assam type tea. Dilapidated china hybrid tea plantations have been successfully revived to the point of maximum productivity.

8. Tea Plucking (R.K. Sud)

Plucking, apparently refers to removal of the apical shoots as a crop. In fact, it is a much wider and deeper subject, having implications of productivity and quality of crop, garden and factory management, marketing and economics and social dimensions. Principals of the plucking are based on physiology of tea bush. Establishment of plucking table, tipping measures, system of plucking etc. are guided by photosynthetic efficiency of the leaves and source-sink relationship. Pruning, tipping and plucking are inter-related operations in tea and severity of one is compensated by leniency of the other to maintain the productivity. A comprehensive information on shoot growth, tipping measures, systems and standards of plucking in relation to physiological, environmental and biochemical parameters have been discussed. Keeping the chinary planting material of Kangra/Darjeeling gardens on the forefront, guidelines have been given in the chapter for management of different flushes and harvesting quality crop.

9. Mechanization in Tea (K.K. Singh)

Due to significant rise in socio–economic status, per capita income and education status, the farm workers are not readily available for tea plantations. Continuous labour wage hike is another deterrent in the economics of tea plantation, compelling to look for alternatives. The mechanization of farm operations in tea plantations across the tea growing countries has evolved as a way to tackle the labour shortage particularly for harvesting of leaves. Different machines of the varying capacities and types have been developed and are in use. New models are being developed to meet the specific needs. In this chapter, tea farm operations that can be mechanized *viz.*, fertilizer application, chemical spraying, mechanical weed control, pruning and skiffing, plucking, bush uprooting and other operations have been discussed to mitigate labour shortage and make tea growing a sustainable venture.

10. Plant Nutrition (R.D. Singh)

Fertilizer is an expensive input in tea production, accounting up to 15-25% of total cost of cultivation. This chapter deals with

15. Pesticide Residues in Tea (Yogesh B. Pakade and Adarsh Shanker)

Tea is an important export commodity and keeping it completely free from pesticide residues is desired. However, it is impractical in view of the serious pest and disease problems where chemical pesticides have to be applied for their management. As tea is subjected to infusion prior to consumption, pesticide residues in tea and its transfer in brew must be monitored prior to marketing. Various international authorities have regulated the use of pesticides by fixing maximum residues levels (MRLs) for commercial purposes. The factors involved in the degradation of pesticides on tea plant are discussed in this chapter. This chapter also explores the fate of pesticide during manufacturing processes as well as during infusion of tea. Alternative extraction methodologies are discussed for pesticide residue analysis in tea which play an important role.

16. Organic Tea Farming (R.D. Singh)

In tea plantations, use of agrochemicals is substantially higher than field crops. Consequently, threat from residual effects of pesticides applied in tea is also high, especially in made tea. This chapter deals with the present status and scope of organic tea farming involving cultural practices like planting disease free seed-stocks, using organic mulch materials, organic manures and bio-fertilizers, recycling farm residues/wastes, and plant protection by bio-pesticides.

17. Plant Physiological Approaches for Enhancing Productivity in Tea (S.K. Vats and P.K. Nagar)

The physiological basis of growth and yield in tea depends on the potential of photosynthetic process, and the rate of transfer of photosynthates to different organs. Both the processes that also decide the operational physiological range are dependent on the genotypic and environmental factors. Often, the interplay of environmental parameters become location specific and assumes greater management role. For instance, shade that also influences canopy temperature and to some extent soil moisture has a greater significance in tropical than temperate areas. Optimal performance in tea, therefore do not depend on light, temperature or soil moisture conditions *per se* but the interaction of it. The role of nutrients, irrigation during dry period, and application of growth regulators have quite well been exploited to improve yield and tackle winter dormancy in tea in areas away from equator. Good Management Practices, thus need to account for these.

18. Tea Biochemistry (Ashu Gulati)

The chapter concentrates on the nature of the non-volatile compounds such as polyphenols and catechins, and their oxidized products *viz.*, theaflavins and thearubigins, theanine, lipids, caffeine and the factors affecting their concentration in the tea shoots. In tea, volatile components though present in minute quantities, yet play pivotal role in the flavoury teas. These volatile compounds are either liberated from their glycosidic precursors by hydrolysis during manufacture, or are formed by thermal degradation or oxidative degradation of fatty acids and carotenoids. The hydrolysis of volatile flavour component precursors is mediated by β-glucosidases.

19. Tea Manufacture (Ashu Gulati)

Teas are classified into three main categories according to the degree of fermentation during manufacturing: non-fermented (green and white teas), partially fermented (oolong tea), and fully fermented (black tea). Activation or non-activation of the enzyme polyphenol oxidase present in tea leaves is the key player for processing of these tea types. Genetic material of the shoot, the environmental conditions under which it grows, and process conditions during manufacturing determine quality of tea. In traditional factories, human judgement plays a vital role in the standardization of process parameters in the made tea. A preconditioning machine developed by CSIR-IHBT based on the concept of physical and chemical wither in black tea manufacture has demonstrated considerable reduction in withering time with improvement in tea quality. Recently, fuzzy logic based system being tried to automate the tea processing. Chemical sensors using metal oxides are being developed for optimizing the fermentation step.

20. Diversified Tea Products (H.P. Singh *et al.*)

The natural nutrceutical compounds in tea leaves rich in antioxidant activities are the host of these attributes. There is a spurt in research activities the world over for the development of various diversified tea products. CSIR-IHBT has also developed some tea based products such as ready to drink black and green tea concentrates, tea wines, tea polyphenols free from caffeine, amino acids and soluble sugars, high grade bio-processed theaflavins which have potential in the world market. Modern trend exhibits further scope for the formulations of more exotic, fruity flavours, rich in nutrients, herbal and ethnic teas for value addition and fortification of human health.

21. Some Recent Innovations in Tea Biotechnology (Anil Sood *et al.*)

There was an urgent need to characterize the existing tea germplasm and molecular markers are developed to categorize the accessions. For effective multiplication of superior clones, protocols for micropropagation are now available. Major strides are made towards improvement of tea quality and yield by dissecting the catehcins biosynthesis pathway, winter dormancy and drought responses. Plant improvement using molecular markers and also by genetic manipulations and transformations are discussed and now standardised for application in tea.

This book is a source of latest information on tea. All the chapters have been well compiled with dedication and sincerity and authors own the authenticity of the information. I hope the book will be useful to variety of users including students, teachers, researchers, planners and policy makers as well as the tea information curious people. The editorial board is indebted to Prof. Sudhir K. Sopory, Vice Chancellor of Jawaharlal Nehru University, New Delhi and Research Council Chairman of the CSIR-IHBT who was the motivating force behind the publication of this book. The commitment and hard work of Sh. Mukhtiar Singh, Library Officer of CSIR-IHBT in compilation of the book deserves special mention for which the editors are deeply indebted. Thanks are also due to Dr. Rakesh Kumar for reviewing of the text of the chapters on tea husbandry. Authors have taken the liberty to access to the research findings and other required information from India and abroad, which have been duly referred.

(P.S. Ahuja)
Chief Editor

CONTENTS

1

HISTORY OF TEA IN INDIA

Brajinder Singh and R.K. Sud*

CSIR-Institute of Himalayan Bioresource Technology,
Palampur (H.P.) - 176 061, India. *E-mail: rksud@ihbt.res.in

INTRODUCTION

According to Lu Yu, author of the world's famous book on tea, "Ch'a Ching" written in 8th century AD, tea was first discovered by the emperor Shen Nung of China in 2737 BC. According to another story in 542 AD, Bodhidharma while traveling in China used to pray and meditate without sleep, but fell asleep while meditating in Shaoline temple in Loyang, now known as Honan province. When he woke up he cut off his eye lids and threw them which immediately took root and the world's first tea plant grew from them. Confucius had mentioned about this in his writing about tea. Regarding use of tea buds, the Han emperor An Ti (206 - 221 AD) defined ming as buds taken from the plant t'u which is now known as tea or ch'a. Tea had become China's universal drink during T'ang dynasty (618 - 907 AD). Lu Yu mentioned about purple tea, black tea and brick tea. During Sung dynasty (960 - 1280 AD) from brick tea further improvement was made and powdered tea became popular and brands like "Grey Eyebrows", "Falcon's Talon", "Ear of Corn", "Sparrow's Tongue" were popular. During Ming Dynasty (1368 – 1644 AD), tea drinking reached new heights of refinement and attained the present day form. Stories of use of monkeys to pluck leaves were known to Ming dynasty. During Lu Yu's time, China became the greatest empire in the world which entered in trade with Persia in West and Japan in East. By the early part of 9th century, tea ceremony or Cha-no-yu was well established in Japan. In 17th century, tea which was imported from China to the West became quite popular and by the downfall of Manchu rule, the British planted tea in India and Ceylon and turned out as producers of tea themselves.

By 2000, tea was grown on about 2.5 million hectares of land in Asia (89 per cent of global tea cultivated areas) and Africa (8 per cent). Most of the growth was due to the increase in productivity rather than an expansion in area. India accounts for about 29 per cent of global production; China, 23 per cent; Sri Lanka, 9 per cent; Kenya, 8 per cent; and Indonesia, 6 per cent. Black tea accounts for more than 70 per cent of world tea production and green tea, 22 per cent. Over the past decade, production of black tea has increased at 1 per cent per annum, while that for green tea at 2.5 per cent. Tea is consumed in both producing and importing countries. India is the largest producer and consumer of tea and accounts for more than 20 per cent of global tea consumption, while the United Kingdom, the Russian Federation, the CIS countries, Pakistan and Japan each account for about 5 per cent of the total tea consumed globally. Over the past decade, world tea consumption has increased by 1 per cent annually. World tea exports have grown by 2.2 per cent annually between 1970 and 2000, from 674 000 tonnes to 1.29 million tonnes. Growth in export values, however, was 4.6 per cent annually during this period, rising from US$ 573 million to US$ 2.3 billion in 2000. Sri Lanka, the largest exporter, accounts for about 20 per cent of world exports, Kenya (18 per cent), China (17 per cent), India (16 per cent) and Indonesia (8 per cent). World net-imports of tea grew at 1.9 per cent between 1970 and 2000, to 1.2 million tonnes. Major importers continued to be the United Kingdom, the Russian Federation, Pakistan, the United States and Egypt.

TEA IN ASSAM

Dr. Robert Fortune introduced tea in India in 1774 when East India Company wanted to establish business in this country. However, Major Robert Bruce discovered wild tea in Assam in 1823 near Rangpur (present Sibsagar) during his visit on a trading mission (Ukers 1935). His brother, Mr. C.A. Bruce, in charge of the British Gunboat division in the war with the Burmese occupying Assam in 1824, took some tea plants and seeds from Singpow Chief and planted them in his garden at Sadiya and some were sent to Commissioner Jenkins at Guwahati. A few leaves of these plants were sent to Dr. N.

Wallich, the then Superintendent of Botanical Gardens, for identification. The leaves were found to be of tea family but of different species. In 1834 the then Governor General of India Lord William Bentinck appointed a Tea Committee to advise on feasibility of commercial tea cultivation in India under which its secretary Mr. G.J. Gordon, visited and procured tea seeds, plants and workers from China. In response to the circular, the Commissioner of Assam, Major F. Jenkins, recommended tea cultivation in Assam where it was growing wild in the forest. He also collected complete specimens of the local plants and forwarded them to the Botanical Gardens in Calcutta. This time Dr. Wallich had no difficulty in identifying the specimens as tea, and the plants were not different from the tea plant of China. Based on this information the Tea Committee recommended the plantation of indigenous plant under proper management for commercial purpose.

A Scientific Commission was constituted in 1835 comprising Dr. N. Wallich, Dr. W. Griffith and Dr. J. McClelland to report on the Indian indigenous tea plants and to select localities for starting experimental tea gardens. In 1836 the Commission guided by Mr. C.A. Bruce, visited tracts at the foothills of the Naga and Patkai as well as to a few in the river valleys where the indigenous tea plant was growing in clumps. Having seen the tea bushes Dr. Wallich expressed the view that there was no need any more to import tea seeds from China, while Mr. Griffith favoured import of the China seeds because a wild plant is not likely to give as good a produce as one that has been cultivated for centuries. It was finally decided that the China plant and not the degraded Assam plant should be used for the Government experiments. The Commission failed to come to a general agreement regarding the most favourable localities for establishing experimental gardens. Dr. Wallich favoured the Himalayan region while the other two favoured Upper Assam where wild tea existed. So Mr. Gordon was sent again to China in 1836 and for many years China tea seed was imported regularly into India. From these seeds, nurseries were raised in the government Botanical Gardens in Calcutta and the plants were sent to Upper Assam, Dehradun, Kumaon and the Nilgiri hills.

Nurseries at Saikhowa near Saidya in upper Assam failed; hence, surviving plants were shifted to a new site near Chabua about 25 kilometers east of Dibrugarh. At the same time tea seedlings were planted near Bhimtal and Almora. Later on, experimental gardens were successfully established with China plants in Kumaon, Garhwal and Kangra districts on the Himalayan foothills. Of the plants sent to south India, a few survived in Nilgiris and a small lot in Wynaad on the western coast. Apart from establishing experimental plots of tea with the China plants and seeds, Mr. C.A. Bruce who was then appointed as the Superintendent of Government tea plantations and later on awarded the English Society of Arts Medal, presented through the Agricultural and Horticultural Society of Bengal, for his contribution in the discovery of the Assam tea plant, raised nurseries of the indigenous tea plant also. He also explored a large part of the territory from Sadiya to Gabru Purbat in upper Assam and discovered numerous tea tracts inside forests. Some of these tea tracts were cleared and the leaves gathered from the bushes were manufactured with the help of workmen brought from China and samples of tea from the indigenous plants were sent to Calcutta in 1836 which received favourable comments, whereupon an invoice of eight chests of Assam tea was forwarded to London in 1838, which was auctioned on 10 January 1839. This was a momentous occasion because not only did it establish the worth of the Assam tea plant but determined the future course of tea cultivation throughout the world, as today, more tea is made from the Assam type of plants than from the China type. Later on Major Jenkins and Captain Charlton also received a medal each from the Agricultural and Horticultural Society of Bengal for the discovery tea plants in Assam. According to another opinion (Baildon 1877), the tea plant of Assam was discovered by a local Assamese nobleman Maniram Dewan, who later worked in the Assam Company for some time and brought the plant to the notice of Robert Bruce during his visit to Rangpur in 1823. It was a Singphow Chief who supplied tea plants and seeds to Mr. C.A. Bruce. Another Singphow Chief prepared 35 out of the 130 chests of tea, which Mr. Bruce sent to Calcutta in 1841. This clearly shows that the Singphows must have been familiar with the plant and were making and drinking tea from antiquity.

TEA IN DARJEELING

In 1841, Dr. Campbell, Superintendent Darjeeling brought China tea seeds from Kumaon and planted near his residence in his garden at Beechwood, Darjeeling, 2000 m above sea level. As a result of its success the government in 1847 decided to establish tea nurseries in this area. Dr. Withcombe, Mr. James Grant, of the Civil Service, and Captain Samler also tried tea cultivation during this period and got success. In 1848 Dr. Hooker planted tea successfully, at Lebong, a place thousand feet below the Darjeeling elevation, and by 1852 several plantations both of Assam and China types were established by Mr. Martin at Kurseong and Pankhabari. By this time first commercial tea estates like Tukvar, Steinthal and Aloobari were also established which used seeds raised plants in the government nurseries. According to Darjeeling Gazetteer, Alubari and Lebongspur Tea Gardens were established by the Kurseong & Darjeeling Tea Company and Darjeeling Land Mortgage Bank. By 1857 six nurseries with a ton of seeds were raised.

Dhutaria garden was started by Dr. Brougham in 1859 and between 1860 and 1864 four gardens at Ging, Ambutia, Takdah and Phubsering were established by the Darjeeling Tea Company and the gardens at Tukver and Badamtam by the Lebong Tea Company. Similarly other gardens like Makaibari, Pandam and Steinthal Tea Estates were also established during this period. The first factory to be mechanized was Tukvar in 1870. By 1874, 113 tea gardens with 71,000 kg production were established. The Darjeeling Consolidated Tea Co. was eastablished in 1896. For meeting labour requirements for tea plantation Dr. Campbell arranged it through a Nepal native Sri Dakman Rai who helped him by immigrating laborers from Nepal and in return Sri Rai was given the grants of freehold lands for tea cultivation presently known as Saurene, Phuguri and Samripani which were planted during 1878-1883.

Tea processing consisted of withering tea leaf on bamboo racks, rolling withered shoots manually by feet and hands till rollers came to the gardens and drying crudely in big cauldrons which gradually changed to chullis. Sorting was used to be done manually by winnowing made tea using fans. Packing was done in wooden boxes made locally.

For soil fertility management, manuring was done by using farm yard manure only as chemical fertilizers were not available at that time. Red Spider (Sinduray) and Blister Blight (Phokay) were known pests at that time and were treated by using sulfur and wood ash.

There were only Pankhabari Road to Kurseong and Old Military road to Jorebunglow in Darjeeling and none of the plantations had motorable roads. Ponies were the only transport available for managers and assistants in the garden. With the improvements of roads, the transports changed to motor cycles and then to jeeps and later cars and now even trucks go to the gardens. A railroad with three sections was constructed in 1881 which solved the transport difficulties of the tea gardens. The Gailla Khola line ran along the bank of the Teesta river which connects the tea gardens lying around Geille Khola, Kalimpong sub-division and Sikkim. The line was damaged beyond repair during the disaster of 1950 and is no longer in existence.

By 2000, there were 80 tea gardens in Darjeeling having 17,760 ha area and producing about 10 million kg tea. Darjeeling constitutes nearly 5 per cent of the total area in India and 2 per cent of total production. Production in Darjeeling hills is only 540 kg made tea per ha.

TEA IN SOUTH INDIA

South Indian tea has been known for over a century, ever since the first estates were planted there during 1860's. As per Hudson *et al.* (1997), preliminary studies on tea in Nilgiris were done by Dr Christie in 1832 and tea planting started in 1859 in Nilgiris, 1875 in Central Travancore, 1878 in Kannan Devan Hills, 1889 in Wynaad and in 1897 in Anamallais. Initially China type tea and their hybrids were planted in the south India but subsequently assam types occupied the prime place over years (Barua 1989) due to their batter adaptation and productivity.

The hills of the Western Ghat range, which runs down the west coast of the great Indian peninsular, were the main home of the South Indian tea gardens. Areas under tea in these hills spread over the states of Kerala and Tamilnadu in south India, with some tea being grown in the state of Karnataka. The total area under tea in these states is about 88,000 hectares. All the tea is grown on hilly terrain (which provides the good drainage that tea requires), at altitudes ranging from 2,000 feet to more than 6,000 feet above sea level. Hence, climatic conditions also vary considerably. These are all monsoon areas, with the south west monsoon, June to September, being the dominant rainfall period. However, some areas have stronger North East monsoon, November to January (Daruvala 2012).

Tea plantations in Tamil Nadu were mainly concentrated in Nilgiri district due to favourable agroclimatic conditions. Originally the district was a coffee growing region and tea was introduced around 1835 following the recommendations of a committee appointed by the Madras Government (Tea Board 1980). Presently the Nilgiri district is far more populated and has some of the oldest tea plantations in south India, some of it over a hundred years old.

TEA IN UTTARAKHAND

Bishop Haber mentioned about wild growing tea in Kumaon area in Uttarakhand in 1824. However, Dr. Rayle, superintendent of government Botanical Garden Saharanpur confirmed that the plants described as tea by Haber, were actually *Osyris nepalenis* and not the tea plants. In 1835, first tea nursery was established at Lakheme-shvar in Almora and Bharatpur in Bhimtal. In 1880, there were 63 tea gardens having 18,637 acres under tea. The area was scattered around planes of Dehradun and hills of Kumaon in the private sector. The tea processing factories were not established in the hilly region due to lack of roads and electricity, and tea was mostly manually processed. During the independence movement, the British growers left the region. The locals had no affinity with tea. Considering the plantations being the property of the British, the locals destroyed some of the gardens. By the end of 1990, the entire industry of Uttarakhand except in Dehradun area had become abandoned and neglected.

The revival of the tea was started in 1986 with the opening of regional office of Tea Board. But the Board's office could not contribute to the industry as it was located at Lucknow, far away from the plantation area. There were only 5 tea gardens existing in languishing condition in the state. These gardens were more than 100 years old and have become unviable. Three gardens namely Arcadia Tea Estate, Hurbanswala Tea Estate and East Hopetown Estate have gone into organic production of tea since 1998 duly certified by an international agency based at Netherlands. The health beneficial qualities of tea in general and that of organic green tea in particular are becoming well known.

Uttaranchal Tea Development Project launched in March 1994 by the Hill Development Department of the Government of Uttar Pradesh, and continued by Government of Uttarakhand for rural development. Project sanctioned for 8-year period with an outlay of Rs. 21.77 crores under Rozgar Yojna, to plant tea in 800 ha. With the support of CSIR-Institute of Himalayan Bioresource Technology (CSIR-IHBT) fresh surveys were conducted in the region for earma-rking the suitable areas for new plantations, manpower trained on tea husbandry practices at Himachal Pradesh and established tea nurseries.

Project envisaged establishment of tea plantations in four districts of Uttarakhand *viz.*, Champawat, Nainital, and Bageshwar in

Kumaun region, and Chamoli in Garwhal region. Nurseries were established in Kumaun in December 1994 and one in Garhwal in October 1995, with tea seeds and tea cuttings procured from CSIR-IHBT, Palampur (Himachal Pradesh) and Darjeeling. Transplanting commenced with these saplings in July 1995 at Champawat, Ghorakhal, and Kausani in Kumaun, and in April 1996 at Nauti in Garhwal. About 220 ha were planted in Kumaun, including 200 ha in Kausani Tea Estate in District Bageshwar; 80 ha planted in Nanda Devi Tea Estate in District Chamoli in Garhwal. Mostly uncultivated land belonging to farmers' and Civil Forest taken on 30 years lease were planted and the tea plot-holding farmers and Civil Forest-owning villagers were offered job in their own tea farms on daily wage basis. A tea factory was established at Kausani Tea Estate in 2002 by a private company in joint collaboration with the state government.

Tea plantations in Uttarakhand extend at an altitude of 1200 to 1900 m above mean sea level and 29-31° North latitude with annual rainfall of 2000 mm adequately received in winter months. The targeted produce in the state was predominantly flavoury black orthodox tea akin to Darjeeling tea. Well known Darjeeling flavoury planting materials *viz.*, clones Tukdah 78, Tukdah 383, Phoobsering 312, Ambari AV 2/Balai, and Rungli Rungliot 17/144 in 60 per cent, biclonal seedstock TS 379 in 20 per cent, and strong-liquoring high-yielding biclonal seedstock TS 449, TS 520, and clone UPASI 9 in the remaining 20 per cent, of each full-fledged 200 ha tea estate.

Green leaf supplied to the tea factory @ Rs. 13.00 per kg for 75 per cent fine crop and average sale price of Kausani tea in 2002, including Calcutta auction and local sale was Rs. 154/kg. Value added tea packed in wooden chestlets of 100 g and 50 g were sold through outlets in cities such as Delhi and Bangalore.

TEA IN KANGRA VALLEY

The credit of introduction of tea in the Kangra valley of Himachal Pradesh goes to Dr. Jameson, the then Superintendent of the Botanical Garden, North West Provinces, Saharanpur. In 1849, Dr. Jameson conducted the feasibility survey of the valley and found it suitable for tea cultivation. In the same year, he brought China tea plants from the nurseries at Almora and Dehradun and planted them in three Government gardens at Kangra, Nagrota and Bhawarna, located at elevations of 750, 870 and 960 m above msl, respectively. Despite having suffered a good deal during transit, the plants showed excellent growth at two places - Nagrota and Bhawarna, but at

Kangra, the plants did not thrive owing to high temperature and scarcity of irrigation water. The success at two places encouraged the Government to go ahead for the establishment of tea industry in the valley. The first Government plantation at commercial scale was set up in 1852 at Holta, near Palampur at an elevation of 1260 m above msl.

Successful launching of the tea plants in the Kangra valley led to the introduction of private enterprise and capital investment. Many outsiders, both native and European, approached the Government for procuring land for tea cultivation. The demand for land multiplied with the success of Holta tea garden. But no land had been reserved by the Government. The arable land was already under agricultural crops, only waste and forest land was available which was under the control of the *zamindars*. In 1860 the Government deputed an officer, Major Paske for six months to facilitate the transfer of waste lands by negotiating between the *zamindars* and the intending purchasers and was successful in transferring 2,596 acres of land. In 1862, Mr. Egorton, the Deputy Commissioner of Kangra, convinced the *zamindars* to hand over larger part of the forest land by offering relaxation in the forest rules, under special power bestowed on him. The zamindars surendered 2,547 acres of land which was auctioned in 1863. In the meanwhile, in 1861-62, some enthusiastic Europeans namely Mr. Duff, Captain Fitzerland, Messrs. Shaw and Lennox gained the confidence of the local people and acquired waste and cultivated lands suitable for tea by private purchase. The Government farm at Holta was also purchased by the European, Major Strutt, in 1866.

The Government garden at Holta was the only estate up to 1860. With the introduction of private enterprise and capital, the area under tea in 1867-68 increased to 2,635 acres, represented by 19 estates. This area was only one-third of the total area acquired by the planters for tea cultivation. In 1872, area under tea reached to 3,292 acres, which was comprised of 28 estates of size 10 acres or more 13 owned by the Europeans and 15 by native proprietors, and 29 small plantations each of size below 10 acres. A decade later, in 1882-83, the actual area under tea increased to 7,994 acres, of which 4,647 acres belonged to 44 estates owned by Europeans and 1,500 acres of 20 estates owned by the native proprietors, besides 1,817 acres in 1,500 small plots belonging to peasant proprietors and 30 acres of land around the bungalows at Dharamshala. In next decade, though total area under tea had risen to 9,537 acres, the area owned by Europeans was reduced to 3,943 acres as 10 European-owned estates were sold to the natives, whereas area owned by the natives increased to 5,594 acres, which also included 613 acres sold by the Europeans. Of the remaining 34 European estates, 3 were owned by large companies registered under Limited Liability Act, two were large unregistered companies and remaining were small estates each owned one or more. As regard to area-wise distribution of 9,537 acres in the Kangra valley, 8,047 acres were in tehsil Palampur, 1,400 acres in tehsil Kangra, 89 acres in tehsil Kullu and one acre in tehsil Nurpur. In the beginning of twentieth century, about 700 acres of new land was brought under cultivation but a similar area of old and unprofitable land went out of cultivation. During this period, area owned by the Europeans had marginally decreased to 3,492 acres where as, area under the native proprietors had risen to 6,200 acres, thus the sum total area reached at 9,692 acres. Later statistics of 1957-58 reported area of tea at 10,000 acres in Kangra and 1,000 acres in near by areas. This rise in area was probably due to participation of more number of small planters in tea cultivation. Tea Board has reported the tea area in Himachal Pradesh (which includes district Kangra and Mandi) to be at 10,454 acres for the last quarter.

The Government plantation at Holta produced 29,312 lbs of made tea in 1860, eight years after its establishment. With rapid extension of area due to large scale entry of private enterprise, the production in 1868 reached to 2,41,332 lbs with an average yield of 92 lbs/acre. The production of tea in the year 1872 increased to 4,28,655 lbs, expressing an average yield of 130 lbs/acre, range of which was 31 to 298 lbs/acre. In 1883, the production of Kangra tea rose to 8,99,957 lbs, showing an average yield of 194 lbs/acre. The yield of the Europeans-plantations was much better than that of the natives. In 1892, the output from gardens owned by the Europeans was a little over one million pounds, with an all round yield of 254 lbs/acre. The average production of whole district increased from 1,799,603 lbs/annum in 1890-94 to an estimated quantity of 20,000,00 lbs in 1918.

Area under tea increased rapidly to about 8,000 acres during 1860's, but thereafter the acrage of large estates and rapid extension diminished. From 1868 onwards, the consistent increase in production with passage of time was attributed not only to the increased area but also to some improvements in the system of cultivation, consequent upon the experience gained. By 1872, the planters realised that small areas intensively cultivated were more economical in management and most profitable in result as compared to less elaborating farming on a large scale. The advantages of manure in boosting productivity were well known, and hence its use was to maximum extent. One planter, Captain Harrison of Bundla Tea Estate developed the concept of 'intermediate planting' where the vacancies were filled up and the lines of tea were converted into continuous hedge rows. This system was found very advantageous and it doubled the productivity. Another planter, Captain Fitzgerland of Baijnath raised the yield of one plot of his estate to 600 lbs/acre by way of high cultivation and better method of plucking and pruning. During 1957-58, the production of tea in Kangra was 2,434,000 lbs while in near by areas was 115,000 lbs. Total production of tea in 1972 was 12,58,255 kg but it declined to 5,25,655 kg in 1974 and thereafter improved slightly and fluctuated between 5,50,935 and 8,72,899 kg till 1980.

Kangra tea industry occupied prime position with respect to its quality from the last quarter of nineteenth century to beginning of twentieth century. Tea made in Kangra during this period was comparable with that on any part of India. The mention of quality of Kangra tea in the Gazetteer of Kangra district (1882-83) is like this "The tea now made is probably superior to that produced in any other part of India. The demand for it has been steadily increasing and much is now bought up by natives for export *via* Peshawar to Kabul and Central Asia". The gold and silver medals won by the Kangra tea in London and Amsterdam markets in the late nineteenth century (1886 to 1895) bear testimony of its quality at international level. The tea made in the hot weather used to be second to none and was sold as well as any.

There were four markets available for sale of Kangra tea, *viz.*, the London market, the European market, the local market and the Central Asian market. Prior to 1880's decade the Kangra teas were little known in the London market, as these were received in little quantities, difficult to scatter among different brokers and agents. But with the passage of time, the production of both black and green teas increased and reached to 10,00,000 lbs and 70,000 lbs per annum, respectively.

About 5,00,000 lbs of black tea was sold in the retail market and rest was sent to Calcutta and London for auction through brokers and agents. The green tea was sold to tea merchants or brokers in Palampur who further supplied it to Amritsar for onward sale to middle-east. Native merchants from Amritsar and nearby areas were regular in visiting all tea plantations of the valley during the season for purchase of large quantities of the coarser black and green teas, the latter essentially for the Central Asian markets. The merchants used to predict the demand of different categories of tea and book their requirements at fixed rates which were to be manufactured in the ensuing season. The city commanded every trade route of the day e.g. connecting Jammu and Kashmir to Ladakh and the Eastern Provinces of Central Asia, as well as to the route to Peshawar through Kabul and further to the great markets of Herat, Khira, Bokhara, and Samarkand. It also commanded the Indus valley route. Its exports supplied the Sindh merchants who traded to Balan Pass with Kohat and Herat. The teas were carried from Amritsar to Karachi to meet the vast trade of Bluchistan and of parts in Persian Gulf. During late nineteenth century, the main tea trade was with Mashad, Central Asia through Nushki-Siestan trade route. During this period, the European planters developed in large scale the retail trade or direct dealing between the growers and the retailing merchants and even the consumers. The tea in small quantities was also exported to Europe, America and Australia. The Kangra valley during 1920's produced nearly half the green tea manufactured in India, and it was exported to Afghanistan and Iran.

But uncertainty of market of green and black teas used to prevail even during these periods. This uncertainty of tea market always prevented the tea planters from their whole hearted participation in proper cultivation and improving productivity of tea leaf. The demand of black tea in London and Calcutta markets was better only if the quality was good. The quality usually suffered due to aberrant weather conditions like being too dry, too wet or too cold for tea.

Both black and green teas were manufactured in the Kangra valley. Black tea was manufactured by the Europeans only, while green tea by the native proprietors because the green tea manufacturing process required very less machinery and also the market for green tea trade was available at Amritsar. The process of black tea manufacture was not different in basic steps from today's manufacturing procedure: young leaves plucked by hands, brought to factories in baskets and then spread out on round bamboo trays for "withering". Next day the withered leaves were subjected to "rolling' in the rolling machines propelled by steam or water power for about one hour. The capacity of each machine at one time was 100 to 200 lbs of withered leaves. The rolled leaves were then exposed for "fermentation" for about 4-6 hours. The fermented leaves were "rerolled" for about half-an-hour, and then passed into drying machine called 'Siroccos' for "drying" or "firing". Alternatively, the drying was done over the charcoal fires in grates. This completes the manufacturing process. The dried tea was then subjected to "shifting" and "packing". For green tea manufacture, the leaves were "scorched" in heated pans and "rolled" off at once by hand then "cooked" upon in the same pans. The leaves then artificially "coloured" with soft stone. For this, the soft stone was powdered and sprinkled into tea and both rubbed together in the pans. Predominating kinds of black teas were Pekoe, Pekoe Suchong, Coarse tea and Fannings; while in green teas Hyson, Young hyson and Coarse grades were popular types. The good quality teas were packed in lead and wooden cases while coarse grades in coarse bags.

During last quarter of nineteenth century, the cost of production of black tea was about 4 annas/lb (one anna = one quarter of a rupee), while the price realized in different markets was averaged around 6 annas/lb, leaving a profit margin of about 2 annas/lb to the grower. In case of green tea, the cost of production was about 3 annas/lb where hired labour was employed, and value of green tea was fixed at about 4 annas/lb, giving an average profit of 1 anna/lb to the grower. The best sale of Kangra tea in London market during 1892 was of 4053 lbs of tea from Kangra valley Tea Company Ltd. which was sold for an average price of 1s $4^{1}/_{4}$ d/lb followed by 41,848 lbs of tea of the same company at an average of $10^{3}/_{8}$ d/lb. The cost at London auction was $8^{1}/_{2}$ d/lb. The Company during the same year sold 24,453 lbs of tea in Calcutta market at an average price of $9^{3}/_{8}$ annas/lb when the cost was 7 anna/lb, leaving a profit of $2^{3}/_{8}$ annas/lb. Later, in February 1896, Bundla Tea Company had the honour to fetch best price for its tea in Mincing Lane which was sold for 3s $2^{1}/_{2}$ d/lb (British coin s (Shilling) = 1/20 of a pound = 12 d (*denarius* penny).

The Kangra tea industry made material progress until late nineteenth century. Area under cultivation considerably increased in both the European as well as the native owned plantations, the quantity of tea turned out for market rose steadily. The availability of labour for tea gardens and factories was in plenty. Some of the larger European planters introduced machinery for black tea manufacturing. The green tea manufactured by the native proprietors always availed easy market. But later on strengthening of tea industry in other parts of India and Ceylon affected the tea industry of Kangra. During last decade of nineteenth century, the output of tea in India increased from 124 to 175 mlbs and that in Ceylon from 72 to 150 mlbs with the result that inspite of increased consumption, the world tea stock rose by 30 per cent and consequently the prices steadily fell down. As a result, the prices of tea came down. Introduction of tea in Java also gave a jolt to Kangra tea industry. In addition to these, imposition of heavy duty by Russia and Amir of Kabul also affected the green tea trade in Central Asia. The retail trade of Kangra tea fell from 5,00,000 lbs to 2,50,000 lbs due to these factors. The problems faced by the Kangra tea industry during late nineteenth century were high lightened by the daily "*Englishman*" in its issue of April 7, 1898 as follows:

"The decline of Kangra tea begins due to high freight charges by Railways, tax imposition by Afgan, discrimination in the revenue charged on tea fields diversified from agriculture, lack of rail link between Palampur and Pathankot, pending repair of Chakki and Matore bridges and absence of British Counselor at Nasirbad (Seistan) and the Bir-j-ang for Central Asia trade. Further Russia has also become a major threat by its entry in tea trade".

Cultivation of only China tea in the Kangra valley was also identified as one of the causes of lower returns. Besides many external factors, the availability of labour had also started posing problems. The downward progress of the industry which set in the later part of nineteenth century, was arrested during 1902 by better management of tea, like judicious system of plucking, abandoning poor and inferior land, manufacture of green tea by the European planters, which they had taken consequent upon the invention of the machinery for its manufacture and existence of its market to the tune of 70 million lbs and also partly by prevalence of favourable climatic conditions.

In the beginning of twentieth century, the Government imposed export tax on all tea exported from India at the rate of 1/4 pie/lb. The money thus collected was handed over to Indian Tea Cess Committee, and was utilized in opening up new markets and expanding tea trade and relieve black tea market for the purpose of balancing the market. The Committee declared an incentive of 6 pies/lb on all green teas exported from the country. United States was a major market for green tea during that time.

But the worst was in store for the Kangra tea industry. Before the industry could restart on a firm footing, it was shattered to the worst extent by a devastating earthquake on April 4, 1905. The entire Kangra valley was rocked. About 20,200 human beings were killed, majority of which were belonging to Kangra and Palampur Tehsils. Numerous houses were demolished. The majority of tea manufacturing units were razed to the ground. Some tea areas were destroyed and many were abandoned. As an outcome of this, the Europeans, who pioneered the production and manufacturing of the Kangra tea, sold their plantations at cheaper prices to the local and left the valley on the ground of safety. The new entrants were unaware of the technical know how of production and processing expertise. They exploited the potential of tea plants fully without proper maintenance, and the result was a continuous decline of output of their plantations with the passage of time. The quality of tea also deteriorated slowly. Most of the planters had their own processing units, where hand rolling of tea for the manufacture of both green and black tea was done, bearing a few exceptions. Such teas lacked good quality and hence, fetched poor prices. Regarding manpower, there was an acute shortage of labour and after the outbreak of the world war in 1914, the young men from the valley joined the military service for better salary and hence, availability of labour was further affected adversely. Fragmentation of land holding was another irreparable set back. The net resultant effect, ultimately, was the loss of interest in tea plantations.

Kangra tea was lacking competition with other tea growing areas of India like Assam, Darjeeling, and Dehradun etc. The reasons identified were:

- Calcutta market was far off and hence, Kangra tea growers had to pay more freight as compared to Darjeeling and Assam growers.
- Tea made from age old Chinery plants in Kangra was not able to compete with that of Assam and Darjeeling because of poor liquor and physical appearance.
- Climate and soil of Kangra was not congenial from flavour point of tea as was the case of Darjeeling teas.

In comparison to Dehradun tea industry, the Kangra tea industry was not strong enough because labour in Dehradun was cheaper, land was levelled and efficiency of labour was better, and as a result of these factors, cost of Kangra tea was 50 per cent higher as compared to Dehradun. Besides, Dehradun had railway line, making it easy to book teas to Calcutta.

A survey of tea resources of Himachal Pradesh conducted by CSIR Complex in 1984 revealed that the total area which was reported as 4,183 ha since 1961 by the Tea Board has shrunk to 2,063 ha, evidently due to uneconomic returns. Of this area, 90 per cent was in Kangra district, while remaining in Mandi. Total number of holdings was 1,660, 82 per cent of which were below 1 ha but produced only 7 per cent of the total crop. Against this, 9 per cent of the total holdings were more than 2 ha which cultivated 70 per cent of total area and accounted for 88 per cent of total production. Only 3 holdings were over 80 ha, representing 17 per cent of the total area. Regarding income, only 8 holdings were managed for 100 per cent dependence, while 1,001 were providing nothing to the owners and rest were contributing between 10 to 50 per cent of the family income. Average yield in the region in 1,980 was 132 kg/ha which was lowest in the country. In general, there was lack of interest in the growers due to non-remunerative returns. The factors identified for this poor performance were the lack of technical know how and requisite inputs, fragmentation of the holdings, century old plantations, low plant population with plenty of vacancies, absentee landlordism etc.

Lack of proper manufacturing facilities and crop husbandry practices had lead to the decline in quality of made tea. Sale of Kangra tea had become difficult due to the fear of prosecution as the tea of Kangra was not conforming to the standards fixed under All India Pure Food and Adulteration Act. In order to save the existence of the tea industry, the Government in 1974 was forced to set up lower standards for Kangra tea.

The first collaborative effort was made by the planters of Kangra by forming Kangra Tea Association in January, 1898. The conditions laid down for its membership disfavoured many of the native planters to participate. All office bearers of the association were Europeans only. The native growers had voting right only in matters of green tea. The association was affiliated to Indian Tea Association (ITA) in April, 1898. Transportation was the major problem during that time. The association requested the Financial Commissioner in 1898 to make Palampur-Baijnath cart road. It also submitted a proposal to lay down Palampur-Pathankot railway track.

After the departure of most of the European planters, the planters' association was formed by the native growers in 1924. Later, in 1930, the name of the association was changed to Kangra Valley Tea Planters' Association and main objectives framed were to improve the methods of tea culture and disseminate such knowledge amongst tea growers. It directed the efforts to improve the quality and economics of tea, and warned the growers against manufacture of substandard teas. The Association approached the State and Central Governments from time to time for reduction in cess, restriction of green tea supply from Assam to Amritsar, exemption from excise duty and sale tax, exemption from Plantation Labour Act 1951, etc. It also demanded representation of Associations' nominee in Indian Tea Licensing Committee. It put forth the need of central modern tea factories in different zones. In order to improve productivity, it urged the Government to involve Indian Tea Association for studying the conditions of Kangra tea plantations and recommend measures to renovate them. After enforcement of the Pure Food Adulteration Act, 1955, the Association demanded for separate standards for Kangra tea as the existing standards did not conform to Kangra tea. Many of the demands were accepted and executed by the Government in due course of time.

Some of the decisions taken by the Association proved hard for the Kangra tea industry. In 1899, when ITA decided to launch a tea research facility in the north east India, it invited the Kangra Tea Planters' Association to join hands in this facility and contribute towards expenditure, but the Association declined the offer, the after effects of which are evident only today. In 1938, the Co-operative Department proposed to organize tea industry on cooperative basis but the Association did not accept its proposal. Later, in 1950 Labour Commissioner, Punjab offered to provide interest free loans for equipments and machinery from the Government with the condition that the planters should consolidate their holdings by clubbing smaller holdings and form a sort of Company, and thereafter, submit their requests for loan. But this proposal also could not materialize. During 1930's, Indian Tea Licensing Committee allotted a quota to the planters to export tea. It was a good opportunity for the Kangra planters to manufacture exportable tea by improving the standards of tea manufacture and field management; but instead, they sold their quota rights to the planters of other areas and missed the opportunity.

In 1962 The Kangra Tea Planters' Supply and Marketing Cooperative Industrial Society Limited was formed. This Society opened up an auction house at Amritsar to promote the sale of Kangra Tea. It was duly recognized by the Tea Board of India. The Society also arranged the supply of inputs to the Kangra growers at that time. To some extent, the society succeeded to relieve the growers' from direct exploitation of tea brokers and middle men. But presently the Society has fallen sick and its contribution is nearly negligible.

In 1979, the State Government appointed Technical Officer Tea in the Department of Industries to provide technical know-how to the tea industry. Restricted by the paucity of extension staff and lack of infrastructure, this department also could not contribute much to save sinking tea industry. However, in the past it did provide subsidized

With the synthesis of all these factors, Kangra tea can regain its past glory and more.

Tea Map of India (Tea Board 2012)

REFERENCES

Baildon S (1877) *Tea in Assam*, A pamphlet on the origin, culture, and manufacture of tea in Assam, W. Newman & Co., Calcutta.

Barua BN (1989) Science and practice in tea culture. Tea Research Association, Calcutta, pp 32-34.

Chakrabarty DN (1990) Kangra revival. *Tea Time* 1(3): 14.

Daruvala D (2012) Organic Tea From South India, http://www. teaand-coffee.net /0400/tea.htm)

Dhar SM (1992) Kangra tea. *Tea Today* 13(2): 6-13.

Dudeja Vijay (1992) Kangra: the tea valley of Dhauladhar - the white mountain range. *Tea Today* 13(2): 1-5.

Gazetteer of the Kangra district, Kangra proper - 1903-04. Punjab Govt., India, pp 118-127.

Gazetteer of the Kangra district, Kangra proper - 1983-84. Punjab Govt. India, pp 160 & 185-193.

Hudson JB, Durairaj J, Muraleedharan N, Dhanakumar VG (1997) Guidelines on tea culture in south India. UPASI Krishi Vigyan Kendra, Coonoor – 643101, The Nilgiris, Tamilnadu, pp 2.

Jain NK (1984) Tea in Himachal: Overcoming technological Problems. *Capital* 195 (4818): 22-25.

Jain NK (1986) Tea industry in Himachal Pradesh in the national context. *Tea Today* 7(2): 9-15.

Jain NK (2012) Saga of Kangra tea development 1984-1991. *International Journal of Tea Science* 8(1): 99-112.

Jain NK, Tamang MB, Chakrabarty DN (1991) Transfer of tea agrotechnology to small holder of Kangra Valley in India. Paper presented in the International meeting on tea information at Tea Research Instt., Chinese academy of Agrl. Sciences at Hangzhou, China on Sept. 2-6.

Maitland Derik (1982) 5000 years of tea - a pictorial companion. Gallery Books, 112 Madison Avenue, New York 10016.

Proceedings of the meetings of Kangra Valley Tea Planters' Association from 1898 to 1970 (Unpublished).

Randhawa MS (1962) Agriculture and animal husbandry in India. Rev. Ed. ICAR, New Delhi, pp 224.

Randhawa MS (1983) A history of agriculture in India. First Printing. ICAR, New Delhi, pp 158-159.

Tea Board (1980) Report on techno economic survey of Nilgiris Tea Industry. Tea Board of India, 14 Brabourne Road, Colcutta-700 001, September, pp 130.

Tea Board (2012). http://teaboard.gov.in/images/teamap.

The Imperial Gazetteer of India (1908) Vol. XIV. Today and Tomorrow's Printers and Publishers New Delhi, pp 390-397.

Ukers WH (1935). All About Tea Vol. I. Tea and Coffee Trade Journal Co., New York.

Uttarakhand Chai Vikash Pariyojna, A Perview. Garhwal Mandal Vikas Nigam Ltd. Dehradun.

www.fmltea.com

www.thunderbolttea.com

2

GENETIC IMPROVEMENT OF TEA

S. Singh, Ashok Kumar, S. Karthigeyan and P.S. Ahuja

CSIR-Institute of Himalayan Bioresource Technology,
Palampur (H.P.) - 176 061, India. E-mail: sanatsujat@ihbt.res.in

INTRODUCTION

Botanically, *Camellia sinensis* (L.) O. Kuntze is a woody evergreen plant of the genus *Camellia* in the Theaceae family, native to southern China. The family Theaceae consists of 30 genera and about 500 species of which tea is the most important plant. Even though some taxonomical problems with the dual nomenclature of tea i.e., *Thea* or *Camellia* exist, today it is botanically referred to as *Camellia sinensis* (L.) O. Kuntze - the China type (2n = 30), *Camellia assamica* (Masters) Wight - the Assam type and *Camellia assamica* ssp. *lasiocalyx* (Planch ex Watt) Wight – the Cambod type. All the three types of tea hybridize with each other resulting in many intermediate hybrid types. The growth habit of tea is in form of shrub or tree, but cultivated tea is maintained as a low bush in a continuous phase of vegetative growth. The leaves of tea plants are simple, alternate and serrate. Flowers are bisexual and regular with sepals 5-6 and petals 5-8. Stamens are many and anthers are two celled. Ovary is superior with 2-4 locules and ovules are 2-4, rarely solitary and with axile placentation. Fruit is a capsule. The commercial part of the plant is the young shoot with apical bud and two juvenile leaves (Barua 1989).

ORIGIN AND DISTRIBUTION

Tea plants are native to East and South Asia and probably originated around the point of confluence of the lands of northeast India, north Burma, southwest China, and Tibet. Tea-drinking originated in China and despite its extensive use in China and later in Japan, it did not spread to other parts of the world until about the middle of the seventeenth century. Opening of the trade route to India and the East facilitated large-scale trading of tea between Europe and the East. Tea shipped to Europe came from China which was the sole supplier of the commodity till the middle of the nineteenth century. However, with the introduction of Assam tea from India in 1839, the future course of tea cultivation changed throughout the world. Tea cultivation spread from India to other countries in Africa and Sri Lanka, through seeds and propagules and has become an important export commodity. Tea was introduced into Japan from China by the Buddhist monks and the custom of drinking tea became popular in Japan by fourteenth century. In India tea was introduced from China in the middle of the nineteenth century and for many years China tea seed was imported regularly into India and plants were grown in Upper Assam, Dehradun, Kangra, Kumaon and the Nilgiri hills. Almost at the same time, with the discovery and introduction of Assam type of tea plants in India, the tea industry got great impetus and tea cultivation spread to different parts of the world. The China type of tea spread to cold climatic regions (Japan, Korea, Turkey, Georgia and foot hills of Himalayan regions in India) while Assam type of tea was very successful in the tropical and equatorial regions of the world. Assam type of tea spread to Sri Lanka, Indonesia, Brazil and number of African countries. In Vietnam, lowland growing tea is var. *cambodiensis*, and highland growing tea is shan type of var. *assamica* (Yamaguchi *et al.* 1999). In Kenya tea was introduced by the Caine brothers (British colonial farmers) who imported dark-leafed "Manipuri" hybrid seed from Assam in 1904 and 1905 to establish a plantation in Limuru, Central Kenya (Singh 1979).

The distribution of tea from its origin to the main tea producing countries had previously been studied using morphological traits (Yamaguchi *et al.* 1999), and terpene index (Takeo *et al.* 1992). Recently, many studies in elucidating the differentiation of tea germplasm in different countries have been developed by applying molecular marker approaches. These studies not only provided the information of genetic relationships between tea germplasm, but also offered molecular evidence to support the assumption and suggestion of the introduction and spread of tea in certain countries. RAPD marker analysis showed that Korean tea has undergone little genetic diversification after being introduced from China (Yamaguchi *et al.* 1999). A comparison of PAL-RFLP DNA diversity analysis between Japanese and Chinese-Korean tea germplasm clarified that limited populations of tea from China have been introduced into Japan. The distribution of tea in Japan first resulted from the limited populations and then evolved into the local tea varieties. Finally, more and more Japanese cultivars were produced by selections from local tea plant and crossing (Matsumoto *et al.* 2002). Similarly, the tea plants in Korea were largely from China and partially from Japan 50 to 100 years ago (Matsumoto *et al.* 2004). RAPD analysis provided the molecular evidence that both the tea cultivars in Korea and the dominant Japanese cultivar Yabukita had close relationship with 'Jiukengzhong', an endogenous cultivar in Zhejiang Province, China, by the co-segregation markers among them. AFLP principal coordinate analysis supported the suggestion that the Kenyan clones have been derived from collections made in the Assam region, India (Paul *et al.* 1997). Details are summarized in Table 2.1.

Table 2.1. Molecular makers studies on distribution of tea in other part of world

Species	Trait/Marker	Reference
Camellia sinensis	Morphological traits	Yamaguchi *et al.* 1999
Camellia sinensis	Terpene index	Takeo *et al.* 1992
Camellia sinensis	RAPD	Yamaguchi *et al.* 1999
Camellia sinensis	PAL-RFLP DNA	Matsumoto 2001
Camellia sinensis	AFLP	Paul *et al.* 1997

GERMPLASM RESOURCES AND CONSERVATION

Germplasm Resources

The germplasm resources of tea consist of cultivated species or varieties, non-cultivated or wild species and genera, weedy relatives, old seed *jats* or land races, improved clones and seed stocks and breeding lines/hybrids. In the hybrids, considerable species introgression has also taken place. Though Yunnan province in south-west China is considered to be the center of origin of *Camellia*, where maximum genetic diversity and most of the species are found, and north-east India bordering Burma is the origin of Assam type plant, where maximum diversity in tea with respect to genetical and morphological characters are found. Besides *Camellia sinensis*, several other *Camellia* species, *viz.*, *C. caudate*, *C. caduca*, *C. irrawadiensis*, *C. taliensis*, *C. kissi*, *C. drupifera* and related genera like *Eurya*, *Pyrenaria*, *Schima* and *Gordonia* are also endemic to this area. Most of these are found in hilly regions of north-east India. Presently, world-wide over 600 cultivated varieties are available (Mondal *et al.* 2004), of which many have unique traits such as high caffeine content, blister blight disease tolerant etc. (Table 2.2).

Table 2.2. Description of tea cultivars with special characters

S. No.	Special characters	Clone	Originator
1.	Wind tolerance	UPASI-2, UPASI-10	UPASI-TRF, India
2	Drought resistance	UPASI-9	UPASI-TRF, India
3.	Frost resistance /tolerance	BS-26	HPKV-TES, India
4.	Smallest leaf	CH-1	IHBT, India
5	Biggest leaf	Betjan	Betjan T.E, India
6	Blister blight tolerance	TRI-2043, DT-1	TRI, Sri Lanka
7	High pubescence content	TRI-2043	TRI, Sri Lanka
8.	High anthocyanin pigmentation	TRI-2025	TRI, Sri Lanka
9.	High tolerance to pH	TN-14-3	TRF, Kenya
10.	Poor fermenter	12/2	TRF, Kenya
11.	Mite tolerance	7/9	TRF, Kenya
12.	Scale insect tolerant	TN 14-3	TRF, Kenya
13.	High polyphenol content (53.7%)	Luxi white tea	TRI, China
14.	High amino acid content (6.5%)	Anji white tea	TRI, China
15	Low caffeine content (0.14%),	Guangdong tea	TRI, China
16.	High caffeine content (6.96%)	Wild tea at Yunnan	TRI, China
17.	Water logged tolerant	TV-9	TES, India

Conservation of germplasm resources

In recent past the genetic diversity is being narrowed down at an alarming rate due to uprooting of old seed grown plantations and replanting thcm with a few recently released, improved, high yielding clones/cultivars. The situation in worsened by deforestation and human activity for other agricultural production, shifting cultivation and industrialization in all tea growing countries of the world. Therefore it is highly desirable to conserve the diverse genetic resources of the tea for further improvement of crop before they are lost forever. With an objective to conserve the tea germplasm resources, China and India, respectively, have made extensive germplasm collections from the wild as well as different parts of the world for posterity and future use. In China, the China National Germplasm Tea Repositories (CNGTR), including Hangzhou Tea Repository in the Tea Research Institute Chinese Academy of Agricultural Sciences (TRICAAS) and Menghai Tea Repository Branch in the TRI Yunnan Academy of Agricultural Sciences (TRIYAAS) were established as national level permanent *ex-situ* conservation repositories in 1990. About 2,665 germplasm accessions, including wild tea plants, landraces, improved clones, strains, genetic materials and others, from the tea growing provinces of China and several other countries around the world had been collected and preserved in the CNGTR by the end of 2003 (Chen *et al.* 2004).

Collection of tea germplasm was started at Tocklai Experimental Station, Jorhat in the early part of the 20th century soon after the discovery of the indigenous tea plant in north-east India. Further, to maintain a heterogeneous gene pool at the national level, regional germplasm collection centers were established in predominant tea-growing areas. Tea germplasm is now maintained at the national level in north-east India at the Tea Research Association, Tocklai Experimental Station, Jorhat, Assam; in south India at the United Planters' Association of Southern India (UPASI), Valparai, Coimbatore; and in north-west India at the CSIR-Institute of Himalayan Bioresource Technology (CSIR-IHBT), Palampur. It is estimated that presently more than 2,532 accessions are held by Tocklai Experimental Station (Singh 1999). Large numbers of commercially important accessions are also available at UPASI and germplasm repository of CSIR-IHBT. The number apparently is quite impressive but when one critically considers the number of collections for a species like tea which is self-incompatible, out crossing, and the rate at which the germplasm is being lost, and then one realizes that the representation is far from adequate. Considering the facts, a selection scheme was started during 1971-72 to collect and conserve the valuable tea genotypes from old seed grown sections in commercial tea plantation, which may be considered as goldmine of tea germplasm. This is a unique scheme where the entire work force from the top manager to the plucker at the ground level participates in clonal selection (Kaundun and Matsumoto 2002, 2003).

PROBLEMS OF TEA PRODUCTION AND BREEDING

The tea industry has shown significant and sustained progress over the years but the world tea production is beset with numerous problems which need to be taken care of. The increase in the productivity was brought about by increasing the area under clonal cultivars. Plant improvement in tea has been brought about in the past by utilizing naturally occurring genetic variations available in the open cross-pollinated populations for selection and development of better planting material which are being propagated as clones. Studies on genealogy suggest that many of the commercial tea clones are related (Wachira *et al.* 2001). Risks of biotic and abiotic stresses associated with widespread cultivation of vegetative clones and populations with narrow genetic base are high, and huge losses may result in event that such constraints happen to arise due to genetic vulnerability. There has been a thrust in the past few years to rationalize the breeding strategy in an effort to buffer the existing and future tea germplasm against the emergence of such risk factors by deliberately crossing disparate parents through intra-specific and/or interspecific hybridization (Anonymous 2006). Further improvements in yield and quality are possible by devising effective breeding strategies based on the knowledge of botany, available germplasm resources and inheritance of different yield and quality contributing characters. However, tea breeding is a lengthy process because of the following factors (Mondal *et al.* 2004):

i. perennial nature
ii. long gestation periods
iii. high inbreeding depression
iv. self-incompatibility
v. low variability for biotic stresses
vi. low variability for abiotic stresses
vii. lack of distinct selection criteria
viii. low success rate of hand pollination
ix. long duration for seed maturation (12–18 months)
x. clonal difference of flowering time and fruit bearing

Therefore, tea breeding has not reached the stage of applying selection pressure through repeated genetic recombination processes.

Table 2.3. Global tea production, productivity and area under cultivation (FAO 2010)

Rank (on production basis)	Country	Area harvested (ha)	Production (metric tons)	Productivity (kg/ha)
1	China	1419530	1467467	1033.77
2	India	583000	991180	1700.137
3	Kenya	171900	399000	2321.117
4	Sri Lanka	218300	282300	1293.175
5	Turkey	75851	235000	3098.179
6	Vietnam	113200	198466	1753.233
7	Iran	19473	165717	8510.091
8	Indonesia	124573	150342	1206.859
9	Argentina	37221	88574	2379.678
10	Japan	46800	85000	1816.239
11	Thailand	19459	67241	3455.522
12	Bangladesh	59500	60000	1008.403
13	Malawi	22500	51589	2292.844
14	Uganda	27100	40800	1505.535
15	Myanmar	76800	32400	421.875
16	Tanzania	18000	32000	1777.778
17	Rwanda	13549	22249	1642.114
18	Zimbabwe	8600	21000	2441.86
19	Nepal	17127	16607	969.6386
20	Mozambique	7100	15800	2225.352

BREEDING OBJECTIVES

Improvements in yield and cup characters of made tea are the two major objectives of tea breeding. However, the breeding objectives vary greatly from one geographic region to the other (Table 2.4). Cold tolerance is primary breeding objective in countries like Japan, China and Georgia where low temperature injury is of common occurrence. Resistance to diseases and pests is an important breeding objective in different regions of the world. Development of drought tolerant cultivars is of major importance in areas where tea is growing under rain-fed conditions. Reduction of winter dormancy, resistance to hail (in hilly regions) and water-logging conditions (low lying plains of Assam) are some of the region-specific breeding objectives. Other breeding objectives are development of varieties for organic tea cultivation and selection of cultivars for diverse product lines such as low caffeine and high catechin content (Mondal 2009).

Table 2.4. Region-specific breeding objectives in tea

S. No.	Objectives	Importance	Region
1.	Improving quality	Directly linked to the profitability	Black tea producing countries (India, East Africa, Sri Lanka, Bangladesh, and Indonesia)
2.	Increasing yield	Horizontal increase of production by extension planting is limited worldwide	Worldwide
3.	Drought tolerance	Reduces productivity and occurs in all tea-growing regions in the world	Worldwide where tea grows as a rain-fed crop
4.	Reduce winter dormancy	No leaf production during winter months	India, Japan and China
5.	Hail resistance	Causes economic loss as young leaves are mostly affected	Hilly regions of the tea producing countries
6.	Water-logging tolerance	Reduce productivity during rainy season	Northeast India
7.	Cold hardiness	Reduced productivity during winter due to snow	Japan, China, and Georgia
8.	Resistance to diseases such as blister blight, stem canker, etc.	Blister blight causes severe damage as only young leaves are infected	India, Sri Lanka, Indonesia, and Japan
9.	Resistance to pests and nematodes	Most important biotic stresses as all cause severe damage to the leaves	Worldwide
10.a.	Suitability to type of manufacturing	For matching the customer's demand as well as better recovery percentage in made tea	Black tea producing countries
10.b.	Suitability to type of manufacturing	For matching the customer's demand regarding low caffeine and high catechin content	Green tea producing countries
11.	Low input responsive clone	Required for organic tea farming	Organic tea
12.	Long internode length	Suitable for mechanical harvest	Worldwide

INHERITANCE OF YIELD AND QUALITY TRAITS

Genetic improvements in tea have been greatly restricted due to lack of information on inheritance of different characters. Study of inheritance in tea is complicated by a number of factors such as out-breeding nature, long life cycle and polygenic inheritance of the traits of interest. In a cross involving China and Assam type parents, the leaf size in hybrids was observed to be quantitative trait with additive gene action as observed in backcross generations (Bezbaruah 1971). One of the important characters long internode of young shoot was observed to be completely dominant over short internode (Wu and Shyu 1966). Observations on the inheritance of yield in biclonal crosses showed that positive correlation existed between the mean yield of parents and progenies in crosses involving widely different clones, but the relationship was broken down when morphologically similar clones were used. Considerable amount of heterotic response in yield, ranging from 21 to 85% over the mid-parental value, was observed (Bezbaruah 1971, 1974). Contrarily, Wu and Shyu (1966) and Bezbaruah (1971) reported absence of any significant differences in yield and quality of reciprocal hybrids from crosses between morphologically different varieties. The inheritance of fermentation ability (polyphenol oxidase activity) of various clones was studied by Toyao (1982). He observed that the normal plants with PPO were dominant over non-fermenter parents. Density of leaf hair and distribution of Calcium oxalate crystals inherited quantitatively (Wight 1958, Wu *et al.* 1958, Wu 1964). However, for inheritance of leaf colour, pose and pubescence, no regular pattern could be explained (Visser 1969).

BREEDING APPROACHES FOR GENETIC IMPROVEMENT OF TEA

Selection

Seed progenies may be used as a source material for selection of vegetative clones but for large scale commercial cultivation, the clone must have good yield and other desirable attributes over the existing standard clones. Therefore, selection is done among the progenies of high yielding clones used as parents in breeding programme.

Selection for Yield: Selection for yield can be achieved by visual selection of bush having different parameters like, bush size and stem numbers, bud break on pruned stick, evenness of the bush, density of plucking points, production of dry matter, resistance to biotic and abiotic factor, genotype x environment interaction and stability of yield.

Visual Selection: It has been observed in the past that in commercial plantations about 10 percent of the bushes produce only 2 percent of the total crop and about 0.5 percent bushes produce as much as or more than 300 percent of the yield of an average bush (Wight 1939). Therefore, with the poorly available literature on genetic study of tea, selection of bushes is really a hard task. Mother bushes could be selected initially on the basis of visual examination of their exomorphic characters as well as certain biometric parameters (Satyanarayana and Sharma 1982). Leaf characteristics *viz.,* size, shape, pose, surface, colour etc. are used for selection of mother bushes for vegetatively propagated clones or as progenitors for breeding better seed plants with characters for high yield. Although yield comparison of the mother bushes is the recommended procedure of screening out the high yielders for final selection of clones, Visser (1961) observed that all the high yielding mother bushes may not be genetically high yielders and significant correlation may not exist between the individual yield of mother bushes and that of their vegetative offspring. Since size of leaf is related with weight of shoot and cost of plucking the same has been used as selection criterion. However, leaf area alone may not be a helpful indicator in predicting yield because light interception depending on leaf posture and its conversion at different profiles influence productivity. LAI rather than total leaf area may possibly be a better selection criterion.

Bush size and stem numbers: Large sized bush with more number of uniform thick stems per unit area and their possible uniform distribution over the bush area is a good selection criterion. Barua and Dutta (1971) established significant correlations between yield and bush area, and between number of shoots and bush area in Assam.

Bud break on pruned stick: The number of buds ready to open (break) on a pruned stick determines the number of primaries which gives out new growth. The evenness of bud-break and number of primaries are considered as useful selection criteria for the yield. The main branch and the primaries developing from them after a prune should have straight or erect growth in order to be high yielders.

Evenness of the bush and density of plucking points: All the buds in a pruned bush should grow simultaneously with good vigour and not turn into *banjhi*. Frequency of occurrence of *banjhi* shoots and time length for opening of successive leaves is also considered as important parameter of yield. Also, the yield potential of a mother bush can quickly be judged knowing the density of plucking points. It is the most important criterion for quick selection or rejection of mother bushes by visual observation. This is an estimate of number of shoots actually harvested from the bush on its plucking table. Tea bushes with higher density of plucking points per unit area spreading to the periphery from the centre should be considered as a selection criterion.

Production of dry matter: Productivity in tea is mainly related to the production of dry matter and its partitioning to that part of the plant which makes up the commercial yield (Magambo and Cannell 1981). The clone with high yield of green leaf may not always necessarily project the high content of dry matter since clonal difference exists in partitioning of dry matter and hence, this aspect is also one of the effective criteria for selection.

Genotype x Environment interaction: Considerable genotypic differences in response to environmental change were reported by Wickramaratne (1981a). Clones of Assam type were found to possess general adaptation whereas clones of Indo-china origin were specifically adapted to low yielding environments. Rono *et al.* (1993) also reported highly significant clone x site interactions regarding growth rate and yield of tea with temperature and altitude. Environment plays a major role in the yield and quality of tea. Tea clones developed or selected should be tested in different environments to assess the G x E interaction over the yield, quality, abiotic and biotic susceptibility (Wickramaratne 1981a). Tea yield varies with altitude, climate, edaphic and genetic makeup and also influenced by seasons. The effect of environment is quite profound and each factor contribution to the total variability of yield is not known. Even within locations effect of micro-climate (shade/partial shade/no shade; aspect; slope; soil heterogeneity; access to irrigation source) may influence the physiological responses of the plants within the clone resulting in wide variations which need to be assessed through long term phenotyping of different clones. Accurate phenotyping of plant responses to various biotic and abiotic factors prevalent in a region can assist in predicting the stability of a potential clone. Based on growth habit, tea is a tree which is maintained as a bush by continuous pruning cycles, therefore, effects of the age of the plant and time of pruning also need to be accounted for during assessment of variations in phenotypic responses of plants within a clone. Although, development of phenotyping facilities for tea under open field conditions, especially in the hilly regions is a big challenge, yet efficient phenotyping of germplasm resources is the key to sustainability of the breeding programmes and future tea improvement.

Multi-locational trials with a different range of environments will be highly helpful to assess the selected clone for high yield and quality parameters. Considering the above parameters, selection of mother bush by visual observation from the existing population is most important for production of improved seed and clonal varieties.

Kamunya *et al.* (2010) undertook genome mapping in tea and tested QTLs for yield in clones from a cross between clones TRFCA SFS150 and AHP S15/10. The 42 clones were tested in two distinct tea-growing regions in Kenya. Twenty-three putative QTLs were detected in the study, 16 in five different linkage groups for one location, two in two groups for second location, and the rest were associated with unassigned markers. No QTL was detected at both sites, which showed strong genotype × site interaction (G×E) but highly effective within-site heritability (h^2 generally>0.7). Problems of overestimated and spurious QTL effects arising from the smallness of the population may be mitigated by generally high within-site heritability. At least two unassigned markers associated with yield at the first location over the whole study period, suggest potential as candidate markers for site-specific marker-assisted selections.

Tolerance to Abiotic Stresses

In the tea fields, occurrence of adverse environment factors like drought, cold and soil factors (environmental stress) are very common. Resistance against these factors should be taken into consideration at the time of field selection.

Drought stress: Xylem water potential measurements alone, or in combination with some measures of stomatal resistance, could be used to select clones for drought tolerance at an appropriate stage in the selection programme. The other important phenomenon observed in drought tolerant clones are characterized by the partitioning of a greater proportion of dry matter to the root system and frame. Accordingly, germplasm resources need to be evaluated for rate of root growth and root depth variability so as to select for high root to shoot ratio in the genotypes. To overcome the problem of drought, UPASI from south India developed the technique of grafting high yielding clone on to the root stock of drought hardy clone and so far, seven graft combinations have been released to the industry. Drought tolerance in tea has been studied in Kenya using six prominent clones and the results indicate that declining soil water content (SWC) reduced both growth and content of polyphenols in tea. Tolerant clones maintained a high polyphenol content at low SWC, and also showed less fluctuation in phenolics when subjected to changes in SWC. There was significant correlation of total polyphenol content with shoot growth and water stress index (WSI) of tea and a linear relationship ($r^2 = 0.97$) between SWC for tea and WSI as well as shoot polyphenol content (Cheruiyot *et al.* 2007).

Jeyaramraja *et al.* (2003) studied the photoassimilatory and photorespiratory behaviour of certain drought tolerant and susceptible tea clones. Net photosynthetic rate in the older leaves was higher in the drought tolerant clones of tea while liberation of the fixed 14C in light from the older leaves was higher in the drought susceptible clones. The tolerant clones translocated more photosynthates to the crop shoots (three leaves and a bud) from the older leaf than the susceptible clones. Concentrations of RuBP carboxylase (RuBPC) or oxygenase (RuBPO) had no relationship with the drought tolerant nature of tea clones but their ratio correlated with the same. Tolerant tea clones had higher catalase activity that could scavenge the hydrogen peroxide formed in the photorespiratory pathway and thereby reduced photorespiration rate. The ratio of RuBPC/RuBPO had a positive correlation with net photosynthetic rate and catalase activity. Negative correlation between RuBPC/RuBPO and photorespiration rate and between catalase activity and RuBPO activity was established.

Using differential display of mRNA, Sharma and Kumar (2005) identified three drought-modulated expressed sequence tags (ESTs). Northern and BLAST analysis revealed that clone *dr1* (drought responsive), induced only by drought but not by ABA, showed significant scores with PR-5 (pathogenesis related) family of PR-protein gene. Another clone *dr2*, repressed by drought but not by ABA, had nucleotide repeats for polyasparate that are also present in chicken calsequestrin-like mRNA. Clone *dr3*, responded similarly to clone *dr2* but did not show significant homology with the reported genes, hence appears to be novel. Identification of these ESTs is an initial step to clone the full length genes and their promoters. EST markers are detected as size difference in amplified product and inherited in a co-dominant manner. These markers are good for mapping, discovery of genes associated with specific QTLs to understand a specific trait.

Cold stress: Regarding response of tea to cold stresses observations on 816 cultivars of native and exotic collections of tea, Toyao *et al.* (1988) concluded that there were no significant differences in cold hardiness of Japanese wild tea among the different sites of collection. However, the plants from China were also resistant to freezing. The Taiwan cultivars were resistant but Taiwan wild collections were extremely susceptible indicating that the latter belong to var. Assamica. Hardiness scores of Indian plants range from 7-10 except Darjeeling plants whose score ranged from 3 to 6. The plants of Sri Lanka, Burma and other South-east Asian countries were less hardy belonging to var. Assamica.

Table 2.5. Grouping of tea cultivars according to frost tolerance studies at CSIR-IHBT (Sharma *et al.* 2000)

S. No.	Category	Tolerance (%)	Cultivar
1	Tolerant	> 95	BS -53
2	Moderately Tolerant	81-94	BS-8, BS-46, BS-12, BS-76, BS-98, UPASI-11, BS-22, ASHA, BS-55, BS-85 and BS-51
3	Susceptible	41-80	BS-42, BS-16, Kangra Jat, TV-20, TRA/AV-2 Balli, BS-64, BS-72, BS-70, BS-65, UPASI-2, BS-97, BS-48, BS-96, TRA/T-78, BS-61, TRA/T- 383, UPASI-8, BS-54, BS-29, TV-16, TRA/B-668, UPASI-6, BS-24, TRA/P 312, TV-17, BS-39, BS-83, BS-44, BS-47, TG-270/2/13, TRI-2024, TV-12, BS-9, RYDAK-1, TV-26, TV-2, UPASI-10, BS-68, UPASI-15, CR-6017, UPASI-9, TRA/T 145, TRA/HV-39, BS-26 and BS-49
4	Highly Susceptible	< 40	TV-18, TRI 2025, TV 23, UPASI-3, TS-449, TV-19, TV-5 AND TV-25

During winter season tea remains dormant in China, Japan and different regions of India for a period of 3 to 4 months and for such regions, early flushing is considered to be a useful selection criterion as the early flushers are quick growers. However, late flushing characteristic could be ideal if selection for frost hardiness is required. The tea populations spread over 2,312 ha, in Kangra valley of

Himachal Pradesh experiences long winters with moderate to heavy frost. The efforts to breed tea cultivars with extended flushing periods, early as well as prolonged flushing were undertaken at CSIR-IHBT, Palampur (1290 m MSL; 76°33" E; 32°6' N) in India (Sharma *et al.* 2000). It was observed that frost tolerance of genotypes in replicated trials indicated significant differences among 44 China, 18 Assam and 3 Cambod types (Table 2.5). Frost tolerance was highest in China followed by Assam and Cambod tea.

Resistance to biotic stresses

Limited progress has been made in tea breeding for resistance to pest and diseases because of the complexity in studying the genetics as well as resistance mechanism involved in it. Production of tea is hindered by various biotic factors.

Disease Resistance: Among the biotic stresses, the major foliar fungal disease of tea-the blister blight caused by the pathogen *Exobasidium vexans* Wasee is concerned as the most serious disease of tea (Arulpragasam 1992). Blister blight inflicted severe crop loss in countries like Sri Lanka, Indonesia, etc. One third of yield loss as well as quality deterioration is noticed even below 35% of disease threshold level (Gulati *et al.* 1993). By spray application of protectant and eradicant fungicides, the disease has been restrained within the economic threshold level. Even though the nature and basis of resistance in tea clones to this disease is not known, certain clones of tea plants in south India were known to manifest resistance to blister blight. Debnath and Paul (1994) observed no correlation between anatomical and morphological characters of 17 clones with disease severity. However, pre and post inflectional biochemical and physical changes in the host plants play a vital role in influencing the mechanism that imparts resistance to the disease. Jayaramraja *et al.* (2006) reported that SA-6 is the most resistant clone to blister blight infection while the clone TEX-34 is susceptible to blister blight disease. Higher amount of epicuticular wax, thickened epidermis, upper and lower cuticle, stomatal frequency on both the surfaces, compactness of the palisade cells might have cumulative effects in providing resistance against blister blight infection in SA-6. Further, chitinase assay as well as western blotting studies confirmed that the constitutive level of chitinase expression was higher in the resistant clone when compared with the susceptible tea clone. Increased content of (-)- epicatechin and lower concentration of (-)-epigallocatechin was also observed in tea cultivars resistant to blister blight leaf disease in comparison to susceptible cultivars (Nimal Punyasiri *et al.* 2005). The resistance of TRI 2043, the purple green leafed cultivar, is attributed to the additional catechin source provided by the presence of high levels of anthocyanins.

***Insect pest Resistance*:** Large number of insect-pests have been reported to cause economic losses in tea. In Japan, mulberry scale has been especially rampant in recent years. With clear varietal differences in resistance to many of the prevalent diseases and pests, they are possible candidates for marker assisted selection (MAS). MAS for mulberry scale resistance has been practiced in Japan (Tanaka and Taniguchi 2003). Mulberry scale lives inside the tree crown where pesticides do not penetrate. The window for control is limited to the larval period, making it a typical tea pest highly difficult to control. Damage from mulberry scale has been steadily increasing over the last decade. Verification of mulberry scale resistance, however, is extremely complex, and MAS using MSR-1 was expected to be a very valuable means for control. The use of MAS with MSR-1, however, needed to overcome a major hurdle. The range of applicable cross-breeding combinations. There were many segregating populations in offspring of scale resistant 'Sayamakaori', but a marker which could be used for both the parent and the offspring was needed. Very rare allele markers linked to coupling were critical for solving this problem. Fortunately, MSR-1 was found at a locus densely populated by many markers, resulting in many linked DNA markers, from which markers closely linked to coupling but rarely detected in most other parents were selected. These established markers were not affected by crossing parents and could be used for selection of offspring as long as the linkage remained intact. This technological solution has made MAS available for breeding mulberry scale resistant tea in Japan.

Selection for Quality

Quality is the overall value of a tea and is an important criterion in selecting a potential clone and is an inherent character of a clone. Quality of tea is something that stimulates the taste buds of the tongue and has the effect of enhancing the value of aroma. But quality differs from one kind of tea to another. The China type of tea plant has a quality different from that of an Assam/Cambod type. Quality is higher in comparison to strength in the China type tea, while high strength relative to quality characterizes the Assam type. Quality of tea clones varies with geographical location, season(s) in an year and over years (Yamanishi *et al.* 1968). Selection with pubescent (hairiness on the undersurface of young leaves and terminal bud) flush generally produces tea with better quality than one with glabrous (non-hairy) flush irrespective of leaf colour (Wight and Barua 1954).

Venkataramani and Padmanaban (1964) from south India observed significant correlation between the ordered arrangement of pubescence and quality of orthodox tea. CTC method of manufacture of tea in Assam did not show correlation between pubescence and quality (Anonymous 1964). Wight and Gilchrist (1961) observed that the CTC roller which does more mechanical damage to the leaf tissues might produce undesirable aromas during cut-leaf manufacture. The appearance of the tea made by these two methods markedly differ- 'tippy' in the orthodox method and 'wolly' in the CTC method but liquor characters of the teas were good. The leaf hairy portion contains large number of polyphenols, amino acids, organic acids, caffeine and carbonyl compounds (aldehydes and ketones) and very few of these compounds were traced in the non-hairy portion of the leaf.

Wellensiek (1947), Wight *et al.* (1963) and Venkataramani and Padmanabhan (1964) observed that leaf colour was found to be correlated with quality of the processed tea. In general, poor quality of tea is associated with dark green and pale green leaves. Light coloured leaves generally produce better teas than the dark coloured ones but, whether the leaf colour is solely responsible to influence the quality is still not known clearly. Occurrence of anthocyanin pigments in the leaf petiole is a characteristic of the southern form of tea. Some low optimum, pigmentation of tea leaf petiole is beneficial and above this low optimum it is detrimental to quality (Wight and Barua 1954). Phloem index (the frequency of calcium oxalate crystals in the leaf petiole parenchymatous cells) is a good indication of leaf quality. Plants with higher phloem index tend to be of good quality in comparison with lower phloem index. Since, large variations due to sampling and environmental factors are involved in the determination of phloem index, its use is restricted (Visser 1969). Chemical parameters like theaflavin, thearubigin, caffeine, aluminium and volatile flavoury compounds (VFC) are useful in certain levels for selecting quality of black tea. Tea rich in catechins will be considered high quality for black tea, but not so for green tea. Flavour in tea is made up of numerous aromas that contribute to its general taste and character. So, aroma is also an important tea quality parameter. Chloroform test (Sanderson 1963) for fermentation ability of the bushes is also a useful indicator for quality.

Lopez *et al.* (2005) studied variations in the substrate level and enzyme activity of prominent south Indian tea germplasm. The content of polyphenols, catechins (substrates) and polyphenol oxidase (PPO) showed variation, which influenced the final black tea quality. The enzyme PPO was observed to occur in tea shoots and catalyzed the reactions between catechins to form theaflavins in the presence of oxygen. The catechins mainly epicatechin (EC), epigallo catechin gallate (EGCG), epicatechin gallate (ECG) and epigallo catechin (EGC) get mixed with PPO during the oxidation process to form quality constituents like theaflavins (TF), thearubigins (TR) and high polymerized substances (HPS). Theaflavins and their fractions such as simple theaflavin, theaflavin monogallate (TFMG), theaflavin digallate (TFDG) in black tea are the essential quality constituents that are responsible for the liquor characteristics where as TR and high polymerized substances impart colour to the liquor. As oxidation of macerated leaves proceed through different stages of tea manufacture, a decline in PPO activity, polyphenol and catechin contents were observed. Data revealed that the oxidation reaction was faster during the initial stages of oxidation. During the period, oxygen consumption was higher and declined thereafter. Ratio between the enzyme (PPO) and its substrate (catechins) were used to characterize the quality potential of tea clones. An attempt has also been made to categorize prominent tea clones as high, moderate and average quality clones based on their enzyme substrate ratio. Theaflavin content (oxidation product) of different tea clones suggests that the ratio between PPO and catechins forms an important criterion which determines black tea quality. Results obtained were compared with standard clones of known high quality (CR-6017) and moderate quality (SA-6). The parameter is important for the selection of high quality clones in the breeding material. Kottur *et al.* (2010) attributed the tea quality to various forms of catechins and phenylalanine ammonia lyase activity.

Ogino *et al.* (2009) observed two tea plants (designated as Cafless 1 and 2) with low caffeine (0.2% or less of caffeine per dry matter weight considered as caffeine-less) in a population derived from natural crossings of an interspecific hybrid 'Cha Chuukanbohon Nou 6' (*Camellia taliensis* x *C. sinensis*). The plants showed low caffeine content and high theobromine, which is a precursor of caffeine, whereas all tested tea (*C. sinensis*) cultivars including 'Cha Chuukanbohon Nou 6' were observed high caffeine/low theobromine contents. HPLC component analysis showed that 'Taliensis-akame' (*C. taliensis*) which is the seed parent of 'Cha Chuukanbohon Nou 6', showed low caffeine and high theobromine content and may be the caffeine-less character donor. Pedigree analysis using SSR markers suggested parent-offspring relationship between plants related to caffeine-less tea breeding lines. Therefore, it is considered that caffeine-less character could be inherited recessively from 'Taliensis-akame' via 'Cha Chuukanbohon Nou 6'. Caffeine-less trait can be introgressed to other tea cultivars using caffeine-less plants in breeding programs.

Hybridization

Tea is a cross pollinated plant and the progenies obtained are highly heterogeneous in nature. Upon selfing of the clones it generally does not set seeds or in very low proportions with reduced germin-ability. Development of pure lines in perennial crop like tea is time consuming and associated with so many practical difficulties. Self and cross pollination experiments carried out in 40 years at Tocklai gives a clear picture of the extent of pollination behaviour in all three commercial tea plants and their progenies. The cause of incompatibility is considered to be a post-fertilization barrier restricting seed set. Depending on climate, method of propagation of the seed bearer and, to a certain extent, the kind of plant, tea is a long duration crop. It takes 5 to 10 years from one generation to another. In breeding improved strains of seed, the selection of plants to be used as parents is most important and its genetic constituents should be well known. In contrast, clonal selection and multiplication is comparatively easy which does not require any knowledge of plants genetic makeup. Better understanding of the factors involved in flowering and fruit-set in tea will be helpful in tea breeding programme.

Intra-specific hybridization

Selection of Parents: Certain criteria have been set for selecting parents in a breeding programme of tea. Based on vigor, uniformity and quality of the progeny, generative clones are identified which implies that a progeny test is required to evaluate the potential of generative clones (parent genotypes). Further, the selected generative clones should be similar in phenotype and good seed bearers. Time of tea flowering is determined by the geographical location. In North-West Himalayan region flowering starts in September and continues till December with the peak in October. Whereas in south India, flowering commences early and continues for a longer period. Maximum fertilization occurs during the peak flowering period. There is considerable variation between different seed trees of each race of tea in flowering time and it may extend beyond December. Shorter flowering season was noticed for Assam type while China type has an intermediate length of flowering. Time of flowering of the parental clones in a clonal seed *bari* (orchard) should synchronize with each other and the seed yield should be optimum to maximum. Hence time of flowering needs special attention in selecting clones for establishing clonal seed baries.

Hybridization Procedure: Hybridization between genetically diverse parents introduces improved hybrid vigor which often reflected higher yield in the progenies and it also helps in broadening the genetic base as genes from plants with different genetic linkages could be brought together. Pollen grains of China, Assam and Cambod types are all morphologically similar and tea pollen loses viability rapidly when stored at high temperature. Tea plants under natural conditions set very low seeds when compared to hand pollination. This is because the large size and sticky nature of the pollen grains and their aggregation into clumps reduces suitability for wind pollination. Insects like syrphids are found on and around the flowering trees and are common pollinators. Although tea flowers secrete large amount of honey in the disc, since it is contaminated with polyphenols, the bees and wasps are not attracted and the small flies have a very narrow flying range. Thus the production of hybrid seeds could be improved considerably by artificial pollination.

To hybridize selected parents, suitable flowering branches are enclosed in fine muslin cloth or polythene bags after removing all the opened flowers, immature fruits and very small flower buds and only optimum sized buds about to open in one or two days are retained. Emasculation in these buds is done by removing the whorl of petals and cutting off the stamens below the anther lobes with a scissors or by forceps. Pollen grains are collected from bagged flowers to eliminate chances of contamination. Pollen grains are collected in a small petridish, kept in sunlight for 20-30 minutes and applied to the receptive stigma of the suitable female parent. Bagging and tagging are done with cross details and date of crossing. Anthesis of tea flowers is uncertain. Stigma becomes receptive 24 hours before anthesis and remains so for 24 to 48 hours until corolla withers. Syngamy and triple fusion occur simultaneously after fertilization and the endosperm is nuclear in initial stage which afterwards changes to cellular. The fruits (capsules) attain physiological maturity in 10 to 11 months, which may contain one to three round or semi-round shaped

the partial sequences of functional genes, including some important secondary metabolism and stress/defense related genes have been generated (Chen *et al.* 2005b). These ESTs generated, will provide useful DNA sequence information to efficiently isolate and identify the functional genes of the tea. Consequently, it is necessary to increase input and promptly deepen the functional genome research, in order to reveal specific genes of tea plant, to better understand the mechanism of genetic variation on the whole genome level for tea breeding.

Table 2.6. A brief description of works of different isolation and characterization of genes (cDNA cloning) in tea

Species	Objectives	Remarks	Reference
C. japonica	Isolation and charecterization of NADH-dehydrogenase subunit F gene of choloroplast	EMBL accession number AF 130216 for complete sequence of 2221 bp	Kim *et al.* 1999
C. japonica	Isolation and charecterization of rbc-L gene	EMBL accession number L 12602 for complete sequence of 1428 bp	Kron and Chase 1993
C. sinensis	Isolation and cherecterization of PAL gene From cv. Yabukita	EMBL accession number D 26596 for complete sequence of 2330 bp	Matsumoto *et al.* 1994
C. sinensis	Isolation and charecterization of rbc-L gene	EMBL accession number X 69732	Savolainen 1994

Genetic transformation

Several transgenic plants have been produced in a wide range of crops, however, due to low transformation efficiency as well as difficult regeneration system production of transgenic tea is difficult. This may be perhaps due to the presence of high levels of polyphenols in tea explants, which has germicidal property against *Agrobacterium tumefaciens* (Biao *et al.* 1998). Several workers attempted genetic transformation of tea through *Agrobacterium tumefaciens* by using different explants such as *in vitro* leaves, somatic embryos etc. (Matsumoto and Fukui 1998, 1999, Biao *et al.* 1998, Mondal *et al.* 1999, Luo and Liang 2000). While stable transformed callus could only be achieved from leaf explants (Matsumoto and Fukui 1998, 1999, Biao *et al.* 1998, Mondal *et al.* 1999, Siswanto and Chaidamsari 1999), however, using somatic embryos as explants healthy transgenic plants could be produced (Mondal *et al.* 2001). Mondal *et al.* (2001) found that pre-culturing of explants and wounding of somatic embryos had no effect on the transformation efficiency whereas, bacterial growth phase (O.D =0.60), cell density (109/ml), cocultivation periods (5 days) and pH of co-cultivation medium (5.6) enhanced the transformation efficiency significantly. Luo and Liang (2000) constructed a vector containing *Bt* gene which was subsequently transformed to tea. The pGA471 plasmid containing Bt gene cry-IA(c) was digested with Hind III and Bgl II and inserted into the vector pCAMBIA2301. However, no transgenic plants were produced.

Mohanpuria *et al.* (2011) produced genetically transformed tea with low-caffeine content using RNAi technology. The caffeine biosynthetic pathway in tea has been proposed to involve three N-methyltransferases such as xanthosine methyltransferase, 7-N-methylxanthine methyltransferase and 3,7-dimethylxanthine methyltransferase. Last two steps of caffeine biosynthesis in tea have been known to be catalyzed by a bifunctional enzyme known as caffeine synthase. To suppress the caffeine synthesis in tea cv. Kangra jat, a partial fragment of caffeine synthase (CS) was isolated from the same cultivar and used to design RNAi construct (pFGC1008-CS). Somatic embryos were transformed with the developed construct using biolistic method. Transformed somatic embryos showed reduction in the levels of CS transcript expression as well as in caffeine content. Transgenic plants showed a significant suppression of CS transcript expression and also showed a reduction of 44–61% in caffeine and 46–67% in theobromine contents as compared to the controls. These results suggest that the RNAi construct developed here using a single partial fragment of CS gene reduced the expression of the targeted endogenous gene significantly.

FUTURE STRATEGIES

With present advances in tea breeding, from selection to hybridization and molecular breeding, further focus on the following aspects is required for giving impetus to the ongoing tea breeding programs in the world:

(a) A massive tea germplasm appraisal and evaluation project around the world, through a common 'Tea germplasm characterization consortium' is required, which already exists for several similar crops (Mondal *et al.* 2004).

(b) Functional markers associated with the traits of interest in tea, such as EST-SSR (Jin *et al.* 2006, Zhao *et al.* 2008) and SNP (Single Nucleotide Polymorphism) need to be studied in detail.

(c) Development of high density genetic linkage map is required for analyzing the important QTLs to facilitate the marker-assisted selection in tea breeding programs.

(d) Early selective markers at the nursery stage for various biotic and abiotic stresses should be generated to revolutionize tea breeding program which suffers from lack of selection criteria and long gestation periods (Ni *et al.* 2008). An important observation in drought tolerant clones is the greater proportion of dry matter in the root system as compared to the leaf biomass. Therefore, selections need to be based on higher rate of root growth and root depth in the genotypes and consequently select for drought tolerance.

(e) The genetics of different traits in tea are still poorly understood and it causes significant negative impact on tea breeding. Thereby, the techniques and approaches of modern molecular biology should be employed to enhance the genetic understanding of important agronomic traits, quality, metabolism and resistance. It will help to guide tea breeding, to enhance the precision and to improve the efficiency (Chen *et al.* 2007).

(f) Cross-pollinated nature of the species is an impediment to efficient use of molecular markers. This limitation can be overcome if doubled haploid plants are developed in tea which will be 100% homozygous genotypes. Development of haploid and doubled haploid genotypes in tea will allow us to understand the genetic basis of inheritance of different traits and also facilitate development of genetic map in tea.

(g) By undertaking distant hybridization we can broaden the genetic base and improve the effectiveness of breeding.

(h) Redefining parameters constituting ideotype of tea plant with increased harvest index is required. This may be achieved by selecting genotypes on the basis of high shoot density, high photosynthetic efficiency and low photorespiration rate (Singh 1999). Also, plucking ability in tea is highly influenced by the internode length with greater preference by the tea pluckers for genotypes with large internode length due to ease of plucking, particularly in China type tea so as to avoid the occurrence of loose leaf which lowers the quality of plucked crop. Long internode length is an important parameter for easy harvest using mechanical harvesters and cut the labour costs for shoot plucking, which account for up to 60% of the field input costs.

(i) Chemical quality parameters for black, green and oolong forms of tea need to be defined for meeting the consumer demands. There is a need to conduct multiple regression analysis of all chemical parameters on tasters' evaluation and tea prices so as to determine the quality aspects of tea.

(j) Tea is a perennial plant species and has high degree of phenotypic plasticity as compared to annuals. Phenotypic responses of the plants to variations in weather and climatic conditions need to be evaluated over a long term to make effective selections. Therefore, development of phenotyping facilities for efficient evaluation of plant responses to changes in biotic and abiotic factors prevailing over a region will pave the way for predicting the stability of a clone and also facilitate accurate organization of the germplasm for sustainable appraisal and conservation programs of tea genetic resources worldwide.

REFERENCES

Ackerman WL (1971) Genetical and cytogenetical studies with *Camellia* and related genera. *Tech. Bull. ARS*, USDA No. 1427, pp 115.

Anonymous (1964) Report TRA, TES for 1964.

Anonymous (2006) Annual Technical Report. Tea Research Foundation of Kenya, Tea Board of Kenya.

Arulpragasam PV (1992) Disease contrl in Asia. In: Wilson KC, Cliffor MN (eds.) *Tea cultivation to consumption.* Chapman and Hall, London, UK, pp 353-374.

Balasaravanan T, Pius PK, Raj Kumar R, Muraleedharan N, Shasany AK (2003) Genetic diversity among south Indian tea germplasm (*Camellia sinensis* and *C. assamica* spp. *lasiocalyx*) using AFLP markers. *Plant Science.* 165: 365-372.

Barua DN (1989) Science and Practice of tea culture. Tea Research Association Calcutta – Jorhat, pp 29-34.

Barua DN and Dutta KN (1971) Distribution of shoots on the plucking surface of a tea bush and its relation to spacing. *Two and A Bud.* 18: 8-11.

Batschauer A, Ehmann B and Schaefer E (1991) Cloning and characterization of a chalcone synthase gene from mustard and its light dependent expression. *Plant Mol. Biol.* 16: 175-185.

Bezbaruah HP (1971) Cytological investigation in family Theaceae I. *Caryologia*. 24: 421-426.

Bezbaruah HP (1974) Tea Breeding: a review. *Indian J Genet Pl Br*. 34: 89-100.

Bezbaruah HP (1987) Use of interspecific hybrids in tea breeding. *Two and a Bud*. 34: 1-4.

Biao X, Kuboi T, Xu J and Yongyan B (1998) Effect of polyphenol compounds on tea plant transformation. *Am Soc Plant Physiol Abstr.* pp 314.

Cabrera C, Artacho R and Gimenez R (2006) Beneficial effects of green tea – a review. *J Am Coll Nutr*. 25 (2): 79-99.

Chen L, Gao QK, Chen DM and Xu CJ (2005a) The use of RAPD markers for detecting genetic diversity, relationship and molecular identification of Chinese elite tea genetic resources (*Camellia sinensis* (L.) O. Kuntze) preserved in a tea germplasm repository. Biodivers Conserv. 14(6): 1433-1444.

Chen L, Yamaguchi S (2002) Genetic diversity and phylogeny of tea plant (*Camellia sinensis*) and its related species and varieties in the section Thea genus *Camellia* determined by randomly amplified polymorphic DNA analysis. *J Hort Sci Biotech*. 77(6): 729-732.

Chen L, Yang YJ and Yu FL (2004) Tea germplasm research in China: Recent progresses and prospects. *J Plant Genet Resour*. 5(4): 389-392.

Chen L, Yang YJ, Yu FL, Gao QK and Chen DM (1998) Genetic diversity of 15 tea (*Camellia sinensis* (L.) O. Kuntze) cultivars using RAPD markers. J Tea Sci. 18(1): 21-27. (in Chinese)

Chen L, Zhao LP and Gao QK (2005b) Generation and analysis of expressed sequence tags from the tender shoots cDNA library of tea plant (*Camellia sinensis*). *Plant Sci*. 168(2): 359-363.

Chen L, Zhou Z and Yang Y (2007) Genetic improvement and breeding of tea plant (*Camellia sinensis*) in China: from individual selection to hybridization and molecular breeding. *Euphytica*. 154: 239-248.

Chen Z and Liao H (1982) Obtaining plantlet through anther culture of tea plants. *Zhongguo Chaye*. 4: 6-7.

Chen Z and Liao H (1983) A success in bringing out tea plants from the anthers. *China Tea*. 5: 6-7.

Cheruiyot EK, Mumera LM, Ngetich WL, Hassanali A and Wachira F (2007) Polyphenols as potential indicators for drought tolerance in tea (*Camellia sinensis* L.) *Biosci. Biotechnol. Biochem.* 71 (9): 2190-2197.

Choudhuri TC (1979) Studies on the morphology and cytology of the progenies of triploid tea. Ph.D. Thesis. Assam Agriculture University, Jorhat.

Das SC (1992) Non-conventional techniques of regenerating polyploids in tea. Proc. 31st Tocklai Conf., TRA, Jorhat, pp 26-30.

Debnath S, and Paul AK (1994) Susceptibility of tea cultivars to blister blight disease and some of their anatomical and morphological characters. *Two and a Bud.* 41: 48-49.

Duan HX, Shao WF, Wang PS, Xu M, Pang RH, Zhang YP and Cui WR (2004) Study on the genetic diversity of peculiar tea germplasm resource in Yunnan by RAPD. *J Yunnan Agric Univ.* 19(3): 246-254. (in Chinese)

FAO (2009) Food and Agiculture Oraganization of the United Nations – Production FAOSTAT.

Feinbaum RL, Storz G and Ausubel FM (1991) High intensity and blue light regulated expression of chimeric chalcone synthase genes in transgenic *Arabidopsis thaliana* plants. *Mol. Gen. Genet.* 226: 449-456.

Ghosh-Hazra N (2001) Advances in selection and breeding of tea-a review. *J. Plant Crop.* 29(3): 1-17.

Gulati A, Gulati A, Ravindranath SD and Chakrabarthy DN (1993) Economic yield losses caused by *Exobasidium vexens* in tea plantations. *Indian Phyto. Pathol.* 46: 155-159.

Hackett CA, Wachira FN, Paul S, Powell W and Waugh R (2000) Construction of a genetic linkage map for *Camellia sinensis* (tea). *Heredity.* 85(4): 346-355.

Hamrick JL (1990) Isozymes and the analysis of genetic structure in plant populations. In: Soltis ED, Soltis PS (eds) *Isozymes in Plant Biology.* Chapman and Hall, London, UK, pp 87-105.

Hou YJ, He Q, Liang GL, Li PW, Peng P and Deng M (2006) ISSR analysis of the hybrid of the descendants of tea *camellias*. *J Southwest Agric Univ (Natural Science).* 28(2): 267-270. (in Chinese)

Huang FP, Liang YR, Lu JL and Chen RB (2006a) Genetic mapping of first generation of backcross in tea by RAPD and ISSR markers. *J Tea Sci.* 26(3): 171-176. (in Chinese)

Huang FP, Liang YR, Lu JL, Chen RB and Mamati G (2004) Evaluation of genetic diversity in Oolong tea germplasm by AFLP fingerprinting. *J Tea Sci.* 24 (3): 183-189. (in Chinese)

Huang JA, Li JX, Huang YH, Luo JW, Gong ZH and Liu ZH (2005) Construction of AFLP molecular markers linkage map in tea plant. *J Tea Sci.* 25(1): 7-15. (in Chinese)

Huang JA, Li JX, Huang YH, Luo JW, Gong ZH and Liu ZH (2006b) Genetic diversity of tea (*Camellia sinensis* (L.) O. Kuntze) cultivars revealed by AFLP analysis. *Acta Horticulturae Sinica.* 33 (2): 317-322. (in Chinese)

Il'yasova D, Martı´n C and Sandler RS (2003) Tea intake and risk of colon cancer in African-Americans and Whites: North Carolina colon cancer study. *Cancer Causes Control.* 14: 676-772.

Janaki-Ammal EK (1952) Chromosome relationship in cultivated species of *Camellia. Amer. Camellia Yrbk.* 76: 106-114.

Jayaramraja PR, Pius PK, Manian S and Nithya Meenakshi S (2006) Certain factors associated with blister blight resistance in *Camellia sinensis* (L.) O. Kuntze. *Physiol. Molec. Pl. Patho.* 67: 291-295.

Jeyaramraja PR, Rajkumar R, Pius PK and Thomas J (2003) Photo-assimilatory and photo-respiratory behavior of certain drought tolerant and susceptible tea clones. *Photosynthetica.* 41 (4): 579-582.

Jha TB, Jha S and Sen SK (1992) Somatic embryogenesis from immature cotyledon of an elite Darjeeling tea clone. *Plant Sci.* 84: 209-213.

Jin JQ, Cui HR, Chen WY, Lu MZ, Yao YL, Xin Y, Gong XC (2006) Data mining for SSRs in ESTs and development of EST-SSR marker in tea plant (*Camellia sinensis*). *J Tea Sci.* 26 (1): 17-23. (in Chinese).

Kamunya SM, Wachira FN, Pathak RS, Korir R, SharmaV, KumarR, Bhardwaj P, Chalo R, Ahuja PS and Sharma RK (2010) Genomic mapping and testing for quantitative trait loci in tea (*Camellia sinensis* (L.) O. Kuntze). *Tree Genet Genomes.* 6: 915-929.

Karasawa K (1932) On triploid. *Thea Bot. Mag.* 46: 458-460.

Karasawa K (1935) On the somatic number of triploid tea. *Jap. J. Genet.* 11: 320-324.

Kato M, Mizuno K, Crozier A, Fujimura T and Ashihara A (2000) Caffeine synthase gene from tea leaves. *Nature.* 406: 956-957.

Katsuo K (1969) Anther culture in tea plant (a preliminary report) *Stud tea.* 4: 31.

Kaundun SS and Matsumoto S (2002) Heterologous nuclear and chloroplast microsatellite amplification and variation in tea, *Camellia sinensis. Genome.* 45: 1041-1048.

Kaundun SS and Matsumoto S (2003) Development of CAPS markers based on three key genes of the phenylpropanoid pathway in tea, *Camellia sinensis* (L.) O. Kuntze, and differentiation between *assamica* and *sinensis* varieties. *Theor Appl Genet.* 106: 375-383.

Kaundun SS and Matsumoto S (2004) PCR-based amplicon length polymorphisms (ALPs) at microsatellite loci and indels from non-coding DNA regions of cloned genes as a means of authenticating commercial Japanese green teas. *J Sci Food Agric.* 84: 895-902.

Kaundun SS and Park YG (2002) Genetic structure of six Korean tea populations as revealed by RAPD-PCR markers. *Crop Sci.* 42: 594-601.

Kaundun SS, Zhyvoloup A and Young-Goo P (2000) Evaluation of the genetic diversity among elite tea (*Camellia sinensis* var. *sinensis*) accessions using RAPD markers. *Euphytica*. 115(1): 7-16.

Kim KJ, Jansen RK and Olmstead RC (1999) Submitted to EMBL. Accession No. AF 130216.

Kottur G, Venkatesan S, Kumar RSS and Murugesan S (2010) Diversity among various forms of catechins and its synthesizing enzyme (phenylalanine ammonia lyase) in relation to quality of black tea (*Camellia spp.*) *J Sci Food Agric*. 90: 1533-1537.

Kron KA and Chase MW (1993) Systematics of the Ericaceae, Empetraceae, Epacridaceae and allied taxa based upon rbcL sequence data. *Ann Mo Bot Gad*. 80: 128.

Kulasegram S (1980) Technical development in tea production. *Tea Quart*. 49: 157-183.

Laskar MA, Das SC, Deka PC and Bhattacharjee H (1993) Tissue culture of tea: Anther culture for haploid plant production. In "Tea Tech. 1993". Proc. Intern. Symp. Tea Science Human Health. Calcutta, pp 194-201.

Lee SH, Choi HS, Kim RH, Lee HY and No IS (1995) Identification of Korean wild tea plants and Japanese green tea germplasm using RAPD markers. *J Korean Tea Society*. 1(1): 129-148.

Lopez SJ, Thomas J, Pius PK, RajKumar R, Muraleedharan N (2005) A reliable technique to identify superior quality clones from tea germplasm. *Food Chem*. 91: 771-778.

Luo JW, Shi ZP, Shen CW, Liu CL, Gong ZH and Huang YH (2002) Studies on genetic relationships of tea cultivars (*Camellia sinensis* (L.) O. Kuntze) by RAPD analysis. *J Tea Sci*. 22(2): 140-146. (in Chinese)

Luo YY and Liang YR (2000) Studies on the construction of Bt gene expression vector and its transformation in tea plant. *J Tea Sci*. 20(2): 141-147.

Magambo MJS and Cannell MGR (1981) Dry matter production and partition in relation to yield of tea. *Expt Agric*. 17: 33-38.

Matsumoto S (2001) Analysis of the differentiation of Japanese green tea cultivars using DNA markers. In: *Proceedings of 2001 International Conference on O-Cha (tea) Culture and Science* (Session II). October 5–8, Shizuoka, Japan, pp J13-J16.

Matsumoto S and Fukui M (1998) *Agrobacterium tumefaciens* mediated gene transfer in tea plant (*Camellia sinensis*) cells. *Jpn Agri Res Quart*. 32: 287-291.

Matsumoto S and Fukui M (1999) Effect of Acetosyringone application on *Agrobacterium*-mediated transfer in tea plant (*Camellia sinensis*). *Bull Natl Res Ins Veg Orn Tea, Shizuoka, Japan*. 14: 9-15.

Wachira FN (1990) Desirable tea plants: an overview of a search for markers. *Tea*. 11: 42-48.

Wachira FN (1991) Newly identified Kenyan polyploid tea strains. *Tea*. 12: 10-13.

Wachira FN, Tanaka J and Takeda Y (2001) Genetic variation and differentiation in tea (*Camellia sinensis*) germplasm revealed by RAPD and AFLP variation. *J Hort Sci Biot*. 76(5): 557-563.

Wachira FN, Waugh R, Hackett CA and Powell W (1995) Detection of genetic diversity in tea (*Camellia sinensis*) using. RAPD markers. *Genome*. 38: 201-210.

Wellensiek SJ (1947) Breeding of tea. In: Zillink T and Harlem Z (eds) *Foundation of general plant breeding*. 2nd ed., pp 305-347.

Wickremaratne MRT (1981a) Genotype x Environment interaction in tea and their implications in tea breeding and selections. *J Agric Sci*. 69: 471-478.

Wickremaratne MRT (1981b) Variation in some leaf in tea (*Camellia sinensis* L.) and their use in the identification of clones. *Tea*. 50: 183-189.

Wight W (1939) Rep. Indian Tea Assoc. Sci. Deptt. Tocklai.

Wight W (1958) Agrotype concept in chemical analysis of tea. *Nature*. 181: 1355.

Wight W and Barua PK (1954) Morphological basis of quality in tea. *Nature*. 173: 630-631.

Wight W and Gilchrist RC (1961) Concerning the quality and morphology of tea. Rep. ITA, TES for 1959, 69-86.

Wight W, Gilchrist RCJH and Wight J (1963) Note on the colour and quality of tea leaf. *Emp J Expt Agric*. 31: 124-126.

Willson KC (1999) Coffee, Cocoa and Tea. New York, CABI Publishing.

Wu CD and Wei GX (2002) Tea as a functional food for oral health. *Nutrition*. 18: 443-444.

Wu CT (1964) Studies on the heredity, variation and morphology of pubescence on young tea shoots. *J Agric Assoc China*. 47:11-22.

Wu CT and Shyu IS (1966) Studies on the utilization of heterosis in tea. *J Agric Assoc China*. 55:1-26.

Wu CT, Wu HK and Fong CH (1958) The distribution of plucking leaf hair of tea varieties and its correlation with yield and quality. *J Agric Assoc China*. 24: 78-82.

Yamaguchi S, Matsumoto S and Tanaka JI (1999) Genetic dispersal of tea plant. In: Jain NK (eds) *Global Advances in Tea Science*, pp. 413-426.

Yamanishi S, Amma S and Yamada T (1968) Studies on the effect of temperature and induction of polyploidy in the tea plant. *Acta Biol Expt Sinica*. 7: 11-18.

Yao MZ, Chen L, Liang YR (2008) Genetic diversity among tea cultivars from China, Japan and Kenya revealed by ISSR markers and its implication for parental selection in tea breeding programs. *Plant Breed.* 127 (2): 166-172.

Yao MZ, Huang HT, Yu JZ and Chen L (2005) Analysis on applicability of ISSR in molecular identification and relationship investigation of tea cultivars. *J Tea Sci.* 25(2): 153-157 (in Chinese).

Zhao LP, Liu Z, Chen L, Yao MZ and Wang XC (2008) Generation and characterization of 24 novel EST derived microsatellites from tea plant (*Camellia sinensis*) and cross-species amplification in its closely related species and varieties. *Conserv Genet,* 9(5): 1327-1331. DOI: 10.1007/s10592-007-9476-y.

Zuo Y, Chen H, Deng Y (2002) Simultaneous determination of catechins, caffeine and gallic acids in green, Oolong, black and *Pu-erh* teas using HPLC with a photodiode array detector. *Talanta.* 57: 307-316.

3

TEA APPLIED GENOMICS: UTILIZING GENETIC VARIABILITY AND CURRENT GENOMIC TOOLS TO ENHANCE PRODUCTIVITY

R.K. Sharma, Pankaj Bhardwaj, Rahul Kumar, Himanshu Sharma and P.S. Ahuja*

CSIR-Institute of Himalayan Bioresource Technology,
Palampur (H.P.) 176 061, India.*E-mail:ramsharma@ihbt.res.in

BACKGROUND

Genetic variability forms the basis for crop evolution. Variability provides a chance to adaptability and is important for the survival of a species. Because of strong commercial pressures, breeders are forced to draw from a narrow range of tested elite germplasm during the development of modern hybrids. Although the substitution of landraces by elite cultivars is generally admitted to cause a loss of diversity (Pollack 2003). Nevertheless, it has still to be proven that plant breeding inevitably leads to the loss of genetic diversity. In this context, previous studies have shown that breeding mainly leads to qualitative rather than quantitative shifts in registered cultivars (Le Clerc *et al.* 2006). Genetic uniformity in crops is undesirable because it tends to make a crop vulnerable to epidemic and abiotic stresses. Many wild relatives of crop plants or land races contain genes which confer resistance to biotic stresses such as pests and diseases, and tolerance to abiotic stresses such as drought, cold and salinity. When such traits are incorporated into economically important varieties, yield losses can be avoided. A pre-requisite for improving the overall plant characteristics is an understanding and management of the natural variation present within the domestic cultivars, landraces and wild relatives of a plant species. Also, for the successful management and sustainable conservation of a plant species, an accurate assessment of genetic diversity is important. Assessment and analysis of the genetic diversity provides an insight regarding genetic relationships among individuals as well as levels and structure of genetic variation. In particular, knowledge of population genetic structure provides a historical perspective of evolutionary changes that characterize a species and allow us to predict how populations will respond to future events of natural and artificial origin (Wallace 2002).

Tea is one of the oldest, most widely consumed, and least expensive natural beverages, known world over for its heritage brew, with various flavors and antioxidant properties. Owing to most heavily consumed beverages across the globe, tea becomes an important revenue source for many countries in the world including India. Further, plantation of tea provides an important agro-based, eco-friendly, employment generating and export oriented industries in all the tea growing countries. The earliest utilization of tea was for medicine (Zhuang 1995, Weisburger 1997). Compounds such as polyphenols, minerals (mainly selenium and fluorine), tea pigments (mainly theaflavins, thearubgins, β-carotene), tea polysaccharides, amino acids [mainly theanine and gamma-aminobutyric acid (GABA)], and caffeine provides protection from irradiation, antimicrobials, enhancement of the immune function, cellular physiology and metabolism modulation, anti-carcinogenesis, regulation of blood pressure, blood sugar and fat levels. Tea has been domesticated and grown over nearly two centuries but the place of origin is still speculative. However, due to specific soil and climate requirements, tea production is confined to particular regions of Asia (India, China, Sri Lanka, Japan, Vietnam, and Indonesia), Africa (Kenya, Uganda, and Malawi), and to a small extent Latin America (Argentina). Annual world tea production grew by more than 3% in 2006, to reach an estimated 3.6 million metric tons (FAO 2008). An increase was also recorded in India, the second largest producer and consumer, where harvests were also 3% higher, totaling (945,000 metric tons) in the same year. Nonetheless, India ranks first as a producer, consumer, and exporter of black tea.

The genus *Camellia* has been described as containing 82 species (Sealy 1958), but other taxonomic classification systems have also been proposed in the recent past. For example, Chang (1981, 1998) recognized 284 species, and Ming (2000) identified 119 species in *Camellia* complex. Of these, only three namely China type, *Camellia sinensis* (L.) O.Kuntze; Assam type, *C. assamica* ssp. *assamica* (Masters); and Cambod type, *C. assamica* ssp. *lasiocalyx* (Planch MS have contributed significantly to the entire genetic pool of cultivated tea worldwide. Because of inter-specific compatibility, these plants breed with related wild *Camellia* species, namely *C. flava, C. petelotii* Wight (1962), and *C. lutescens* (Sharma and Venkataramani 1974). Additionally, *C. irrawadiensis* and *C. taliensis*, which have morphological proximity to tea also contributed to the cultivated tea gene pool (Banerjee, 1992). Therefore, cultivated tea is highly heterogeneous, having several intergrades, introgressants and putative hybrids. These can be categorized from China types through intermediate to Assam types, and hence the existence of true China, Assam, or Cambod type of tea is doubtful (Visser 1969).

In India, tea cultivation has started with introduction of tea germplasm from China in the year 1836. Subsequently, its spontaneous large-scale hybridization with indigenous tea accessions and wild relatives from a wide range of geographic distribution in India led to hybrid swarms with tremendous genetic variability. In order to maintain a heterogeneous gene pool at the national level, regional germplasm collection centers were established in predominant tea-growing areas. Region wise, tea germplasm is now maintained at Tea Research Association, Tocklai Experimental Station, Jorhat, Assam (north eastern Indian region), United Planters Association of Southern India (UPASI), Valparai, Coimbatore (The Southern Indian region) and in CSIR-Institute of Himalayan Bioresource Technology (CSIR-IHBT), Palampur (The north-western Indian region). It is estimated that presently more than 2532 accessions are held by Tocklai Experimental Station (Singh 1999). Large numbers of commercially important accessions are also available at UPASI and IHBT. There are ~200 Darjeeling tea collections are also available at Darjeeling Tea Research and Development Centre, Kurseong, Darjeeling, India. In late 90's, Tata Tea Limited has established tea collection centre at Munnar, which was later on succeeded by The Kanan Devan Hills Plantations Company Private Limited (KDHP) in 2005. Additional, 90 tea collections collected from north western region are also available with Tea Breeding Centre, Himachal Pradesh Agricultural University, Palampur.

Future of tea industry depend upon the availability of high yielding, quality tea clones with greater tolerance to pest, diseases and environmental stresses. Genetic improvement of tea requires extensive survey for identification, characterization, evaluation, domestication, maintenance and utilization of germplasm for development of superior plant material. Conventional breeding programme in tea, however, limited by long gestation period, out-breeding nature and self-incompatibility. Status of germplasm and application of different marker technologies in tea have been summarized and well documented (Sui *et al.* 2008, Bandyopadhyay 2011). This monograph links status of emerging genomic information in recent past that can expedite the genetic improvement in tea and hence increase the productivity. This will provide a background for future possibilities of modern tea breeding together with current efforts for development of sequence based markers such as microsatellite markers and links genetic diversity inferences of existing gene pools for identification of diverse parental groups and developing efficient phenotying tools to support operational breeding. Insights of preliminary attempts on quantitative trait locus (QTL) mapping in tea are also reviewed along with the future prospective of association genetics to dissect quantitative traits.

POPULATION AND GENETIC DIVERSITY

Amount and distribution of the genetic variation existing in the gene pool is the key for the success of any breeding programme and sustainability of the crop. In general, our knowledge about the global genetic diversity in tea has been very low. Moreover, diversity status of related *camellia* species is also limited. In the past decades, morphological and biochemical descriptors have been routinely used for characterization of tea populations. Descriptors are valuable tools for identification of tea at the varietal level, however, unable to distinguish inter- and intra-varietal variations. Further, due to out-breed nature, tea is a plant with many overlapping morphological, biochemical and physiological abilities (Purseglove 1968, Wickremasinge 1979) and may not account for all the diversity in the species. Therefore, recent attention has been shifted to the use of different molecular markers for determination of genetic diversity and establishing genetic relationship among different tea populations. The first report on utilization of molecular markers in tea came from efforts by Wachira *et al.* (1995), where in RAPD markers were used for assessment of genetic diversity in Kenyan tea clones and recorded a moderate to high level of genetic diversity in the Cambodian and China populations, respectively. Moderate to high level of genetic

belonging to different morphotypes (Assam type, Assam hybrid, China hybrid and Cambod) were distributed in all the groups (Raina *et al.* 2012). A set of one hundred five core collections identified that captured 98% diversity and represent all the major groups have been identified in this study. These accessions, therefore, can be utilized to enlarge the diversity of cultivated tea clones in tea breeding. However, the breeding of tea cultivars needs a sustainable effort from the preservation of tea germplasm to the development of superior varietal material through wide genetic crosses. Overall status on genetic diversity has been given in Table 3.1.

Table 3.1. Tea genetic diversity status

Species	Techn-ique	Size of Popu-lation	Description of Polymorphism	Remarks	Reference
C. sinensis	AFLP	32	10 primer combination and 73 polymorphic loci.	More polymorphism within the popul-ation (79%) rather than between the populations (21%)	Paul *et al.* 1997
C. sinensis	RAPD	38	21 primers and 157 polymorphic loci	70% of variation were detected with in the population	Wachira *et al.* 1995
C. sinensis	RAPD	27	17 primers and 58 polymorphic loci	3 primers sufficient to distinguish all the accessions	Kaundun *et al.* 2000
C. sinensis	RAPD and AFLP	48	37 RAPD primers and 6 AFLP pri-mer combinations produced 266 polymorphic bands	AMOVA revealed a significant structuring among the population	Wachira *et.al.* 2001
C. sinensis	RAPD	15 genetic resources	20 RAPD primers amplified 129 polymorphic bands in all the tested resources.	94.2% genetic diversity	Chen *et.al.* 2005
C. sinensis	RAPD	23 related species	15 primers and 107 polymorphic loci	95.3% genetic diversity	Chen *et al.* 2002
C. sinensis	CAPS	50	7 primer comb-inations, 31 poly-morphic alleles	83% of genetic div-ersity within the verities	Kaundun and Matsumoto 2003 a

C. sinensis	RAPD	90 (6 populations)	12 primer, 25 polymorphic loci	84% genetic diversity within population and 16% between population	Kaundun and Park 2002
C. sinensis	RFLP	297	PAL cDNA probe	2 tea groups were identified	Matsumoto *et al.* 2004
C. sinensis	ALP	44	12 primer combinations	The markers devised here were sufficient to differentiate all the teas	Kaundun and Matsumoto *et al.* 2004
C. sinensis, C. assamica and C. cambod	AFLP	49	3 primer combination and 1555	94% maximum similarity between Assam and Cambod and 85% maximum similarity between china and Cambod	Balasarvanan *et al.* 2003
C. sinensis	AFLP	29		Genetic similarity ranged between 68% to 92%.	Mishra and Sen Mandi 2004
C. sinensis	ISSR	48	21 primers and 381 polymorphic loci	92.07% genetic diversity with population and 7% genetic diversity between population	Yao *et al.* 2008
C. sinensis, C. assamica and C. assamica spp lsiocalayx	AFLP	123	7 AFLP primer combinations	Overall genetic diversity 49 %; no significant difference in the tea population growing at the national level.	Sharma *et al.* 2010 (outcome of the national tea network programme coordinated by CSIR-IHBT)
C. sinensis, C. assamica and C. assamica spp lsiocalayx	AFLP	1644	7 AFLP primer combinations	PCoA and NJ analysis clustered 1,644 tea accessions and clones into six major groups with one group in each, constituted mostly by China hybrid, Assam China hybrid and Assam hybrid morphot-ypes, of distinct genetic identity, therefore, recorded highly heterogenous nature of tea	Raina *et al.*2012 (outcome of the national tea network programme coordinated by CSIR-IHBT

ASSESSMENT OF GENE FLOW AMONG THE GOBAL TEA CULTYIVATIONS OTHER THAN CENTRE OF ORIGIN

Morphological traits were the first markers to be utilized to assess the possible amount of gene flow from the centre of origin (Takeo *et al.* 1992, Yamaguchi *et al.* 1999). However, considering overlapping morphological traits, trend has been shifted from morphological markers to more reproducible molecular marker approaches. These studies not only provided the information of genetic relationships between tea germplasm, but also offered molecular evidence to support the assumption and suggestion of the introduction and distribution of diversity to other countries. Tea cultivation in India started in early nineteenth century with seed and plant imported from China. Indigenous tea was also simultaneous discovered from Assam, Manipur and Burma. Owing to outbreed nature, free hybridization occurred between indigenous and China populations resulting a hybrid swarm with tremendous heterogeneity in Indian tea. AFLP based genetic analysis revealed little differentiation among tea populations cultivated at the national level are highly heterogeneous (Sharma *et al.* 2010, Raina *et al.* 2012). RAPD marker based genetic analysis revealed little genetic diversification in Korean tea after being introduced from China (Yamaguchi *et al.* 1999). Comparative study with CAPs markers between Japanese and Chinese-Korean tea germplasm clarified that few Chinese tea populations have had gene flow from high yielding tea clones from Japan. However, genetic variation of tea in Japan has first resulted due to the evolution of local tea varieties followed by intensive selective out-crossing (Matsumoto *et al.* 2002). Similarly, the tea plantations in Korea were largely introduced from existing tea populations of China and Japan in 50 to 100 years ago (Matsumoto *et al.* 2004). Tea, being introduced crop in Sri Lanka from India has very limited genetic diversity. Published literature suggested that tea cultivations in Kenya are also from the extensive gene flow from Assam tea cultivations from north-eastern India (Paul *et al.* 1997). Details are summarized in Table 3.2.

Table 3.2. Molecular makers studies on distribution of tea in other part of world.

Species	Trait/Marker	Reference
C. sinensis	Morphological traits	Yamaguchi *et al.* 1999
C. sinensis	Terpene index	Takeo *et al.* 1992
C. sinensis	RAPD	Yamaguchi *et al.* 1999

C. sinensis	PAL-RFLP DNA	Matsumoto 2002
C. sinensis	AFLP	Paul *et al.* 1997, Sharma *et al.* 2010 Raina *et al.* 2012

DNA FINGERPRINTS FOR UNIQUE IDENTIFICATION

In India, morphological descriptors data have been developed for 1763 tea collection representing the major tea gene banks in India with all the vegetative traits (unpublished data). Of the 12 quantitative morphological characters, leaf size characters (length and width with higher eigen value (-0.737 and 0.334) could able to explain variation in 121 tea accessions (Rajkumar *et al.* 2010). Morphological, phytochemicals and terpenoid index have also been used to characterize tea species. It has been argued that these descriptors may not reflect the correctness of genetic differentiation because most of them are largely affected by environmental and developmental factors. Tea, being a highly heterogygous plant with many overlapping morphological, biochemical and physiological attributes recorded a continuous variation with high degree of plasticity (Purseglove 1968, Wickremasinge 1979). Morphological characteristics have been proven difficult to identify the discrete taxonomic groups in tea. As an efficient, auxiliary means to distinguish intra- or inter-specific variation of tea germplasms, the molecular markers have been proven to be useful to discriminate the tea varieties, cultivars and common adulterants in commercial tea from the market (Wachira *et al.* 1995, 1997, 2001, Kaundun *et al.* 2000,Lai *et al.* 2001, Singh & Singh 2001, Mondal 2002, Kaundun and Matsumoto 2003a, 2003b, Chen and Yamaguchi 2005, Chen *et al.* 2005a, Yao *et al.* 2005). RAPD markers have been routinely used for varietal and inter-specific differentiation of tea (Lee *et al.* 1995, Wachira *et al.* 1997). The cytoplasmic DNA, which is inherited maternally, could be used for to determine the seed parent (Tanaka 2006). The registered parents of two Japanese cultivars, 'Yutakamidori' and 'Meiryoku', were found to be incorrect and genuine parents were identified by RAPD analysis (Tanaka and Yamaguchi 1996). 5S rDNA based RFLP analysis could able to differentiate three varietal types in Indian tea. Nucleotide sequencing of 5S rDNA gene derived amplicons could identify putative SNPs, distinguishing *C. sinensis* (China) tea from the Assam and Cambod (Singh and Ahuja 2006). These studies proved the importance of molecular markers as effective tools to authenticate the released cultivars and their parents, and therefore can be helpful to protect intellectual property rights. DNA finger-

		rces	polymorphic bands in all the tested resources.	evaluate the genetic diversity and relationship, but also to identify tea genetic resources.	
C.japonica *C. sinensis*	RAPD and SSR	20	33 RAPD and 1 SSR primer combination	The hybrids play an important role in genetic breeding for *C. sinensis* and *C. japonica*	Tanaka *et al.* 2003
C. sinensis	RFLP	463	PAL cDNA as probe	Results indicate a process of different-iation from the ancestral material introduced from China to the local tea plants and, culti-vars were produced by selecting from local tea plants and crossing.	Matsumoto *et.al.* 2002
C. sinensis, *C. assamica* and *C. assamica* spp lsiocalayx	5S rDNA RFLP	28	5S rRNA gene primer	5S rDNA gene could be used as a molecu-lar marker to distin-guish *C. sinensis* Chinary tea from the other important types of tea,	Singh and Ahuja 2006

ESTABLISHMENT OF GENETIC UNIFORMITY

In general, conventional tea is cultivated either through seed or cutting derived from the superior plants. Vegetative propagation is a routine practice to get the optimum yield. In India, on an average 50% of tea cultivation is under clonal cultivation. Molecular makers can be routinely utilized to ensure genetic uniformity in vegetative mode of propagation. However, considering limitations in vegetative propagation (slower rates, non-availability of suitable planting material, abiotic stresses and survival rate) micro-propagation appears to be an alternative ideal choice to circumvent the traditional difficulties in conventional propagation (Mondal and Chand 2002, Mondal *et al.* 2004). Genetic stability in the micropropagated plants derived from organized meristem cultures are unquestionable. However, cryptic genetic defects arising via somaclonal variation during the morphogenesis limits the wider utility of micropropagation systems. Molecular markers have been routinely utilized for establishment genetic uniformity in tissue culture raised plants in tea (Devarumath *et al.* 2002, Mondal and Chand 2002, Singh *et al.* 2004, Thomas *et al.* 2006) (Table 3.4).

Table 3.4. Molecular fingerprinting status of tissue culture raised plant in tea.

Species	Technique	Population/ cultivar	Description	Remarks	Reference
C. sinensis	RAPD	T-78/ 18 Plants	Total no of loci 221 Polymorphic loci 24	Polymorphism among four plants showed an identical banding pattern suggesting the occurrence of a single mutation. Results demonstrated that RAPD can be used successfully to determine the genetic instability among micro propagated plants which otherwise were morphologically indistinguishable.	Mondal and Chand 2002
Camellia sinensis Camellia assamica	RAPD ISSR and RFLP	UPASI 26, UPASI 27 and UPASI 3/ 14 plants	14 RFLP Probes 25 RAPD primers 25 ISSR primers	ISSR fingerprinting (12.8%) detected more polymorphic loci than RAPD fingerprinting (4.28%). RFLP detect no polymorphism	Devarumath *et al.* 2002
Camellia sinensis	RAPD	UPASI 9/ 16 Plants	22 primers	Twenty one primers exhibiting amplified products gave monomorphic banding patterns. Only one primer amplified a unique extra band of similar size in four plants.	Singh *et al.* 2004
Camellia sinensis	ISSR	UPASI 10/ 15 Plants	15 primers	Genetic variability of somaclones derived from single line cotyledonary culture ranged from 33.0 to 55.0%.	Thomas *et al.* 2006

SSR MARKER DEVELOPMENT

The ubiquity of microsatellite or simple sequence repeats (SSRs) in eukaryotic genomes and their usefulness as genetic markers has been well established over the last decade (Hamada *et al.* 1982, Tautz and Renz 1984). Microsatellites are mainly characterized by high frequency, co-dominance, multi-allelic nature, reproducibility, extensive genome coverage and ease of detection by polymerase chain reaction with unique primer pairs that flank the repeat motif (Gupta and Varshney 2000). As a result of these characteristics, microsatellite have become the most favored genetic markers for plant breeding and genetic application such as, assessment of genetic diversity, constructing framework genetic maps, marker aided selection and comparative mapping studies. SSR marker development have been started with the identification of fifteen SSR loci (Freeman *et al.* 2004). In another study 11 SSR markers were developed from *Camellia sinensis* with PCR-based isolation of microsatellite arrays (PIMA) based method (Hung *et al.* 2008). Availability and continuous enrichment of expressed sequence tags (ESTs)/ transcriptome sequence database in tea can serve the alternative genomic source for development of microsatellite markers. Twenty four polymorphic SSR markers reported (Zhao *et al.* 2007) from the EST database available for *C. sinensis*. In another study ten markers (EST-SSR and ALP markers) were also developed (Ueno *et al.* 2009). Further, considering the multiple advantages of UGMS markers, 519 unigenes containing 563 SSRs were identified from 4356 unigenes (1614 contigs and 2742 singletons). These microsatellites mainly comprised of di (50.8%), tri (30.8%), tetra (6.6%), penta (7.5%) and few hexa (4.1%) nucleotide repeats. Among the dinucleotide repeats, (GA)n.(TC)n were most abundant (83.6%). Recently, 173 highly polymorphic novel SSR markers were used to investigate the genetic diversity among the tea accessions of different *Camellia* spp. These markers have also recorded a high level of cross transferability (83.6%) to the related species at various levels Functional annotation of the unigenes containing SSRs was done through gene ontology (GO) characterization identified that 60% of them revealed significant sequence similarity with the known/putative proteins of *Arabidopsis thaliana*. Genetic analysis these markers also proved highly heterogeneous nature of Indian tea. Further, test for IAM and SMM models for the UGMS loci showed excess heterozygosity and did not show any bottleneck operating in the tea population (Sharma *et al.* 2009, Sharma *et al.* 2011). Network efforts coordinated by CSIR-IHBT also identified 625 polymorphic microsatellite markers derived from enriched genomic libraries of nine different repeat probes (unpublished). These newly developed markers were tested on a diverse array of tea accessions and provide a tool for understanding demography and population structure in tea. The majority of the SSRs developed were proven to be highly polymorphic both between and within different geographical origins and offered a potential value to investigate the population diversity and genetic origins of tea plant. Overall status of microsatellite markers is given in Table 3.5.

Table 3.5. Status of microsatellite markers development in tea

Species	Method/ Probe	No of markers	Description	Remarks	Reference
C. sinensis	Multiple probe	15	High number of alleles with high PIC	Potential to evaluate the population struct-ure and origin of Tea plant.	Freeman *et al.* 2004
C. sinensis	PCR Ampl-ification of Micro-sate-llite Array	11	Alleles ranged from 4-18	provide a tool for understanding demo-graphy and popul-ation structure in tea	Hung *et al.* 2008
C. sinensis	EST database	10	Numbers of alleles and expected het-erozygosity ranged from 2 to 13 & from 0.28 to 0.90	Marker resource will be useful for analyzing the genetic diversity of *Camellia* species.	Ueno *et al.* 2008
C. sinensis	EST database	24	2-8 poly-morphic alleles with high heter-ozygosity	molecular marker res-ource for management and conservation stu-dies on the tea germ-plasm in the future.	Zhao *et al.* 2008
C. sinensis	EST database	61	2-16 poly-morphic alleles with high heterozy-gosity	Markers recorded highly polymeric, high level of transferability and Gene ontology analysis proved their wider applicability in genetic and function analysis in tea	Sharma *et al.* 2009
C. sinensis	EST database	112	2-16 polymor-phic alleles with high he-terozygosity	Highly polymorphic & heterozygous	Sharma *et al.* 2011
C. sinensis	Enriched genomic libraries derived from 9 rep-eat probes	625	2-8 poly-morphic alleles with high heterozy-gosity	Markers recorded highly polymeric, high level of transferability is proved their wider applicability in genetic analysis in tea	unpublished (tea network group)

Sharma RK, Bhardwaj P, Negi R, Mohapatra R, Ahuja PS (2009) Identification, characterization of unigene derived markers in tea (*Camellia sinensis* (L.) O. Kuntze). *BMC Plant Biol.*9: 53.

Sharma VS, Venkataramani KS (1974) The tea complex. I. Taxonomy of tea clones. *Proc Indian Acad Sci.* 53:178–187.

Singh D, Ahuja PS (2006) 5S rDNA gene diversity in tea (*Camellia sinensis* (L.) O. Kuntze) and its use for variety identification. *Genome.* 49: 91–96.

Singh D, Singh MP (2001) Organization of 5S ribosomal RNA genes in tea (*Camellia sinensis* (L.) O. Kuntze). Genome. 44: 143–146.

Singh ID: Indian tea germplasm and its contribution to the World's tea Industry. *Two & Bud* 1979, **26:**23-26.

Singh M, Bandana, Ahuja PS (1999) Isolation and PCR amplification of genomic DNA from market samples of dry tea. *Plant Mol Biol Rep.* 17: 171–178.

Singh M, Saroop J, Dhiman B (2004) Detection of intra-clonal genetic variability in vegetatively propagated tea using RAPD markers. *Biol Plant.* 48: 113–115.

Sui NI, Yao M, Chen L, Zhao L, Wang X (2008) Germplasm and breeding research of tea plant based on DNA marker approaches. Front. Agric. China, 2(2): 200–207.

Takeo T, You XQ, Wang HF, Kinukasa H, Li MJ, Cheng QK, Wang HS (1992) One speculation on the origin and dispersion of tea plant in China–One speculation based on the chemotaxonomy by using the content-ration of terpen-alcohols found in the tea aroma composition. *J Tea Sci.*12: 81–86.

Tanaka J (1996) RAPD linkage map of tea plant and the possibility of application in tea genetics and breeding. *J Tea Res.* 84: 44–45.

Tanaka J (2006) Study on the utilization of DNA markers in tea breeding. *Bulle Nat Inst Veg Tea Sci.* 5: 113–155.

Tanaka J, Metoku S, Takeda Y (2003) Garden-variety camellia 'Robiraki' derived from crossing between *Camellia japonica* as seed parent and *C. sinensis* as pollen parent. Application of RAPD and SSR marker analysis to tea breeding by interspecific hybridization. *Breed Res.* 5: 149–154.

Tanaka J, Yamaguchi S (1996) Use of RAPD markers for the identification of parentage of tea cultivars. *Bulle Nat Res Ins Veg Orna Plant Tea* .(B) 9: 31–36.

Tautz D, Renz M (1984) Simple sequences are ubiquitous repetitive components of eukaryotic genomes. *Nucl Acids Res.*12: 4127–4138.

Thomas J, Vijayan D, Joshi SD, Lopez SJ, Kumar RR (2006) Genetic integrity of somaclonal variants in tea (*Camellia sinensis* (L.) O. Kuntze) as revealed by inter simple sequence repeats. *J Biotechnol.* 123: 149–154.

Ueno S, Tsumura Y (2009) Development of microsatellite and amplicon length polymorphism markers for *Camellia japonica* L. from tea plant (*Camellia sinensis* (L.) O. Kuntze) expressed sequence tags. *Mol Ecol resources*. 9: 814–816.

Visser T (1969) Tea *Camellia sinensis* (L.) O. Kuntze. In: Veenman H, Zanon N (eds) Perennial crops breeding in the tropics. Wageningen, Netherlands, pp 459–493.

Wachira FN, Powell W, Waugh R (1997) An assessment of genetic diversity among *Camellia sinensis* L. cultivated tea) and its wild relatives based on randomly amplified polymorphic DNA and organelle-specific STS. *Heredity*. 78: 603–611.

Wachira FN, Tanaka J, Takeda Y (2001) Genetic variation and differentiation in tea (*Camellia sinensis*) germplasm revealed by RAPD and AFLP variation. *J Hortic Sci Biotechnol*. 76: 557–563.

Wachira FN, Waugh R, Hackett CA, Powell W (1995) Detection of genetic diversity in tea (*Camellia sinensis*) using RAPD markers. *Genome*. 38: 201–210.

Wallace LE (2002) Examining the effects of fragmentation on genetic variation in *Platanthera leucophaea* (Orchidaceae): inferences from allozyme and random amplified polymorphic DNA markers. *Plant Species Biol*.17: 37–49.

Weisburger, J.H (I997). Tea and health: a historical perspective. *cancer letters*, 114, 315-317.

Wickremasinghe RL (1979) Tea. In: Mark EM, Steward GF (eds) Advances in food research, Academic press, New York. Vol.24. pp 229–286.

Wight W (1962) Tea classification revised. *Curr Sci*. 31: 298–299.

Yamaguchi S, Matsumoto S, Tanaka J (1999) Genetic dispersal of tea plant. In: Jain N K, ed. Global Advances in Tea Science. New Delhi, Aravali Books International (P) Ltd., 413–426.

Yao MZ, Chen L, Liang YR (2008) Genetic diversity among tea cultivars from China, Japan and Kenya revealed by ISSR markers and its implication for parental selection in tea breeding programs. *Plant Breed*. 127: 166–172.

Yao MZ, Huang HT, Yu JZ, Chen L (2005) Analysis of applicability of ISSR on molecular identification and relationship investigation of tea cultivars. *J Tea Sci*. 25 (2):153–157

Zhao LP, Liu Z, Chen EL, Yao EMZ, Wang EXC (2007) Generation and characterization of 24 novel EST derived microsatellites from tea plant (*Camellia sinensis*) and cross-species amplification in its closely related species and varieties. *Conserv Genet*. 5:1327-1331.

Zhuang, X (1995) Development of world tea industry. In: *100 years of world tea industry* (edited by Q. Chen & X Zhuang). pp 1-52. Shangai, China: Shangai Science & Technology Education press.

4

PLANT PROPAGATION

R.K. Sud

CSIR-Institute of Himalayan Bioresource Technology,
Palampur-176 061, H.P., India. E-mail: rksud@ihbt.res.in

INTRODUCTION

The objective of plant propagation is to produce healthy and well grown plants in the shortest possible time with a minimum of losses. The propagation stage is the most vulnerable growth stage in the growth cycle of any plant, and any adverse factor, which operates during the propagation stage, will reduce quality or result in losses (Hort1001 2004).

Raising plants through seeds and nodal cuttings are the common methods of tea propagation followed throughout the tea growing regions. Propagation through seeds is the oldest and easiest way of plant multiplication. During old times, when tea was introduced from China to India and other countries, seeds were imported to raise the plants. Thus, the initial method of tea propagation was only through seed. Regarding vegetative propagation, literature reveals that earliest method of plant multiplication practiced in China and Japan were layering, marcots and wood cuttings (Barua 1989).

An account of progress in the vegetative propagation of tea in Indonesia since 1902 is available in literature (Anonymous 1973). It has been mentioned that for tea nursery initially the callusing beds were prepared under shade of *Tithonia diversifolia*, and the individual nursery beds were shaded with ferns or cut *Imperata cylindrica* grass, and later on plastic sheets were introduced for covering beds. These beds were made under shade trees.

Vegetative propagation of tea with green single leaf nodal cutting was demonstrated by Tunstall (1931) for the first time in Tocklai Experimental Station, Assam and subsequently in Sri Lanka, Java and the former USSR (Barua 1989), but success rate at that time was very low. In south India, the nursery techniques were standardized in 1955 (Hudson *et al.* 2002). The present methods of tea propagations with higher success rate are result of perfection of each aspect of nursery management achieved in due course of time.

In India, tea is cultivated in 5,78,458 ha area, the second highest tea acreage in the world after China (Tea Board of India 2012). It is a common recommendation for tea estates to uproot and replant 2-4% of the plantation with high yielding high quality promising cultivars for sustaining the garden productivity. For a garden with an average population of 8,000 bushes/ha, there would be an annual need of 160-320 nursery plants for replanting. Total annual requirement of nursery plants for the India would be around 86 million numbers for replanting at a minimum rate of 2% of area each year. Besides, the plants are required for establishing new plantations and for infilling and interplanting in vacant patches and low density plantation, respectively particularly in rejuvenated / deep pruned sections.

Guidelines have been laid for new plantation. A tea garden should have both clones as well as seed raised plants in equal proportion. A single clone should not have area more than 10 %. There should be 5-6 clones in each garden. The cultivars should be selected based on their suitability in the locality, need for balancing production and quality etc. These guidelines have been formulated for stable productivity, avoiding narrowing of genetic base and vulnerability of single variety (Bezbaruah and Singh 1992). Therefore, there is continuous need of new cultivars for replanting, new planting and vacancy infilling. Both clonal as well as seedlings are needed to be multiplied to fulfill the requirement.

In north-east and south India, replanting and new plantation have been a common feature of tea industry. Consequently, many garden-owned as well as private tea nurseries exist in these regions.

In Himachal Pradesh, the tea productivity is far below the national average or even that of its counterpart Darjeeling with identical agroclimatic conditions. The long term solution to this problem is new planting (CSIR Complex 1990). So far no serious efforts have been made by tea growers to replant their gardens. The reason could be that china hybrid tea is cultivated in the state, which regains its productivity after rejuvenation prune, and being a multistemmed plant, it is capable to tolerate the shock of rejuvenation pruning. However, the new plants are continuously required for vacancy infilling and interplanting in pruned tea sections. Limited demand for new plantations has also been coming. These demands are mostly met by tea wing of the State Agriculture Department who is raising and supplying seedlings plants at a subsidized price. Tea Federation in association with Tea Board also supplied biclonal stocks and clonal plants during 1997-99 and thereafter discontinued the activity. CSIR Institute of Himalayan Bioresource Technology, Palampur has created awareness about quality planting material amongst tea growers. The Institute supplied tea plants and propagation material (cuttings) of suitable cultivars to the state as well as to Uttarakhand state (CSIR Complex 1990).

REQUIREMENTS

Tea is a perennial woody plant. Raising a full grown nursery plant ready for transplanting in the field is a multistep long duration practice that takes over a year time. Success of tea nursery depends on many factors, management being the key factor. Like any other perennial or horticulture crop, weather, soil and water are major requirements for propagation of tea.

Weather

Growth of nursery tea plant is better in hot and humid climate. According to Burt (2004) a temperature range of 30-35°C is optimum and should never exceed 35°C; the optimum night temperature for nursery plants is 15°C and should not be lower than 5°C; and light intensity should be in the range of 25,000 to 40,000 lux.

In Georgia, Patarava (1987) has studied low and high temperature resistance in 1-2 years tea plants, raised from seeds or cuttings and grown at 20000-lux light intensity at 0 to 50°C and 0 to -15°C. He distinguished five temperature zones for tea plants *viz.*, basal (25 to 30°C), 2 hardening temperatures (5 to 15°C and 35 to 40°C) and 2 lethal temperatures (-3 to -15°C and 45 to 55°C).

Frost is very harmful to tea plants. Relative humidity and tea shoot growth are very closely associated. Relative humidity more than 90% ensures better rate of plant establishment. High and well-distributed rainfall is important for maintenance of soil moisture and atmospheric humidity, both contributing towards better plant growth.

In tea growing areas of Kangra total annual rainfall is 250 cm but 70-80% of the rainfall is received in mid-June to mid-September, thus drought like conditions prevail in the remaining months. Atmospheric humidity also varies considerably between 80-30% based on rainfall pattern. The temperature range is suitable and light intensity can be managed as per requirement.

Soil

Physical, chemical and biological properties of the soil affect establishment and growth of nursery tea plant. The soil media possessing these properties in optimal may vary from clone to clone. Preliminary trials are necessary to establish the best rooting medium for an individual clone in a particular location.

Soil should neither be too clayey or too sandy. Clayey soil causes over callusing and poor rooting whereas sandy soil demands frequent irrigation. In sandy soil, plant roots grow deep and get broken during shifting. If compost is mixed with sand in equal proportions better success of rooted cuttings can be obtained (Chasandoerjat 1970). Sandy loam texture is generally preferred for better success rate. It maintains the proper balance of moisture and air. Top soil collected from jungle, bamboo-*bari*, thatch or any vegetation is suitable for nursery particularly for raising seedlings in sleeves. For vegetative propagation, pure forest soil may not be suitable. UPASI has recommended two separate media in a sleeve for better success (i) rooting medium consisting of subsoil and sand in equal proportion (sandy loam texture) and (ii) growing medium, made of topsoil or jungle soil and sand in 3:1 ratio (clay loam texture) (Hudson *et al.* 2002). The poor subsoil initiates better root growth whereas topsoil promotes shoot growth. In Central Africa, it was observed that tea cuttings planted in bags containing subsoil in upper layer and topsoil in the lower layer had a mean survival percentage of 92%, compared with 80% for those with subsoil alone (Kayange 1972). Only forest topsoil may give poor growth, but when mixed with a large volume of subsoil and covered by a subsoil cap it proved satisfactory (Green 1973). However, the source of the subsoil (depth and site) is also important. Besides, bulk availability of topsoil is also a great problem. Use of subsoil with little modifications or in combination with the topsoil is providing some solution. Bordoloi and Bezbaruah (1983) in Assam found better survival of tea cuttings of three clones after one year in subsoil amended with single super phosphate at 10 g/sleeve (size 15 x 20 cm), as compared to topsoil alone. But subsoil alone resulted in poor shoot and root dry weight. Gabrichidze (1990) also recorded outstanding results by enriching red subsoil with phosphorus and potash, and subsequently top dressing with nitrogen. Kayange (1989) in Central Africa observed that the compost, prepared from tea shoots and leaves, if mixed with subsoil in 3:1, 2:1 or 1:1 gives better results in terms of plant growth. But plant growth on subsoil alone was poorer.

Soil pH is the most important chemical parameter for suitability of nursery soils. The pH should be within 4.5-5.5 range. Little amendment in pH can be done with the application of aluminium sulphate or lime/dolomite. In case soil is strongly acidic (pH is below 4.5), slaked lime @ 2 t/ha should be applied; on the other hand if soil is sub-acidic (pH more than 5.5), aluminium sulphate 2-4% solution needs to be applied @ 1 litre/cubic feet soil (Tocklai 2004). Nursery soil with pH of above 5.5 would prove detrimental for survival of chinery clone cuttings (CSIR Complex 1989). Application of aluminium sulphate @ 600 g m^{-3} of nursery soil favoured root and shoot growth of the cuttings owing to better acidifying effect on soil whereas elemental sulphur @ 300 g m^{-3} was the second best (Hamid *et al.* 2006).

The electrical conductivity, a measure of total soluble salt content, of the soil should be less than 0.05 dS/m, and if more, it can be reduced by leaching out of the salts with water washing (Hudson *et al.* 2002). Availability of plant nutrients particularly nitrogen, phosphorus and potash and organic matter should be high for supporting plant growth.

Biological activity of soil is equally important for raising nursery plants. The soil should contain sufficient amount of organic matter. Plants grown in tea compost had higher root and shoot growth of nursery plants than on topsoil (Kayange 1989). Many researchers have been studying effect of different microbial inoculations on the performance of tea plants. Deori *et al.* (1998) in Assam planted tea cuttings in soil media to which finely ground roots of the weed species *Ageratum conyzoides, Mimosa invisa* and *Borreria hispida*, inoculated with VAM with and without phosphate fertilizer had been added. After 9 months, it was observed that mycorrhizae enhanced the overall growth and dry weight of the tea shoots compared with controls. The effect was more pronounced in cuttings which were supplied rock phosphate. Tea root dry weight was higher in pots treated with *A. conyzoides* roots and rock phosphate. In Himalayan region, *Bacillus subtilis* and *Pseudomonan corrugate* were identified as promising bacterial inoculations for tea plants, which exhibit antifungal and phosphate solublising activity and are capable of surviving at low temperature (Trivedi *et al.* 2003).

Nematode infection in tea nursery soil is a commonly observed problem in all tea growing regions. As a thumb rule, tea nursery soil should not have nematode more than 6 numbers per 10 g soil.

Rao (1976) has discussed in detail nematode infestation problem in tea nurseries and concluded that root-knot nematode (*Meloidogyne* spp.) was the most important parasite in India. Regarding control measure, soil heat treatment and the use of chemicals both before and after planting were recommended. Neog and Bora (1999a) tried neem cake, mustard cake, decaffeinated tea waste and carbofuran 3G against *Meloidogyne incognita* in a tea nursery in Assam. All the doses were effective in respect of increasing plant growth characters and reducing the number of galls, egg masses and final nematode populations as compared to the control. Carbofuran was recorded as the most effective followed by neem cake and mustard cake at 3% (w/w). In other study they (1999b) observed that the physical treatments of soil *viz.*, soil steam sterilization at 15 lb psi for half an hour, sun drying of soil for 10 days, and soil heat treatment in hot pan for 3 hours at 60°C more effective than Carbofuran 3G. No nematode infection was recorded in the steam sterilization treatment. They (1999c) had also screened some of the clones against *Meloidogyne incognita* in Assam.

In Sri Lankan tea nurseries root-lesion nematode, *Pratylenchus loosi* has been the main concern. Its heavy infestation results in severe stunting of nursery plants (Gnanapragasam and Jebamalai 1982). The infested nursery soils when treated with dazomet 98% G and methyl bromide were free of nematodes even after 12 months. Growth of the nursery plants was far better in dazomet treated soil (Sivapalan *et al.* 1980, Gnanapragasam 1982 & 1983). Besides, Fenamiphos (Nemacur-P) 5% G at of 7 g/plant was observed to significantly suppress infestation up to as much as 60% and stimulated plant growth (Gnanapragasam *et al.* 1987).

Khan *et al.* (2003) in Bangladesh recorded 10 genera of plant parasitic nematodes associated with root zone soils of tea seedlings. Nine of them were identified as *Aphelenchoides* sp., *Helicotylenchus dihystera, Hemicriconemoides mangiferae, Hoplolaimus indicus* (*Basirolaimus indicus*), *Pratylenchus loosi, Meloidogyne* spp.,

Tylenchorhynchus annulatus, T. davainei and *Zygotylenchus*. The occurrence of *P. loosi* and *Meloidogyne* spp. was the highest. Their population was highest in the month of May. Population density of the nematodes was positively correlated with soil organic matter and silt content and monthly rainfall recorded during the months of the study.

In Kangra, during 1989-90, stunted plant growth of nursery tea plants at CSIR-IHBT nursery was observed particularly in case of assamica type plants, which was suspected to be due to nematode infection. Application of Furadon @ 2 g per plant to such plants was found effective.

Irrigation

Availability of regular irrigation supply is a key factor for any nursery. A period of 10-18 months, depending on the agro-climatic conditions, is required for raising a plant from seed or cutting. During this period, frequent irrigation is needed for growth of the plants. The water requirement of nursery plant increases with growth of plant because of increased foliage density and root volume. There is no information available in the literature on the water requirement in quantitative terms. The water requirement obviously will vary in different tea regions in accordance with the weather conditions.

Close proximity of nursery to perennial source of irrigation ensures good success provided other requirements are met. But in tea, many times over watering becomes the cause of failure particularly with flooding, because the plant is highly sensitive to water logging and does not tolerate very wet soil conditions. High and well distributed rainfall reduces the burden of frequent irrigation. In case of water scarcity, the irrigation requirement can be minimized by checking evaporation losses using mulching of nursery beds. Use of sprinklers or foggers or auto-misting heavily economizes the water use and results in excellent performance in terms of plant growth by way of increasing humidity, but may hike the cost of their raising particularly in the initial years.

In Kangra, the main source of irrigation is *kuhls*, the diverted flow of water through small channels from the streams and small rivulets. With population pressure the available water is being prioritized for drinking and other uses, and the irrigation *kuhls* are getting dried particularly in summer season when the demand of water is at peak. Digging of shallow bores for water supply mainly for drinking by the state government is now common practice in water scarcity belts. Such bores have also been dug for horticulture crops and can also be practiced for tea nurseries. This may hike the cost in the initially years, but in long run it will ensure better success.

NURSERY STRUCTURE

Location

Tea nursery should be located at lower elevation in a warm-sheltered position for early development of the plants. Availability of perennial source of irrigation should be the main consideration. The place should be easily accessible for supervision, should be sunny and bright and located on raised or well drained land (CSIR Complex Palampur 1987). It should not be prone to hail or high wind velocity. A commercial unit should be well connected to the road for transportation of the soil, plants and farm inputs. Selection of low-lying site is also a practice in tea plantations for its close proximity to water. But some of these sites may have high water table, which restricts or even inhibits development of roots in plants in the nursery beds. Besides, low lying areas frequently form traps for cold air, a condition which is not conducive of plant growth (Kehl 1964). Availability of suitable soil for nursery in the nearby area is another important consideration.

After selection of the site, a deep drain is dug around the purposed nursery area for efficient removal of surface and ground water in the monsoon. A shelterbelt with evergreen trees like *Acacia, Casica, Grevellia* etc. is planted to create a favourable microclimate conducive for better plant growth and protection against wind and storm (Hajra 2001).

Types of nursery structure

Nursery structure is established with a view to create a microclimate conducive for better plant growth. The temperature, humidity and shade are optimized within the nursery structure. Based on the height of shade, structures are of three types, low shade, high shade and over-head shade. Overhead shade structure is erected over the beds to cut down 50% of light. Shading can be provided by locally available thatch or straw spreading, bamboo rith, commercial plastic nets of specified shade or coir mat over the structure.

Low shade: The shade is only 20 cm high from the ground level in case of rooting beds or from the top of sleeve. Very dense shade is provided with bamboo rith mat or straw mats etc. For intercultural operations both sides are lifted, alternatively and there after kept at original position. It is used in areas of high relative humidity. This type of shade was not found suitable in Kangra condition as it lead to very high mortality of the planted cuttings.

***High Shade*:** The shade is provided over the polythene tunneling in the rooting or sleeve staked beds. Height of one side of the shade structure is kept at 90-100 cm and of other at 125 cm. The top is covered with heavy shady material such as bamboo rith or coir mat in slanting position. All three sides are kept covered. This is a low cost temporary structure and is erected as per convenience. The structure was tried in Kangra and results were satisfactory.

***Overhead Shade*:** Shade is provided about 1.8 m above the ground level. The structure may be permanent or temporary and movable depending on the nature of nursery. Angle or bamboo poles 2.4 m tall are used to erect the structure of which 0.6 m is buried in the soil. The poles are placed at a spacing of 4.5 m x 2.7 m and 6 mm thick MS rod or bamboo strips are provided at the top of poles to support the shooting material. Shading material could be rice-straw (commonly used in Assam), coir mat (used in south India) or plastic net, depending on the availability of the material, which may allow only 50% or less light to pass through it in a diffused manner. Coir mats may be given a through spray with 1% copper sulphate solution before using them, for enhancing durability.

An experiment in Kenya involving seven shading regimes (30, 40, 50, 60, 70, 80 and 90%) using UV-stabilized black non-biodegradable net shade material indicated that black artificial shading materials with 40-90% shading capacity are suitable for use in tea nurseries (Wachira and Kamunya 2004).

High shade structure is convenient for the worker to carryout the operation inside the nursery (CSIR Complex Palampur 1987, IHBT 1997). Permanent shade is more economical in long run. In Kangra, this type of tea nurseries were found more convenient and effective for clonal multiplication, perhaps because in comparison to other structures, it creates better growth conditions owing to favourbale temperature and humidity.

Nursery beds and sleeves

***Rooting beds*:** In case of vegetative propagation, tea cuttings are planted in a special prepared bed for rooting. This bed is called rooting bed. Microclimate of the bed is modified to encourage easy initiation of the roots and their further growth and development from the cut end of tea cutting inserted in the soil. For this purpose, the beds are prepared under the shade structure and are covered with polythene tent or polytunnel to maintain the desirable range of temperature and humidity. In regions where atmospheric humidity is very high, say more than 90%, the polythene tent is not required. The soil of the bed is also modified to maintain its desired standard of physical, chemical and biological properties to suit root growth as described in the section 'Soil'.

Presently, beds of any convenient length with width of 1.5 m are prepared in rows keeping a distance of 0.6 m between two neighboring beds. A 15 cm deep and 15 cm wide drain is dug on the upper side of the bed keeping the excavated soil in the bed. The whole area within the beds is dug and leveled maintaining fine tilth. The beds are raised about 15 cm from the ground and final width of 120 cm is kept with slight camber in the centre of the bed surface to avoid water logging. Sufficient sand is added in the bed to give sandy loam texture. About 500 g super phosphate/m^3 of soil is mixed in bed and soil is made compact. The beds are kept ready one month prior to planting and are covered with mulch for controlling weed. Alternatively, pre-emergence herbicide atrazine or simazine @ 5 ml/L may be sprayed after preparation of bed.

***Callusing beds*:** Sometimes, instead of rooting of the cutting, only callusing is required to be done in the beds. In that case very high proportion of sand is mixed in the bed. Tea cuttings are planted at very close distance and are uprooted after the development of the callus. Callused cuttings are better for transportation to distant places than fresh cuttings because of less mortality, and sleeved plants because of low volume. Callusing takes about 3-4 weeks. It may be judged by complete healing of the of the cut end of cutting with mother leaf above the bed. The callused cuttings are transplanted in the sleeves and kept under polytunnel/plastic tent in the nursery for their further growth.

***Seedling beds*:** The beds are demarcated and prepared in the same way as mentioned in 3.3.1. Sandy loam soil is the best for raising seedlings on orthodox beds. Wherever possible, the virgin soil should be used. In absence of virgin soil, uprooted tea area or area with poor soil may be used after the site has been put under green crops like *Crotolaria anagyroides, Mimosa invisa* or grasses like *Tripsecum* *laxum* (Guatemala), Cymbopogon (*Citronella*) etc. for a period of about two years (Tocklai 2004).

Staking beds: These beds are made for staking the sleeves filled with soil. The dimensions and layout of the beds are the same as described in the rooting beds. While preparing the beds, the surface of the bed is cheeled, light hoeing is done and soil is compacted, giving a slight camber to the bed surface. For holding of the stacked sleeves, the borders of the beds are provided with bamboo or iron wire railings. The beds are kept moist but water logging is strictly avoided.

Sleeves

Plastic sleeves of 150 gauge thickness are used. Both ends of the sleeves are kept open to avoid water logging for tea nursery. Size of sleeve is 10 cm lay flat x 25-30 cm length for the cuttings and 12-15 cm layflat x 20-25 cm length for the seedlings. Soil with desirable physical, chemical and biological properties is used for filling the sleeves. Sleeve filling should commence 6-8 weeks before propagation. The soil should be kept in shade and not be allowed to become dry. It should be sieved through wire mesh number 4 to separate pebbles, wood pieces etc. Super phosphate at 500 g/cubic meter soil is thoroughly mixed. While filling soil into sleeves, care is taken to give gentle and uniform compaction with a blunt wooden club. The filled sleeves should not bulge or have gap between the layers of soil. The filled sleeves are stacked on the beds and watered. Mulching of the sleeves before sowing helps maintain the moisture, soil microbial activity and weed control. Alternatively, pre-emergence herbicides, *viz.,* atrazine or simazine or oxyflorfen or pendimethalin may be used for weed control, at least 15 days before sowing/planting.

VEGETATIVE PROPAGATION

Management of mother bushes

Mother bushes are the bushes selected or planted, from where tea cuttings are collected for vegetative propagation. The progenies raised through vegetative propagation of a single mother bush form a clone. A plot of the recommended mother bushes is called multiplication plot. The multiplication plot is maintained separately for each clone. This plot is laid out in shade free area (Anonymous 2004) to induce hardiness in the cuttings. A plot where the new planting material is kept and later on multiplied for establishing multiplication plot is known as nucleus plot. Nucleus plot is established either by new planting or grafting on the existing tea bushes at such a convenient place where constant supervision is possible. The bushes of nucleus plot serve as mother bushes for multiplication plot and are maintained as exhibition plot after the establishment of the multiplication plot. The size of multiplication plots is based on the requirement or demand of cuttings of a particular clone.

(a) (b) (c)

Sleeve filling and arranging, (a) collection and sieving of top soil, (b) filling of sleeves and (c) arranging sleeves in staking beds

Fig. 4.1. Sleeve filling and arranging in the nursery bed

(a) (b) (c)

(d) (e) (f)

Raising of clonal nursery tea plants through vegetative propagation, (a) collection of top and sub soil for sleeve filling, (b) soil filled sleeves kept in the staking beds with provision for erecting polytunnel, (c) collection of shoots for making cuttings, (d) cutting making, (e) covering of the bed after planting of cutting under overhead shade, and (f) raised plants

Fig. 4.2. Vegetative tea nursery

The rooting and growth of nursery plant is clone specific (Ali *et al.* 1969). The rooting performance of tea cuttings having one-quarter, one-half or three-quarters of a leaf was compared with that of cuttings bearing 1 or 2 leaves. Single-leafed cuttings gave the best results in respect of survival, growth and the sprouting of axillary buds, and cuttings with three-quarters of a leaf were the next best, whereas strike rate of one-quarter or one-half leaf cutting was poor (Ali *et al.* 1969).

Time of cutting collection is important for their development into plants. A report from Brazil, indicates that the ideal time to cutting collection was winter, whereas in summer and autumn mortality of the planted cuttings was extremely high (Lima *et al.* 2011). Number of nodes (1-3) of cuttings has slight differences for establishment. The effects of cutting the leaves were not consistent, but root growth was adversely affected. However, field survival of plants raised from 3 nodes had 80% survival, compared with only 30% or less of those with single nodes (Anonymous 1972).

Irradiation of single leaf cuttings with 1000-2000 rad/hour stimulated callus formation and root and general growth in the cuttings (Eliadze 1973).

Experiments have been conducted to evaluate the potential of plucking shoots as an alternative to conventional green stem cuttings for large-scale clonal production in different parts of world. Nyirenda and Mphangwe (1998) in Central Africa found that plucked shoots rooted very well under a polyethylene tunnel, especially when the apical buds had been removed. Misting inside the polyethylene tunnels helped to maintain the turgidity of shoots. Later on Chapotoka *et al.* (2001) also in Central Africa compared the shoot and root developments of clonal tea plants raised from plucking shoots with those raised from conventional mature green stem cuttings. Seven months old plants raised from plucking shoots already exhibited 2 branches per plant, while plants of the same age raised from conventional cuttings had none. Plants from plucking shoots were generally taller, had more branches, and greater shoot and root dry weights compared to those from conventional cuttings. Plants propagated from plucking shoots survived better in the field compared to those from stem cuttings.

Bagrodia and Varghese (2000) in Assam also raised plants with freshly harvested shoots of three leaves and a bud size. They planted cuttings of three different clones in the rooting beds during September and October. The cuttings rooted after 8 weeks. The percentage of rooting differed with clones. These raised plants were transplanted in the field after completing their nursery life. Initial performance of these plants in terms of vigour of growth was satisfactory.

Sanikidze (1972) in Georgia tried softwood cuttings for propagation. These cuttings rooted well in the nursery but there were considerable mortality at transplanting. Therefore, planting of cuttings directly in pots was recommended to reduce transplanting losses.

Planting of cutting: Cuttings are initially planted in the rooting beds. The beds are well compacted and thoroughly irrigated one day before planting. The cuttings are planted at a spacing of 5 cm x 5 cm for accommodating maximum number of cuttings. However, the spacing may be increased for bigger size of mother leaf cuttings. The cuttings are inserted in the bed by holding them at node between thumb and index finger and push the stem into the soil in one motion until the base of leaf remains slightly above the ground. In order to avoid injury to the stem or bark during planting, holes at requisite spacing in the bed can be made with a dibber. The holes should be shorter and smaller than the stem of cutting to prevent loosening of the cutting. The cutting should be inserted in such an angle that the mother leaf stands at upright position and does not touch adjoining cuttings. The cut ends of the cuttings should not be touched with fingers. Also, the mother leaf should not sag on the soil. The big or drooping leaves should be cut to half its length to prevent its sagging. Watering should not be done immediately after planting otherwise the cuttings will be disturbed. After completion of the planting of the cuttings, the bed is covered with polythene sheet and sealed airtight, on all the sides by covering with soil, giving a tunnel shape. The polythene sheet should not be disturbed unless required for intercultural operations and spray of pesticides. It should again be sealed after watering.

After rooting, the cuttings are transplanted into the sleeves with minimum damage to the roots and proper compaction of soil. The size of sleeve or in other words volume of soil affects the growth of tea plant. Rooted cuttings of three different clones after transplanting into well-drained bags containing 2, 5 or 8 kg of dry soil exhibited better growth in terms of root and shoot dry weight in the largest bags (Kayange 1986). Similarly, nursery plants grown in bigger bags (holding 5 or 8 kg of dry soil) became better established than those grown in smaller bags (holding 2 kg of dry soil) (Malenga 1997).

The cuttings can directly be planted in the sleeve. In this case, the cuttings are inserted in the center of the sleeve. Planting in the sleeves is better for the better rooting clones. Direct planting in sleeves cuts down the cost of rooting bed and further transplanting.

McCubbin and Steenkamp (1998) planted single node cuttings initially in shallow (35 ml volume) or deep (56 ml volume) pot trays and later on after 5 months transplanted them to an irrigated field. Plants from the deeper pot trays established better due to more space for root growth, than those from the shallow trays. They concluded that tea field establishment by planting out directly from pot trays may be a viable proposition, enabling reduction of nursery costs. However, management of the field into which the plants are placed is critical.

Fig. 4.3. Development of shoots in the planted cuttings

It has been observed that plants grown in the centre of the bed were taller and more branched than those on the periphery; moreover, there were more vacancies in the latter position (Ndamugoba 1977). Also, the ratio of shoot dry weight:root dry weight for all clonal nursery plants remains in the proportion of 2:1 (Kayange 1986).

Post planting care: The nursery beds are kept weed free. When the cuttings have produced 2-3 leaves on the shoots, foliar spray of the approved insecticides is done to protect them from insect damage. The sprays are repeated at monthly intervals. Copper fungicide may be mixed with the insecticide, if required particularly against blister blight. The beds are to be irrigated when top soil tends to become dry. After removal of the polythene sheet, application of fertilizers, growth regulators etc. may be done as per requirement.

It is a common observation that in spite of an initial high sprouting ability in the nursery, some clones have low survival rates. This observation implies that sprouting is not invariably accompanied by rooting of cuttings in such clones. It was reported from Nigeria that the outstanding performances of most clones at the nursery stage were not maintained in the field with respect to green leaf production (Okelana and Esan 1994).

Application of fertilizers and growth regulators

Response of phosphate for development and growth of the nursery plants is a common observation in different countries. In Georgia, phosphate at 0.15 g/kg substrate during rooting in the beds and subsequent application at 0.2 g and potash at 0.1 g/kg soil with nitrogen at 0.15 g applied 3 times as a top dressing after transplanting in sleeves exhibited the best plant growth and development (Chanukvadze 1978). Aiyadurai and Sivasubramaniam (1977) in Sri Lanka observed that tea plants receiving the soil applications of triple super phosphate or rock phosphate had higher dry matter contents of the roots. Ali and Mannan (1974) in Bangladesh found that tea cuttings grown in sandy loam topsoil when treated with triple super phosphate, showed highest shoot and root growth after one year.

Hamid *et al.* (2000) in Pakistan studied the effect of fertilizer mixture (ammonium sulphate and potassium sulphate at 3.5:1.0 ratio) on growth of tea cuttings in the nursery. The mixture significantly increased the soil acidity as well as root growth, plant height and number of leaves per plant. Ruan *et al.* (2000) had also studied the effect of nitrogen form and phosphorus source on the growth, nutrient uptake and rhizosphere soil property of *Camellia sinensis* L.

Application of zinc to nursery plants, through soil or foliar, had also been reported to improve the plant growth. Ranganathan (1980) found that soil application of zinc sulphate with fertilizer mixture reduced premature leaf fall, flowering and the incidence of zinc deficiency in tea nurseries and improved plant growth. Response of zinc application has also been reported by Hamid *et al.* (2000).

In Sri Lanka, Kulasegaram and Kathiravetpillai (1972) studied the effect of fertilizer (NPKMg mixture) on growth and apical dominance using free growing nursery plants of vigorous clones. Application of fertilizer showed early bud-break and increase in the number and duration of active phases, with a reduction of the dormant phases. These resulted higher dry matter accumulation and more lateral shoot growth than with lower levels. Leaf production on both main and side shoots increased at the higher levels of nutrition.

Tea seedlings raised in association with AM (arbuscular mycorrhizal) fungi generally showed an increased growth and nutritional status over those grown in the absence of AM fungi. The extent of growth and nutritional status enhanced by AM fungi and the mycorrhizal dependency of the host vary with the species of AM fungi. Seedlings inoculated with *Scutellospora calospora* had greater biomass and seedling quality than other mycorrhizal seedlings (Karthikeyan *et al.* 2005).

Influence of growth regulators on nursery plants is inconsistent and clonal specific. In Assam, several growth regulators were tried on one-year old nursery plants of different clones and clonal specific response of zinc sulphate + urea, Cytozyme (a protein digest extract containing growth regulators, protein complexes, enzymes and micronutrients) and tricontanol (a biologically active long-chain aliphatic alcohol) were recorded (Barbora *et al.* 1992). Later, effect of tricontanol in combination with other nutrients was studied in Assam (Baruah *et al.* 1997). When sprayed in combination with zinc sulphate, urea and boric acid, tricontanol increased plant height, number of leaves, stem diameter, leaf area, lateral growth, dry weight of shoot and total dry weight in both free growing and tipped nursery plants. Chlorophyll content of leaves, nitrogen content of shoots, and starch content of roots were increased when tricontanol was sprayed in combination with zinc sulphate and urea. Volume of root increased with tricontanol when applied in combination with urea and boric acid. Tricontanol in combination with other nutrients improved the growth and vigour of nursery plants and alleviated the deleterious effects of tipping.

Investigations of Liu *et al.* (2007) with 5 species of arbuscular mycorrhizal fungi suggest that *Glomus fasciculatum* has high affinity with the tea plants; it resulted in improving the rooting of cuttings and the nursery raising period by growth promotion, and inducing an increase in GABA (gamma-amino butyric acid) content in leaves which leads to high quality. Singh *et al.* (2010) reported significant increase in most of the growth parameters (root/shoot length, dry weight and root/shoot ratio) in AM fungi treated plants, which also showed better tea quality parameters than non-inoculated plants. Significant increase in amino acids and total protein content up to 31% and 100%, over control, were recorded, respectively. Similarly, increase in total polyphenols (15%) and caffeine content (34%) over control was also significant. Enhanced sugar content (maximum 16% in total soluble sugars, maximum 45% in reducing sugars, and maximum 69% in non-reducing sugars) was also recorded.

Use of hormones

Use of hormone for efficient rooting of the cuttings has always been a controversial issue because of the inconsistency in results.

Scarborough (1973) gave 20 hours soak treatment to tea cuttings in IBA at 50 to 150 ppm, and observed 100% rooting after 4 months with the highest rate, whereas soaking in solutions of IAA, IPA, L-NAA or a mixture of IBA and L-NAA, or quick-dipping cuttings in 50% alcohol plus these auxins proved less effective. It was also observed that auxin treatments increased root weight but suppressed shoot development. Sanikidze and Sardzhveladze (1973) in Georgia dipped tea cuttings, taken from different parts of the plant, in IBA or NAA at 15, 30 or 50 mg/l for 8, 16 or 24 hours and observed that average percentage of rooting among the hormone treated cuttings was appreciably lower than among the untreated cuttings. According to them this was attributed to sufficient reserves of endogenous growth stimulators in the physiologically mature shoots and to phytotoxic effect of IBA and NAA. In Pakistan, Noor *et al.* (2002) observed that application of IBA at 8000 ppm gave the best results with regard to percentage rooting, and root and shoot growth, during a period of 6 months from July-January. But when applied in combination with NAA, IBA had an antagonistic effect beside cost factor. Interestingly, Nyirenda and Mphangwe (1998) had observed better results with respect to rooting of young tea shoots when sugar solution was used instead of IBA. The influence of IBA on the number and length of roots in cuttings particularly when collected in winter season was also observed in Brazil (Lima *et al.* 2011).

Rout (2006) also reported the positive response of IBA on rooting as compared to IAA and NAA. At 75 ppm, IBA gave maximum percentage of rooting, number of roots and root length. Further it was reported that IAA-oxidase activity of IBA-treated cuttings increased slightly as compared to control. The activity was found to decrease

Bhattacharya *et al.* 2004). With recent popularity of the improved polyclonal and biclonal seed varieties, interest in these seeds is ever increasing.

BICLONAL AND POLYCLONAL SEEDS

Despite a lot of care and scientific management, the tea productivity fails to increase beyond a certain limit. It is then, that the replanting of existing areas with improved planting materials becomes an absolute necessity. The use of better biclonal and polyclonal seed stocks is an important option for improving tea productivity. While the seeds produced from the crosses of two most superior cultivars are 'biclonal hybrids', the ones obtained from a series of crosses with a number of superior cultivars are the 'polyclonal seeds' (Govindasamy 1995, Bezbaruah 1981). In order to produce such seeds and conduct such hybridizations successfully, tea plants are especially kept aside and allowed to grow into seed bearing trees that are maintained in '*seed baris*'.

Seed *baris* are specially designated areas where tea seed bearing plants are maintained and nurtured. It is a practice in some areas to obtain seeds from trees especially kept aside for this purpose. The seed bearers are allowed to grow in a normal way without following pruning or plucking operations. Irrespective of the region or practice, the seed bearers specifically belong to high *jat*, and are of healthy nature. In a seed *bari*, the selected elite plants are planted in a spacing of 3-4 m x 3-4 m for China type or 5-6 m x 5-6 m for Assam type based on their normal spread; and are usually allowed to grow indefinitely. The capsules grow on the canopy and only the dehisced seeds that drop on to a cleared ground are collected every day.

FLOWERING AND POLLINATION

Flower buds generally develop on seed trees and are directly governed by the phasic shoot growth or in other words the 'flushes' of tea plants (Barua 1970). The flower buds develop from the axils of bud-scales in the lower portion of the bushes. Although at times, they appear to develop from leaf axils, in reality however, they are produced from the axils of bud-scales of short axillary shoots. The scars of caducous bud-scales or cataphylls generally mark the positions of the leaf-less growth of a particular year. The buds remain covered by the bud-scales initially. However, they finally bloom into conspicuous white flowers on extended shoots that arise during the period of active vegetative growth. The whole process from bud to flowers generally takes about 120-150 days (Bezbaruah 1975). Once the flowers bloom, the total flowering period continues for some time but depending on the latitudinal variations, the period of flowering varies considerably in the different regions of the world. Thus while flowering is observed all throughout the year in the regions at or near the equator (Elliot and Whitehead 1926), the period narrows down significantly with every increase in latitude. Distinct variations in the flowering period have also been observed between different cultivars growing in a particular region. The ability to flower early or late may be an individual feature of the plant (Bakhtadze 1931).

Tea shows an appreciable degree of self-sterility but better seed-set is observed only after cross pollination. The tea flowers are reported to be pollinated by the Hymenoptera group of insects comprising of the bees and wasps, and also the Diptera group consisting of insects and small flies with short flying ranges. Although the tea flowers secrete considerable amounts of nectar, the number of insects visiting them is very low. This is because the peak flowering time of tea generally does not coincide with maximum insect activity. Adding to the problem is the fact that the tea pollens are heavy, sticky and tend to remain in clumps. The number of insects available for pollination is also insufficient to bring about complete pollination. Therefore, the percentage of fruit set is poor despite an abundance of flowers. Although pollination in tea is independent of photoperiod, it is highly temperature dependent.

In a detailed study on the floral morphology, pollen biology, stigma receptivity and phenology of flowering and fruit set in the Srilankan clones, Ariyarathna *et al.* (2011) reported unequal reproductive successes among the genotypes. This was attributed to significant variability and asynchrony in pollination and stigma receptivity. However, they observed February and March to be the major flowering period when free crossing between different genotypes occurred. Profuse mature fruits containing an average of two seeds were obtained during February to May, and these matured 8-9 months after pollination. After pollination, more than 1 month was required for zygotic development but early embryonic development continued for 4 months.

In the recent years, the physiological, biochemical and medicinal attributes of the flowers have also drawn considerable attention. While the inhibitory activities of alpha-glucosidases and amylases from flowers was reported by Wang *et al.* (2010), the anti-allergic activities of compounds such as acylated oleanane-type triterpene saponins and sasanquasaponins from the flower buds of a close relative of tea, i.e., *Camellia sasanqua* was demonstrated by Matsuda *et al.* (2010). Later, Joshi *et al.* (2011 a,b) studied the changing fate of flavan-3-ols in the developing flowers of Kangra jat from unopened buds to full bloom stage and also extracted the fresh fruity odour from flowers by super critical fluid extraction. The same year, Han *et al.* (2011) reported a water soluble anti-diabetic agent from the tea flowers.

FRUIT AND SEED DEVELOPMENT

Following pollination and fertilization, the first visible sign of fruit set is indicated by withering of the corolla and the anther lobes from the pedicel after a period of 24 to 48 hours. The style and stigma also undergoes dehiscence, and the ovaries become covered with persistent calyx. Fruit set is however, very poor (Harller 1964) and the fruits or the capsules drop continuously, right from the time of flowering up to the time of fruit maturity. This phenomenon further reduces the number of seed bearing capsules/fruits.

The entire process from flowering to anthesis followed by complete seed development is considerably long and generally takes about 12-18 months (Barua 1970, Bhattacharya *et al.* 2002). While the flower initials develop up to the stage of anthesis in one season, the fertilized ovaries continue to develop until the stage of dehiscence towards the latter part of the following season. The fruits or the capsules are generally light green in colour during the initial phases of growth but dry brown patches appear on the fruit wall at full maturity and this can be taken as an indicator of seed maturity and fruit dehiscence. Faster growth with normal development is observed in fruits with two to three healthy seeds as compared to the ones with only one seed. Mature fruits of about 26.5 mm and about 6.69 g fresh weight usually contain 2-3 healthy seeds.

Seed growth and the time of development vary with genotype (Table 5.1). During the initial stages of seed development, the embryos are tiny, white (less than 0.1 mm) and are enclosed within a brittle white testa. Embryos are completely bathed in transparent liquid endosperm which occupies 98% of the entire volume of seeds during the initial phases. However, with further growth and development of the embryo, the endosperm changes into a jelly like viscous consistency. By the stage of embryo maturation, the endosperm occupies only 50% of the seed volume. During the last phase of seed development, the embryo grows further at the expense of the endosperm and fills up the entire seed volume. At this time, only a thin, fine, papery layer is left attached to the inner wall of the seed coat. The mature seeds are non-endospermic and have a linear thick embryo protected by a thin brittle testa (Figures 5.1 and 5.2).

Table 5.1. Seed development in four different tea types

Clones	Types	Seasons					
		Spring (May-Apr)	**Summer (May-Jun)**	**Monsoon (Jul-Aug)**	**Autumn (Sep-Oct)**	**Winter (Nov - Dec)**	**Late winter (Jan-Feb)**
TG 270/2/B	Assam	Immature	Partially mature	Mature	Aged	-	-
UPASI-9	Assam	Immature	Immature	Mature	Aged	-	-
Tukdah-78	China	Immature	Immature	Mature	Aged	-	-
Kangra *jat*	China	-	Immature	Immature	Partially mature	Mature	Aged

Fig. 5.1. Fruit and seed development in tea

seeds on a tarp or a flat plastic nursery in full sun while keeping them moist with water sprays prior to sowing. Only after the seed coats crack in a day or two, the seeds are sown with their "eyes" (the hilum) in a horizontal position (parallel to the surface of the medium). These are then buried under 1 inch of moistened but coarse vermiculite having good drainage but excellent moisture-holding capacity. The seeds are allowed to germinate under shade (80%) for about 1–2 months.

The newly emerged seedlings are transferred to 30–40% shade only after 3-4 leaves and well established roots develop. These are then nourished with a few granules of slow-release fertilizer such as Nutricote 13-13-13 or 18-6-8 coupled with half strength foliar fertilizer formulated for acid-loving plants with NPK ratio 3:1:1 (e.g., 30-10-10, 21-7-7 and micronutrients). The seedlings (about 1 foot tall) are finally transferred to full sun for a few days and then transplanted in the fields (Sato 2007).

Recent studies have focused on the chemical characterization of the leaves of seedlings. Thus, Chen-Lin *et al.* (2008) showed that the active ingredients such as catechins and purine alkaloids are actually formed in the tender leaves of the seedlings only after germination. However, L-theanine is especially abundant in the roots (13.429 mg/g). Therefore, natural L-theanine can be easily obtained from the tender roots of tea seedlings.

PHYSIOLOGY OF TEA SEEDS

The seeds of tea were first reported to belong to the 'Intermediate' category wherein they were thought to exhibit the characteristic features of both the 'recalcitrant' and 'orthodox' seed types (Bewley and Black 1983, Chin and Roberts 1980). However, in 1993 Berjak and her group showed that tea seeds are actually of 'recalcitrant' type and are highly sensitive to desiccation and chilling. It was further confirmed by a number of other workers that tea seeds are indeed recalcitrant and thus have a high critical moisture content of 28-30% (Barman 1988, Bhattacharjee and Singh 1994, Bhattacharya *et al.* 1996, Bhattacharya *et al.* 2002). While Berjak *et al.* (1993) stated that desiccation up to a minimum moisture level of about -11 MPa or 0.2 to 0.3 g water/g dry matter is suitable, Bhattacharya *et al.* (1996) reported that a water potential of -5 bar is crucial for certain cultivars like TS-450. The effect of rate of dehydration was assessed for embryonic axes from mature tea seeds and the desiccation sensitivity of the axes at different developmental stages was estimated using electrolyte leakage. Axes dried slowly in the seed suffered desiccation damage at higher water contents than rapidly (flash) dried excised axes. Flash drying of isolated axes imposes a stasis on deteriorative reactions that do not occur during slower dehydration process (Berjak *et al.* 1993).

Recently, Chen *et al.* (2011) also showed that desiccation treatment is detrimental to seed germination due to accumulation of H_2O_2 in tea embryos. However, the damage is partly ameliorated by increase in ascorbic acid and catalase. Using physiological and proteomic methods, they further demonstrated that 23 proteins associated with defense response, metabolism and redox status were up-regulated following desiccation. These assuaged desiccation damage to tea seeds partially thereby, resulting in improved germination rates. They recommended induction of desiccation tolerance for improved seed germination.

The moisture content of seeds have also been reported to vary from 30 to 53% (dry weight basis) depending upon the genotype and environmental factors. The rate of loss of moisture was directly related to moisture content reaching an equilibrium point at about 13% (dry weight basis) on the 20th day of storage in open Petri dishes.

STORAGE

Storage of seeds is difficult because the seeds exhibit recalcitrant behaviour and cannot be dried below critical moisture content. Thus several workers attempted different strategies to overcome this problem. While Visser *et al.* (1958) stored their seed samples at 0°C and 100% relative humidity for 7 months, Gomez (1966) stored the seeds at 7°C in polythene bags for more than three months. Bhattacharjee and Singh (1994) further attempted to develop a proper method for bulk storage by treating the seeds with Jalshakti (a starch polymer) @ 0.625 and 1.25% by weight, and storing them in high density polythene lined hessian bags. They attained about 88% germination even after 6 months of storage. They found that the seed viability could also be retained when the seeds are treated with charcoal powder prior to storage in high density polythene lined jute bags. Bhattacharjee *et al.* (1996) also studied the effects of various moist sand preconditioning treatments on extending viability and found that the best treatment was moist sand incubation with PEG at -5 bar water potential plus treatment with 0.05% mercuric chloride. About 70% germination was attained even after seven months of storage. More than 70% germination could also be achieved by storing seeds for nine months to six years by retaining the moisture content at 14% (Hu *et al.* 1993).

In an attempt to cryopreserve the tea seeds, Chaudhury *et al.* (1990) desiccated the seeds to different moisture levels and exposed them to liquid nitrogen. However, they found that the seeds were badly damaged due to ice crystal formation and subsequent membrane damage injury. Usage of cryoprotectants or further rehydration for 24 hours after retrieval from liquid nitrogen did not bring about any improvement in retaining the seed viability. Wesley-Smith *et al.* (1992) used the embryonicaxes of tea for cryopreservation. They studied the interrelationships between water content, rate of freezing to -196°C, thermal properties of water and survival. They used three freezing rates i.e. 10°C/min, 200°C/min and rapid freezing by plunging samples into nitrogen slush. They found that cooling of axes with sufficiently high water content at a freezing rate that prevents the formation of large ice crystals can minimize the damage and enhance the survival rates. Hu *et al.* (1993) further recommended partial drying and post thaw preconditioning to improve the survival and germination of cryo-preserved seeds of tea.

In a recent approach, embryonic axes of tea seeds were encapsulated in MS medium supplemented with 3% Na-alginate, 100 mM $CaCl_2$ and 0.75 M sucrose and desiccated for 1 hours under the laminar air flow prior to cryo-preservation. When these cryopreserved beads were cultured on solid Murashige and Skoog (MS) medium, they showed better revival. However, the non-encapsulated embryonic axes could not survive (Kaviani 2010).

USES

The tea seeds also have commercial utility besides their importance in crop improvement programmes. In China, the seeds collected from wild populations are cold-extracted to yield edible oil and a residual cake or powder. The oil is extensively used in China for cooking, and also for cosmetics. A strong antioxidant activity has also been reported in the methanol extract of tea seeds (Lee 2004). The residue remaining after oil extraction from seeds is available in the form of cake or powder and contains saponins. It is used in organic farming as an insecticide or molluscicide in rice fields, ore-culture, and horticulture. It does not hamper plant growth and is a source of nutrients (Zhongguo *et al.* 1995, Wekumbura *et al.* 2010).

The seed cake is also used as fisicide especially in shrimp farming to eliminate unwanted fishes and harmful insects in the aquaculture ponds. The saponins of tea seeds are available in liquid forms too and have been reported to have remarkable antiseptic, fungicidal, bactericidal and lipid lowering effects.

REFERENCES

Ariyarathna HACK, Gunasekare MTK, Kottawa-Arachchige JD, Paskarathevan R, Ranaweera KK, Ratnayake M, Kumara JBDAP (2011) Morpho-physiological and phenological attributes of reproductive biology of tea (*Camellia sinensis* (L.) O. Kuntze) in Sri Lanka. *Euphytica* 181(2): 203-215.

Barman TS (1988) Maturation, germination and storage of tea seed. Proceedings of 30th Tocklai Conference, Tocklai, Assam, India, pp 102-105.

Barua DN (1989) Science and practice in tea culture. Tea Research Association, Tocklai Experimental Station, Jorhat, Assam, India.

Barua PK (1970) Flowering habit and vegetative behavior in tea (*Camellia sinensis* L.) seed trees in North-East India. *Annals of Botany* 34(316): 721-735.

Berjak P, Vertucci CW, Pammenter NW (1993) Effects of developmental status and dehydration rate on characteristics of water and desiccation sensitivity in recalcitrant seeds of *Camellia sinensis. Seed Science Research* 3: 155-166.

Bewley JD, Black M (1983) Biochemistry of Germination and Growth. In: Bewley JD, Black M (eds) *Physiology and Biochemistry of Seeds in Relation to Germination*. Springer Verlag, New York, Vol. 1, pp 133-176.

Bezbaruah HP (1975) Tea breeding – A Review. *Two and a Bud* 22(2): 123-130.

Bezbaruah HP (1981) Recent advances in tea breeding. Proc. 29th Annual Conference. Tocklai, Assam, India, pp 68-72.

Bhakhtadze KE (1931) The blossoming and fructification of the tea plant under chakva conditions (Russian). Bull. Res. Inst. Tea U.S.S.R. Georgia State Press, Tiflin, No 2.

Bhattacharjee H, Singh ID (1994) Storage of tea seeds. *Two and a Bud* 41(1): 32-34.

Bhattacharya A, Nagar PK (2006) Changes in endogenous cytokinin activity during seed development in tea. *Indian Journal of Plant Physiology* 11(3): 287-290.

Bhattacharya A, Nagar PK, Ahuja PS (2002) Seed development in *Camellia sinensis* (L.) O. Kuntze. *Seed Science Research* 12: 39-46.

Bhattacharya A, Nagar PK, Ahuja PS (2004) Changes in endogenous indole-3-acetic acid and some biochemical parameters during seed development in *Camellia sinensis* (L.) O. Kuntze. *Acta Physiol. Plant.* 26(4): 399-404.

Bhattacharya AK, Rahman F, Basu RN (1996) Preservation of tea (*Camellia sinensis* L.) seed viability. *Seed Research* 22(2): 108-111.

Chaudhury R, Lakhanpaul S, Chandel KPS (1990) Germination and desiccation tolerance of tea (*Camellia sinensis* (L.) O. Kuntze) seeds and feasibility of cryopreservation. *Srilankan Journal of Tea Science* 59(2): 89-94.

Chen L, XiaoChun W, Zheng Zhu Z, Qi C (2008) Distribution of catechins, purine alkaloids and free amino acids in tea seedlings. *Journal of Tea Science* 28(1): 43-49.

Chen Q, Yang LM, Ahmad P, Wan XC, Hu XY (2011) Proteomic profiling and redox status alteration of recalcitrant tea (*Camellia sinensis*) seed in response to desiccation. *Planta* 233(3): 583-592.

Chin HF, Roberts EH (1980) Recalcitrant Crop Seeds. Tropical Press, Malaysia, pp 152.

Elliot EC, Whitehead FJ (1926) Tea planting in Ceylone. Charles Subusinghe and Sons (P) Ltd, Colombo, Sri Lanka, pp 278.

Gomez ASB (1966) Reports of Tea Research Institute, Ceylon 2: 33-34.

Govindasamy SK (1995) Field performance of biclonal seed stocks. *The Planter's Chronicle* June: 257.

Han Q, Yu QY, Shi J, Xiong CY, Ling Z, He PM (2011) Molecular characterization and hypoglycemic activity of a novel water-soluble polysaccharide from tea (*Camellia sinensis*). *Carbohydrate Polymers* 86(2): 797-805.

Harller CR (1964) Tea Culture and Marketing of Tea, 3rd edition, Oxford University, London, pp 70–89.

Hu J, Guo CG, Shi SX (1993) Partial drying and post thaw pre-conditioning improve the survival and germination of cryo-preserved seeds of tea (*Camellia sinensis*). *Plant Genetic Resources Newsletter* 93: 1-4.

Joshi R, Poonam, Gulati A (2011a) Biochemical attributes of tea flowers (*Camellia sinensis*) at different developmental stages in the Kangra region of India. *Scientia Horticult.* 130(1): 266-274.

Joshi R, Poonam, Saini R, Guleria S, Babu GD, Kumari M, Gulati A (2011b) Characterization of volatile components of tea flowers (*Camellia sinensis*) growing in Kangra by GC/MS. *Natural Product Communications* 6(8):1155-1158.

Katsuo K, Toyao T, Kayumi S (1970) The germination of tea seed. *Study of Tea* 39(1&2): 14-19.

Kaviani B (2010) Cryopreservation by encapsulation-dehydration for long-term storage of some important germplasm: seed of Lily [*Lilium ledebourii* (Baker) Bioss.], embryonic axe of Persian lilac (*Melia azedarach* L.), and tea (*Camellia sinensis* L.). *Plant OMICS* 3(6): 177-182.

Koshioka M, Yamagachi S, Nishijima T, Yamazaki, H, Ferraren DO, Mander LM (1993) Endogenous gibberellins in the developing liquid endosperm of tea. *Biosci. Biotech. Biochem.* 57(9): 1586-1588.

Lee CP (2004) Identification of bioactive compounds in oil of tea seed (*Camellia oleifera* Abel) Session 33G, Neutraceuticals and Functional Foods: Lipid and Probiotic functiona Foods. IFT Annual Meeting 2004, July 12-16.

Matsuda H, Nakamura S, Fujimoto K, Moriuchi R, Kimura Y, Ikoma N, Hata Y, Muraoka O, Yoshikawa M (2010) *Chemical and Pharmaceutical Bulletin* 58(12): 1617-1621.

Mohan MS, Siby M, Achuthan R (1994) Influence of the method of collection and storage on the germination of UPASI: BSS-1 seed. Newsletter UPASI Tea Research Institute No. 3(2): 4.

Sato D (2007) Germinating Tea Seeds (*Camellia sinensis*). *Soil and Crop Management* SCM:17.

Sebastiampillai AR, Amandappa TI (1979) The influence of moisture and temperature on the germinability and longevity of tea (*Camellia sinensis*) seeds. *Tea Quart.* 29: 48: 8-20.

Visser T (1969) Tea *Camellia sinensis* (L). O.Kuntze. In: Ferwards FP, Wit F, Veennman H, Zonen Wo (eds) Outline of Perennial Crop Breeding in Tropics, Wageningen, pp. 459-493.

Visser T, Tillekeratne LH, de Waas (1958) Observation on the germination and storage of tea pollens and seeds. *Tea Quart.* 29: 30-35.

Wang YF, Yang ZW, Wei XL (2010) International Journal of Biological Macromolecules 47(4): 534-539.

Wekumbura WGC, Mohotti AJ, Vidhana ALP, Mohotti KM (2010) Comparison of assimilate partitioning in organically and conventionally grown tea. *The Journal of Agricultural Sciences* 5(1): 32-41

Wesley-Smith J, Vertucci CW, Berjak P, Pammenter NW, Crane J (1992) Cryopreservation of desiccation sensitive axes of *Camellia sinesis* in relation to dehydration, freezing rate and the thermal properties of tissue water. *Journal of Plant Physiology* 140(5): 596-604.

Zhongguo J, Sheng C, Xue YJ, Sheng C, Bing ZZ (1995) Molluscicidal activity of seed of *Camellia sinensis*. *Chinese J. Schistosomiasis Contral* 13(2):114-116.

6

YOUNG TEA MANAGEMENT

R.D. Singh

CSIR-Institute of Himalayan Bioresource Technology,
Palampur (H.P.)-176 061, India. E-mail: rdsingh@ihbt.res.in

Tea plant when allowed to grow freely may produce flowers and fruits in due course but may bear few shoots for manufacture of tea. Young plants are therefore, pruned to form a low spreading frame for early canopy cover to facilitate manual or mechanical harvesting of tea shoots to develop an ideal and sturdy bush-frame for sustainable vegetative growth and to provide substantial crop in the formative phase for better economics.

YOUNG TEA RAISING

Pruning: Until 1970's in India young tea plants were not pruned by knife to allow them to get established in main field. In north east India young tea plants were not pruned for at least 2 years in the field until the main stems became 1.5 -2.5 cm thick at collar (ground) level (Barua 1989). The plants were then cut across at 45 -50 cm from the ground and the main stems re-pruned at 15 to 23 cm from the ground. The latter operation is termed as centering which is done at a level above two-three lateral branches. The basic purpose of centering is to promote branching at a low height. The branches were then pruned to attain well-balanced and spreading frame. The plants without branches below the level of centering were most adversely affected by inclement climatic conditions (Barua 1989). In plantations raised by hill-seeding centering was adopted at about 30 cm height from the ground.

Around 1980's according to Barua (1989) young plants in north east India were pruned when the stem diameter at collar level became about 1 cm. Before pruning few random plants were tested for root starch content by a rapid visual method using iodine solution. Pruning was delayed if roots were deficient of starch and the plants were flushing. The plants were cut across at 35-45 cm height and centered immediately afterwards between 10-20 cm preferably above three laterals.

Normally height of the pruning is related to the length of internode. Thus pruning and centering of chinary and china hybrid plants is possible at a height lower than that of assamica type bushes without losing branches on the bush after pruning and centering. Subsequently tipping and plucking can also be done at a lower height in the chinary bushes. Augmenting income from young bushes helps in reducing the payback period which is a key factor in the economics of planting new areas or uprooting and replanting old fields of tea (Awasthi 1981).

The standard procedure at the Tea Research Institute of East Africa is to prune sleeve-grown plants at 15 cm when they attain a height of 30 cm or more leaving at least three leaves below the pruning height (Barua 1989). The laterals growing from the stump are cut across at 28 cm when they are 1 cm thick at that height. Pruning is repeated at 41 cm when majority of the shoots become 1 cm thick at that height. The bushes are then tipped at about 61 cm when three leaves are produced above the pruning measure. Plucking continues until pruning at the end of three-four years.

In Sri Lanka the first pruning in young tea is done at 23 cm height at about 9 months after planting. Later the bush is pruned at 35-38 cm (4-5 months after the first prune) and finally tipped at 45 cm. Plucking continues for about 5 years. In the seventh year final frame-forming prune is done at 45 cm. This indicates that one prune is considered insufficient to produce a well-formed bush.

Hard plucking methods like *janam* plucking or fish leaf plucking after tipping promote growth of more laterals to make up the gaps in the bush-canopy and plant- spread (Barua 1989).

Bending and pegging: Of late bending and pegging operation for raising young tea is gaining importance. In this method the main stem and branches of the young plant are bent towards ground and fixed in that condition with the help of pegs (Barua 1989). Peg can be made of pruned tea stem twig of tree bamboo etc. Success of pegging depends on growth of the main stem which should be ideally 40-60 cm tall at the time of pegging but not too thick and hard (Barua 1989). Thick stem may break on bending and can not be easily fixed by pegging. Thus the stem should not be thicker than 8 to 10 mm at the collar level. Branches should be allowed to grow 40-60 cm tall before bending and pegging. A prostrate and drooping branch should be trained to stand at an angle with the main stem. In a pegged stem new growth takes place from the basal part. Pegging should continue until the plant has developed branches in all directions.

Tipping: Tipping is generally done at 35 to 50 cm height from ground depending upon the type of plant topography and climatic conditions. It is adopted to promote growth of branches or stems evenly on the bush frame (Barua 1989). The thin peripheral branches are tipped higher for allowing them to become thick.

In East Africa pegging after a cut across and decentering provided significantly higher yield than any form of pruning including the Standard TRI pruning even at the end of two pruning cycles (TRIEA Annual Report 1977). Barbora (1984) obtained higher crop in north east India from pegging with or without centering than from any type of pruning during initial 3 years until the bush is given the final frame-forming prune. Pegging without centering yielded more only in the first year than pegging following centering. In the second pruning cycle two pegging treatments and the standard TES method of pruning provided similar yield. These three treatments increased the size of bushes and the number of pruning sticks per unit area to the same extent which was more than that of any other forms of pruning. Pegging proved superior to pruning in Sri Lanka too (Manipura and Yatawatte 1974).

Overall pegging requires skill regular monitoring and corrective actions without which it can not be effective and economical. In a pegged young tea section manuring and weeding should be dome with great care.

Time of pruning/pegging: At active growth stage roots of the plant normally have low content of carbohydrates which may not be sufficient to backup regeneration of shoots after pruning. To overcome the situation number of leaves removed from the plant should not be more than the optimum number of leaves required for carrying out the operation (Barua 1989). The dormant period and cold weather are suitable for pruning young tea provided the soil has some moisture. Thus pruning should not be done during hot and dry weather. Pruning during dormant phase should be completed much before the time of bud-break.

Plant growth regulators as pruning agents: Many plant growth regulators are being used for stimulating branching by suppressing apical dominance. The examples are Ethephon or Ethrel (2-chloroethyl phosphonic acid) CCC or Cycocel (2-chloroethyl trimethyl ammonium chloride) SADH or Alar (deminozide) Goal (oxyfluorfen) Morphactin (chlorflurenol methyl ester) paclobutrazol etc.

Manivel *et al.* (1998) have observed that paclobutrazol sprayed at 500 ppm increased number of laterals in lower region of young tea. Application of polyamines (350 ppm) in young tea improved water use efficiency and consequently induced drought tolerance in the plant.

Application of Ethephon at 1000 ppm and Goal at 500 ppm on young transplants produced more branches than normal pruning and centering (Barua 1989). The factors determining bioefficacy of the chemical pruning agent are kind of chemical formulation spray concentration growth stage of the plant and type of the plant (TES Annual Report 1981 & 1985). The presence of roots and the state of dormancy appear to be prerequisites for the action of the chemical.

Foliar application of GA_3 50 ppm on Tocklai clones TV-7 TV-10 and TV-18 planted in pots provided significantly higher plant height leaf area and girth during initial growth stage. Plant biomass showed increment due to treatment with triacontanol 5 ppm (Deka and Dutta 2001).

MULCHING

In a 4-year-old vegetatively propagated young tea mulching with black plastic peat or tea factory residues provided better shoot and root growth than that of the non-mulched controls (Tabagari and Kopaliani 1985).

In vegetatively and seed-borne tea plants mulching with black plastic tea factory residues or peat improves soil moisture and temperature conditions. Black plastic mulch also provided complete control of weeds (Tabagari and Kopaliani 1984). Kopaliani *et al.* (1991) have observed that in terms of maintenance of rhizospheric temperature (22 to 25°C) mulching with black polyethylene was more effective than mulching with peat or tea factory waste.

Othieno (1975) assessed the surface run-off and soil erosion in a young tea planted on a land with 10% slope and concluded that the run-off and soil erosion was higher in the absence of erosion control measures i.e. in plots subjected to manual weeding and herbicide weeding. Mulching gave the lowest run-off and least soil erosion. Similarly in a 4-year old newly planted tea on land with 25-39% slope soil loss with clean weeding during the first and second year was 89 and 51 t/ha respectively compared with 7 and 1 t/ha when Guatemala (*Tripsacum laxum*) grass mulch was used and it was 5 and 2 t/ha with Mana grass mulching (Sandanam and Rajasingham 1982). Though the difference among the treatments in terms of crop yield was marginal in the initial two years in the second year yield was highest due to Mana grass mulching. In both years soil moisture was highest in the mulched plots.

Type of the mulch material *viz.*, black plastic stone chippings and organic mulches affect soil moisture content and water status of young clonal tea. After 4 years of continuous application the grass mulches had significant influence on water holding capacity of the soil but it induced shallow rooting in tea plants which made the tea more susceptible to drought (Othieno 1980; Othieno and Ahn 1980).

In young tea annual mulching with straw provided higher yield as compared to deep forking (Maehara 1976). Similarly in Assam studies on mulching in young clonal tea with Guatemala grass siam weed (*Eupatorium odoratum*) Indian rhododendron (*Melastoma malabathricum*) water hyacinth (*Eichhornia crassipes*) rice straw and black and white plastic sheets showed that stem girth bush spread volume of feeder roots and number and density of plucking points significantly increased by mulching. Total yield (shoot dry weight) was highest with rice straw mulch followed by Guatemala mulch (Baruah and Sarma 1996).

Studies conducted by Nyasulu (1996) in Mulanje Malawi on effects of stylo (*Stylosanthes*) *Crotalaria* beans (*Phaseolus*) and groundnut as cover crop or Guatemala grass as a mulch revealed that tea yield was highest (1939 kg made tea/ha) with Guatemala grass followed by *Crotalaria* (1528 kg made tea/ha) compared with 1497 kg made tea/ha for non-mulched controls. The other cover crops depressed yields. Bush vigour was increased by Guatemala grass and reduced by beans. Soil K content was greatly increased by Guatemala grass mulch.

Mulching is at times limited by unavailability of suitable biomass such as Mana grass (*Cymbopogon confertiflorous* syn. *C. nardus)* and Guatemala grass. In an attempt to study feasibility of using paddy husk sawdust or tea residues as alternative mulching materials Wijeratne *et al.* (1994) have observed that these alternative materials could be used for desired effect on soil moisture and plant growth. However a thick layer of paddy husk should be avoided as it may cause poor drainage condition.

In Malawi Guatemala grass mulch increased bush area by 19 % compared with the no-mulch control. Tea yield declined due to the cover/intercrops compared with no-mulch treatment although the detrimental effect declined with time being almost nil by the fifth year. The Guatemala grass mulch provided 30% higher tea yield than no-mulch treatment in the first year but the difference between these two treatments declined to almost nil by the fifth year (Nyasulu 1995 and 2000).

In Assam a pot experiment to study the effect of biomass of *Borreria hispida* a weed as partial supplement to inorganic N fertilization on growth of young clonal tea (TV-22) revealed that application of 15 kg N as *Borreria* + 30 kg N as urea/ha significantly increased height number of leaves collar diameter shoot: root ratio fresh weight and dry weight of tea plants (Saloi *et al.* 1997). Plants receiving 15 kg N as tea waste in combination with 30 kg N as urea/ha produced the highest leaf area/plant.

In a 3 year old tea plantation on 36-80% slope in Cachar district of Assam India the soil loss varied from 4-172 t/ha depending on the slope soil type and the extent of vegetative cover (Goswami *et al.* 1997). The trend of soil loss closely followed the rainfall pattern. The soil loss from the plot under strip weeding was about 1.75 times less than that from the plot under 100% weed control.

INTERCROPPING IN YOUNG TEA

Ekanayake (2003) suggested that in young tea during initial 2 years inter-row space can be utilized for raising short-term crops such as pulses (Li *et al.* 2008) and grains. It will augment farm income improve soil fertility and control weeds. Such intercropping has been reported to significantly increase green leaf yield of young tea (Baruah *et al.* 2005). However after initial 3 years the effect of intercrops on tea yield tends to decline (Bore 2005).

In subtropical hilly region straw mulching (Peng *et al.* 2007) and intercropping with white clover in young tea improved physico-chemical attributes of soil (Peng *et al.* 2008). The above two treatments increased moisture content in soil at 0-20 cm depth shortened drought period suppressed weeds reduced insect infestation and enhanced growth of earthworms. Song *et al.* (2006) also reported significant improvement in growth yield and quality of young tea treated with straw mulching and white clover intercropping compared with untreated control.

ROOT GROWTH IN YOUNG TEA

In young tea several adventitious roots form the skeleton of the future root system (Yamashita 1994). Root system of clonal plants is generally shallow unlike that of seedling plants. The fine white roots play an important role in nutrient uptake and always account for more root dry weight than any other root. Bandyopadhyay (1999) has described that under trickle irrigation in Assam cultivars TV-1 and TV-19 exhibited a thick mass of fibrous roots in the top 30 cm of soil. The young white and cream coloured roots were mainly responsible for absorption of nutrients and water. The number of suberised reddish-brown roots was highest in the top 30 cm of the soil. Between 74 and 94% of roots were available within 90 cm of the soil surface. At the surface layer the roots were well distributed within a radial distance of 90 cm.

TILLAGE

In young clonal tea deep hoeing and fork hoeing either once or twice a year during June and December provided highest average annual yield (1821 kg/ha) (Sarkar *et al.* 1983). However care should be taken to level the land after this operation otherwise the depressions in the land may cause water logging and subsequently incidence of root-rot pathogens.

WATER REQUIREMENT

In a pot experiment on water requirement of young tea under greenhouse condition the plants wilted when the soil moisture content declined below 20%. The plants died at soil moisture content of 15% (Darmawijaya *et al.* 1989).

REFERENCES

Awasthi RC (1981) Economics of some major development projects in field. Proceedings of the 29th Conference Tocklai Experimental Station, pp 25-32.

Bandyopadhyay SK (1999) Rooting pattern of tea (*Camellia sinensis*) under trickle irrigation. *Annals of Agricultural Research* 20(4): 497-501.

Barbora BC (1984) Factors contributing towards success of young tea. *Two and A Bud* 31: 32-37.

Barua DN (1989) Science and Practice in Tea Culture. Tea Research Association Calcutta p 509.

Baruah S, Sarma D (1996) Effect of mulching on growth and yield of young tea. *Journal of the Agricultural Science Society of North East India* 9(2): 141-144.

Baruah Swapon, Ahmed Nassimuddin, Saikia Samiran (2005) Intercropping in young tea plantation. *Natural Product Radiance* 4(1): 35-39.

Bore JK (2005) Effects of intercrops on yields of young tea. *Tea* 26(2): 52-56.

Darmawijaya MI, Dachman Nyanjang (1989) Water need of tea plant. *Indonesian Agricultural Research and Development Journal* 11(3): 42-45.

Deka M, Dutta AK (2001) Effect of plant growth regulators on the growth and frame formation of young tea plants belonging to three different origins. *Journal of Plantation Crops* 29(1): 27-31.

Ekanayake PB (2003) Crop diversification and intercropping in tea lands. *Tropical Agricultural Research and Extension* 6: 66-70

Goswami H, Barooah AK, Dutta PK (1997) Soil erosion from young tea fields in Cachar teelas. *Two and a Bud* 44(2): 12-14.

Kopaliani RS, Ugulava VV, Dzhobava TS (1991) Effect of interrow management methods in young tea plantations on soil temperature regime. *Subtropicheskie Kul' tury* 5: 133-136.

Li JianLong, Tu PanFeng, Chen Na, Tang JinChi, Wang XiuRong, Nian Hai, Liao Hong, Yan XiaoLong (2008) Effects of tea intercropping with soybean. *Scientia Agricultura Sinica* 41(7): 2040-2047.

Maehara M (1976) Effect of straw mulch and deep tillage in tea fields. *Japan Agricultural Research Quarterly* 10(3): 132-137.

Manipura WB, Yatawatte ST (1974) Studies on the methods of bringing clonal tea into bearing in the high country of Sri Lanka. II. Further observations on the effects of cutting and bending on branching and the growth of young plants. *Tea Quarterly* 44(1): 37-46.

Manivel L, Kumar RR, Marimuthu S, Venkatesalu V (1998) Plant growth regulators for crop management in south Indian tea. *Journal of Plantation Crops* 26(1): 63-69.

7

TEA PRUNING

R.K. Sud

CSIR-Institute of Himalayan Bioresource Technology,
Palampur (H.P.)-176 061, India. E-mail: rksud@ihbt.res.in

DEFINITION AND IMPORTANCE

A free growing tea plant attains shape of a tree with a height of 8-12 m in case of assam type or to a shrub of 3-5 m top in case of chinary type, and mostly produces flowers and fruits. Such plant has a much larger leaf area (Hadfield 1974) and accumulates greater dry matter (Magambo and Cannel 1981), due to better light interception by the photosynthetic surface as a result of better leaf disposition and canopy structure (Wadasinghe and Senanayake 1990) than a well managed tea bush of same age but no commercial yields are possible in free growing tea plant. Thus, to keep a tea plant in vegetative phase and economically productive, the vertical growth of the plant is kept under check by way of artificially imposed conditions by cutting the main stems and branches of the plant at a particular height in specific intervals. This operation is called "pruning". Pruning is done to maintain the bush at proper plucking height and to retain its ability to produce active shoots. It needs to be repeated at a time when the bush tends to turn from the vegetative phase to reproductive phase. In a pruned and well managed tea bush, the shoot growth is very active, shoot density is less but size is more. With time the growth rate declines, shoot density increases, and size of shoots becomes smaller and smaller, and even many shoots turn *banjhi* and fail to grow. At this stage, the bush enters in reproductive phase through profuse flowering. This stage reaches in 4-6 years depending on type of bush. Chinery bush tends to flower earlier. Therefore, there is need to prune the bushes at periodical interval. Other reason for adopting pruning is that with continuous plucking, there is formation of multi-pronged structure, sort of knots, called crows' feet, which obstruct easy movement of sap, containing nutrients from roots, and photosynthates from maintenance foliage to the growing shoots, resulting in their restricted growth. Hence, these crows' feet need to be removed through pruning for smooth flow of the sap. Even after pruning, with passage of time the height of plucking table continues to rise. The undue rise in the height of the bush affects the production of shoots as well as labours' output and overall productivity. This, therefore, necessitates periodical pruning.

Since the pruning operation results in heavy crop loss during the year of prune, it sometimes is referred as a necessary evil. According to Tocklai (Chakrabarty and Barpujari 1992), the need to maximize productivity has made the field work in tea estates more intensive in character, and since pruning has direct bearing on bush productivity, there should be an accepted pruning policy for the estates which must ensure: (i) keeping the bushes efficient all the time in respect of health and uptake of nutrients, so that maximum number of healthy shoots appear from the pruning sticks and majority of them reach the tipping height in time, (ii) harvesting even crop as much as possible and maintaining acceptable quality throughout the season, (iii) reducing incidence of pest and diseases, and (iv) keeping the bush heights within reasonable limit for efficient plucking.

In the very early years of tea cultivation in China, there was no concept of pruning. The tender shoots, which grow in flushes on the freely growing tea plants, were harvested with considerable care, and as such this harvesting posed lot of difficulties (Ukers 1935). The pruning practice, perhaps, originated with the observations of sickling of wild tea trees in Assam forests by Capt. Robert Bruce around 1850. Mitra (1976) opined that the concept of pruning in tea probably came from horticultural practices followed in certain fruit crops for giving shape to the tree to regenerate new branches. No record exists as to when the practice of annual prune started in India. Pruning is the second most labour intensive process, after by plucking (Moorti *et al.* 1985).

Pruning, tipping and plucking at regular intervals are the operations applied to the tea in a sequence to maintain it in bush form with a flat top surface. These operations are interrelated and severity of one is compensated by the leniency of the other (Jain and Tamang 1990).

Importance of pruning is well understood and commonly followed by the big growers and corporate. But a survey of small tea growers of Assam revealed that there was need for proper understanding of system of pruning as well as other tea husbandry practices for both trained and untrained growers (Baruah *et al.* 1998). However in Sri Lanka, a study indicated that pruning is one of the highest adopted packages besides clone selection, field establishment, and fertilizer application (above 75% level) (Jayamanne *et al.* 2002).

The major issues in pruning are selection of suitable time for the operation and the selection of pruning cycle, i.e. duration in years between two consecutive prunings.

OBJECTIVES OF PRUNING

Keeping in view the above points, the aims of pruning may be summarized as below:

1. Stimulate vigorous shoot system, induce more vegetative growth, suppress reproductive phase.
2. Establish plucking table and maintain convenient height to facilitate plucking operation.
3. Renew maintenance foliage and maintain its sufficient volume to meet the physiological needs of the plant.
4. Rationalizing wood /shoot ratio, remove knots and interfaced branches.
5. Removal of dead, diseased and defunct wood, regenerate the frames, thus help rejuvenate bushes that have crossed the period of maximum productivity.
6. Crop regulation and uniform crop distribution for avoiding crop rush particularly during peak periods
7. Maintain quality in processed tea by encouraging active shoot growth and minimizing formation of *b [illegible] jhi* shoots.
8. Control spread of pests and diseases.

TYPES OF PRUNING

In order to achieve the above objectives three types of pruning systems are practiced from time to time in the life of a tea bush. These are: formative prune, maintenance prune and corrective prune.

Formative prune

Formative prune, also called frame forming prune, is the initial prune where free growing young tea plant is trained and brought up to form table-topped bush (Figure 7.1) by pruning them at increasing predetermined heights (Templer 1975). A newly planted out tea is a single-stemmed and free growing plant. The operation is performed for cutting its vertical growth and removing apical dominance of free growing leader shoot and to convert it into multi-branched or multi-stemmed pluckable bush by encouraging development of new shoots of higher order. Formative pruning prepares the young tea bush frame into an expansive plucking table. For achieving the objective the plant is initially pegged, thumb or lung pruned or decentred. The formative prune results not only in increased branching but also in higher branch density at the pruning level. The summary of the sequence of the formative prune is given below:

i. Thumb or lung pruning the leader stem at 15-20 cm height by breaking the single stem to initiate branching below it.

ii. Decentre the leader stem at 8-15 cm, cut across of the side braches at 30 cm or alternatively peg them.

iii. Tip the new shoots at 55 cm and pluck subsequently and give skiff to level the table during subsequent years.

iv. Cut across at 35 cm and remove crossed branches.

v. Tip at 55 cm and keep on plucking.

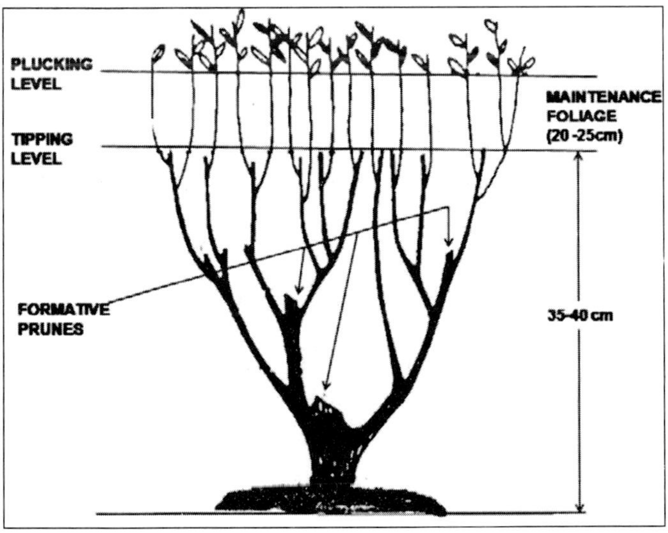

Fig. 7.1. Formative pruning

Types of skiffing

Depending on severity, the skiffing operations have been categorized as 'deep skiff', 'medium skiff', 'light skiff' or 'level-off skiff' (Figure 7.5). Normally the severity of skiffing increases with the passage of time from the last pruning operation (Kulasegaram *et al.*

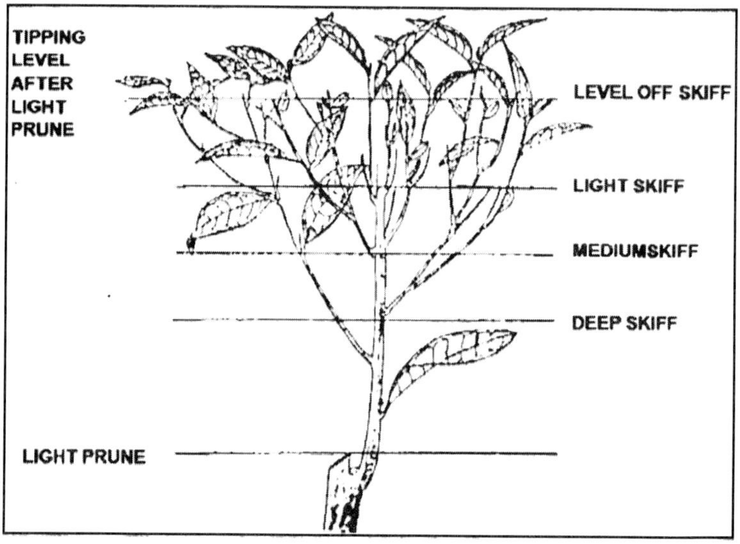

Fig. 7.5. Primary shoot showing different types of skiffing

1975, Jain and Tamang 1990, Barua 1989). Important feature of skiffing is to retain more mature leaves on the bush frame. Skiffing operation ensures early crop as well as total crop during the season, except deep skiff where part of early crop is lost. Investigations have shown that skiffing operation promotes the growth of the top lateral bud (Nakano 1999).

The presence of *banjhi* shoots could adversely affect the growth of new shoots. In Darjeeling, where plucking of *banjhis* between flushes on unpruned or light skiffed tea in the third or fourth year of the cycle is not possible, skiffing can be done to remove the *banjhis* immediately following a round of plucking (Chakrabarty and Barpujari 1992).

***Deep skiff (DS)*:** Deep skiff is severe form of skiffing where cut is applied midway between the current plucking table and last pruning level to remove the congestion of the top hamper and unproductive shoots. DS ensures a fork in most of the primaries, thereby promising higher crop. This skiffing is applied to extend the pruning cycle.

***Medium skiff (MS)*:** Medium skiff is applied to remove the crows' feet which are unproductive and confined at the top of the table. For removing these unproductive twigs, skiffing is usually done at a level of 15 cm above the pruning mark or 5 cm below the tipping level.

Light skiff (LS): In this skiff, a superficial cut is given in the red stems on the plucking table to remove a majority of the previous plucking points and over growth and to reestablish the table.

Level-off skiff (LoS): This is a skiff where cut is made in the green stem on the table with an objective to make the table leveled to the first plucking.

LEVEL AND TIME OF PRUNING AND SKIFFING

Height, timing and methods of pruning and skiffing are the major factors affecting tea yield.

Level of pruning

Height of pruning depends on (i) existing height, (ii) last pruning height, (iii) incidence of pests and diseases, (iv) anticipated crop and market, and (v) existing vacancies in the plantation.

In Kenya, a 10-years cumulative yield data on pruning levels indicated that the lower the pruning height below 60 cm the lower was the yield (Njuguna and Magambo 1990). The amount of crop lost in the first 2 years, as a result of pruning at heights of 15 and 30 cm, and that lost in the first 3 years by pruning at ground level were not recovered by the end of 10 years. Malenga (1997) found that in mature tea deep pruning resulted in severe crop loss, which is restricted during the years of pruning. He further (1999) observed that deep pruning below a height of 30 cm significantly affected bush survival and recovery resulting in lower production in the following season. However, pruning at 40 cm or above showed no crop losses. Hence for assamica tea plantation with healthy bushes, the pruning level should be 40-45 cm, but in special circumstances where stems are badly damaged due to sun-scorch or canker or fire or other reason, more severe pruning would be appropriate in long term.

In studies on the effects of pruning at different levels, Magambo (1980) observed that the time to bud-break for bushes pruned at 60 cm was 21 days, compared with 60 days for the other severe prunings. Branch die-back was greater on bushes pruned at 30 cm height. Yields and surface area/bush after 6 months, following tipping at 10 cm above each pruning height, were greatest with pruning at 60 cm, and declined as the height of the plucking table was lowered.

Hard pruning may be desirable when the plucking table has become too high or the frames of the bushes are diseased. A pruning height of 25-34.6 cm is suggested for large-leaf hybrid and Assam types at mid and high elevations (Tocklai 1972). In south India, rejuvenation pruning of debilitated tea bushes out yielded normally pruned tea from the third year (Ram 1974) or by the end of the following cycle (Thomas 1979), compared with normal pruning.

Marimuthu and Ranganath (1994) assessed the chemical composition of the clonal tea pruned to 45, 60 or 70 cm. The TSS and tea brightness, briskness and body were highest with a pruning height of 60 cm and the flavour index with a height of 70 cm. All the other compositional and quality parameters measured showed an inconsistent response to pruning height.

Regarding root growth, it has been reported that growth of both fibrous as well as main roots, was adversely affected by the severity of the pruning (Esartiya 1972).

For rejuvenation pruning in Himachal Pradesh, CSIR-IHBT has observed significantly higher shoot density and trend of higher yield when bushes were pruned below or at ground level compared with that of pruning at 10 or 20 cm above ground level (Table 7.1, Jain *et al.* 2002).

Table 7.1. Effect of level of rejuvenation pruning on yield of old china hybrid tea bushes under different pruning cycle (Jain *et al.* 1992)

Level of Pruning	Made Tea Yield (kg/ha)	
	HP-UP-DS-UP	HP-UP-UP-DS
Below ground level	2089	2277
At ground level	2201	2027
10 cm above ground level	2129	2056
20 cm above ground level	2330	1854
Mean	2187	2054
LSD	NS	NS

HP: Heavy/Rejuvenation prune, UP: Unprune, DS: Deep skiff

Time of pruning

This has a significant effect on the recovery of the bushes after pruning. In tea growing regions where winter dormancy prevails *viz.*, the regions away from 16° N or S latitude, winter season is the best time of pruning. For example, in north India tea plant does not flush in winter period (November to February) and can be pruned during this period. But in other regions where there is no winter dormancy *viz.*, regions located around equator to 16° N or S latitude, pruning should be done just before the rush period when the starch reserve is enough. After the rush period the starch reserve will be less so pruning should be avoided during this period.

Applying pruning at the beginning of every month from January to December at heights of 30 or 55 cm, Odhiambo and Othieno (1992) in Kenya observed that yields were lower when bushes were pruned to 30 cm. Though initially, October or November pruning gave the highest yields but cumulative yield of the 3 year cycle was more from pruning in December than from pruning in July but these differences could not be attributed to a single seasonal factor. According to Mwakha (1997), the experimental results on timing of pruning in Central Africa remained controversial. However, it is a general recommendation to complete the pruning by May or mid-June. Experiments on delayed pruning (Mkwaila 1994) showed that the recovery period of the bush was so protracted that it covered (i) the entire season resulting in under production, (ii) the response to fertilizer application decreased and there was no response at all when pruning was delayed until August, and (iii) a loss of crop due to thrips infestation. Bore (2003) based on his work on shoot and root TNC (total non-structural carbohydrate) in relation to tea pruning concluded that the best pruning time in Kenya was October, provided drought stress can be avoided. But Nyabundi (2004) suggested that though October pruning with retention of sufficient lungs results in the highest amounts of plant carbohydrate reserves, yet June pruning is more practical because there is little shoot growth and hence less crop to be lost and such plants have adequate reserves to survive the December/February dry season and take advantage of the March/May rains for increased yield.

Spring and winter teas are the most desirable and command the highest prices in Taiwan. Pruning in spring rather than winter increased the yields in the following winter and spring in Taiwan particularly when mechanical plucking was applied at all seasons (Ho *et al.* 1992)

Udayakumar and Muraleedharan (1998) in south India observed that time of rejuvenation pruning affected recovery. Fields pruned in April, May and June took 160, 200 and 230 days to recover, respectively. The yields of rejuvenated fields varied depending on vacancy percentage, timing of pruning and irrigation. It has also been observed that pruning improves yield during the drought season of the following year (Satyanarayana *et al.* 1994).

Considering the starch reserve in roots, November-January is the ideal time for pruning tea bushes in Darjeeling hills (DTRDC 2002). Even within the winter period, the time of different types of pruning is dependent on the severity of pruning. Deeper the pruning level, earlier the operation to be completed. Heavy pruning in mid to high elevations of Darjeeling can be started as early as August provided labour availability is not a constraint and starch reserves in the roots are satisfactory. In low elevation gardens, it should commence in October.

Using an isotope tracing technique, Xei *et al.* (1993) concluded that (i) correct pruning after the first harvest could improve the quality of summer crop, (ii) pruning was best carried out when plant nutrient status is highest and (iii) heavy pruning of shoots after harvesting the spring crop was not beneficial.

Shade may interact with recovery of the pruned bushes. Recovery after cut-across pruning done in November or December was reported to be better and more rapid under shade than with unshaded bushes (Kilavuka 1980).

Table 7.2. Effect of month of rejuvenation pruning on yield of old china hybrid tea bushes under different pruning cycle (Jain *et al.* 1992)

Month of Pruning	Made Tea Yield (kg/ha)	
	HP-UP-DS-UP	HP-UP-UP-DS
November	2647	2277
December	2187	2027
January	2032	2056
February	1881	1854
Mean	2187	2054
LSD	637	411

HP: Heavy/Rejuvenation prune, UP: Unprune, DS: Deep skiff

In Himachal Pradesh, tea was pruned in alternate years during 1962-68, in November, December, January or June. June pruning increased the mean yield by 65% and the yield of first flush by 117%

compared with winter pruning (Mehta 1972). But in the later years June pruned tea bushes were observed to be more susceptible to blister blight disease than those pruned in winter months (Sugha *et al.* 1990). For rejuvenation of old china hybrid tea bushes in Himachal Pradesh, CSIR-IHBT has observed that pruning in November month resulted in higher yield compared to other months from December to February over four-year cycle (Table 7.2, Jain *et al.* 1992).

For precise recommendation of pruning time, it is important that both time and site replicated experiments should be conducted.

Level and time of skiffing

The Japanese workers reported that earlier and deeper the autumn skiffing the earlier the first plucking in the next year, and also the lighter the skiffing the larger the total crop (Yamashita *et al.* 1975). Nakano (1999) concluded that the period from skiffing to the opening time of the first leaf was shortened by skiffing from spring to summer, but it would be lengthened by skiffing from autumn to winter.

Evaluating skiffing at different levels from 10 to 25 cm above the previous pruning height, Waheed *et al.* (1998) in Pakistan observed that leaf yields increase with increasingly severe skiffing. This was due to an increase in the number of shoots/bush and number of plucking points/bush. It was concluded that severe skiffing to 10 cm above the previous pruning height could be used to increase tea yields 3-fold.

For Darjeeling area it has been recommended that in order to keep the crop loss to the minimum, the skiffing should be completed before mid August (DTRDC 2002).

PRE-REQUISITES FOR PRUNING

Enough starch reserve

Root starch reserves play an important role in recovery of tea after pruning until new shoots develop. Earlier studies by Selvendran and Selvendran (1972) on the polysaccharide content of the plant immediately after pruning and during recovery suggest that starch was the main reserve polysaccharide. Root wood contained the highest starch content on a dry weight basis. Recovery from pruning was accompanied by a decrease in the starch content on the root and stem whereas the holocellulose (hemicelluloses A and B and alpha-cellulose)

fugus easily infect the bushes recovering from pruning (Mouli 1993, Premkumar *et al.* 1998, Rajalakshmi 2003), because of both the greater susceptibility of new foliage and increasing blister size from more rapidly expanding lesions (Sugha and Singh 1988). From disease management point of view, correct time of pruning is very crucial (Hudson and Muraleedharan 1998, Baby and Premkumar 1999). Studies in Himachal Pradesh with respect to the effect of pruning time and shade/sun on the development of blister blight of tea revealed that blister blight may be alleviated by avoiding shade and June pruning (Sugha *et al.* 1990).

Otieno (1997) discussed the epidemiology of *Hypoxylon serpens* in tea plantations in Kenya. According to him, soil is a source of inoculum of the pathogen and outbreak of *Hypoxylon* wood rot is encouraged by low pruning beside bark injuries.

For minimizing pesticide usage and thereby the potential risks of residues in made tea, Whittle and Rattan (1995) in Central Africa suggested to follow recommended pruning schedules in association with other cultural practices.

Shot-hole borer *Xyleborus fornicatus,* pest of tea very common in south India and Sri Lanka, which burrows within the stems of the tea plant and cultivates a fungus in galleries within the stems, on which the larvae feed. This weakens the bush and can lead to premature death as a result of wood-rot. Alteration of the pruning cycle, and burning or burying the pruned material are among the important control methods (Seneviratne 1995). Muraleedharan (1995) presented a summary of published suggestions and recommendations for the control of shot-hole borer (*Euwallacea fornicatus*) on tea in India and highlighted pruning height, post-prune spray, manuring of pruned (infested) fields, infestation trend in the pruning cycle.

Gope and Das (1992) in north-east India observed that the China tea was more susceptible to damage of *Brevipalpus phoenicis* than the Assam, and pest infestation was reduced the most when bushes were pruned rather than skiffed. Radhakrishnan (1997) opined that in south India, rejuvenation pruning can be used to combat canker.

On tea bushes, the bottom level harboured significantly higher dense populations of *C. carinatus* mites than that of leaves at the top and middle level, also the density of the mite was greater on bushes in the third and fourth years from pruning than on bushes in the first half of the pruning cycle (Muraleedharan *et al.* 1994). Thus shorter pruning cycle accompanied with removal of bottom leaves may help control the mite.

In Central Africa, young tea plants recovering from pruning are affected by 'hot and cold' disorder where new-shoot leaves get curled and yield is reduced. Rattan (1996) observed that if the pruning time is preponed from June and July (currently recommended time of pruning in Central Africa) to April or May, the percentage of dead or dying plants (assessed in November) as a result of the disorder was reduced to half by early pruning.

POST PRUNING CARE

There are certain operations following the pruning, which are very crucial for the sustained productivity and should be done after pruning is over.

Knife clean out

'Knife clean out' refers to removal of the mature *banjhi* shoots, thin shoots and other unproductive shoots, which always remain under the canopy of the maintenance foliage and never contribute to productivity, with the help of knife. Bush repair and correction of uneven cuts during pruning are also carried out during this operation.

Bush sanitation

Cuts on the bush frame during pruning provide lot of scope for the pathogen penetration. In order to protect the tea bushes against the pathogens, copper fungicide pasting is applied on the cuts. Lime washing of the frame is also carried out for bush sanitation. In case there is infestation of moss or lichens, micronished shell lime at 1:10 is applied on the stems.

Management of pruning litter

In order to utilize its nutrients and to check soil erosion, pruning litter should be buried or arranged in between tea rows as 100 kg of pruning litter is equal to 200 kg of FYM. Retaining pruning litter, particularly in the absence of fertilizer results in enhanced yields (Monkhouse and Lennox 1976). Sri Lankan worker studied the effect of incorporating tea prunings into the soil on some soil characteristics (Weeraratna *et al.* 1977). During the first 2 months after incorporation, prunings decomposed rapidly and resulted in increased

N and K availability and organic matter content of the surface and sub-surface soil. Incorporation reduced leaching of added N and K fertilizers and the addition of N fertilizer reduced K availability in the soil.

PRUNING CYCLE

The time interval in years between two successive prunings is called a pruning cycle. A pruning cycle consists of a prune followed by skiffs in following years until the prune again.

Different types of pruning and skiffing are adopted depending upon the condition of the bush and other external factors. Based on the type of pruning, the re-grown shoots are tipped at different pre-determined heights above the ground level or prune level. Pruning immediately results into a crop loss in the particular section but overall crop is not allowed to suffer by balancing the proportion between the pruned and unpruned sections of the gardens and for this reason it becomes important to follow a set of pruning programme or the "pruning cycle", which may be short or extended covering a period of 2-6 years. Length of pruning cycle depends on many factors *viz.*, growth pattern of tea, cultivar planted, climatic conditions, quality and quantity of the targeted crop, labour availability for plucking, prevalent market conditions etc.

Before the slump in tea market and when the growers were getting prompt leaf payments many tea estates in Kangra valley have adopted rational pruning cycles according to CSIR-IHBT recommendations balancing the prune: unprune percentages around 33:67. The objective is to harvest at least 35% of total crop during first flush, which is most valuable as it can be processed property due to its chemical constituents, weather conditions and labour availability for no tough competition with other crops.

In Darjeeling, a study of pruning cycles of 2, 3, 4 or 6 years at several estates for 12-year period revealed that a 4-year cycle gives highest total yields in the region (Hazarika *et al.* 1985). Since the first and second flush crop fetches better price in this hilly region, most of the gardens tend to keep 80% or more of the areas under unprune or light skiffed (DTRDC 2002). But practice of 8 to 9 years long pruning cycle having more LS, LoS and UP is also not uncommon. However, this may affect the bush structure and productivity in the long run. At mid to high elevation and under very slow growing conditions, a 5-year cycle LP-UP-LS-DS-UP can be considered in the region. For production of quality shoots, a 3-year cycle LP-UP-LS or a 4-year cycle LP-UP-DS-UP in most of the areas of Darjeeling hills may be adopted.

In Sri Lanka, yield continuously increases from year first to third year of pruning cycle, followed by reductions in the fourth year. De Costa *et al.* (2009) elucidated the physiological basis of the yield decline during the fourth year of the pruning cycle. The fourth-year decline in harvest index was brought about by reductions in both shoot number and mean individual shoot weight, which indicate reduction in sink strength. Tea bush showed reduction in light-saturated photosynthetic rate of maintenance foliage during the second half of the pruning cycle, indicating reduced source capacity. Hence, a combined reduction of both sink strength and source capacity during the fourth year could have brought about the significant yield reduction in tea. A significant increase of root starch in the fourth year indicated a down-regulation of physiological activities of the bush towards the end of the pruning cycle. There is need to further investigate the mechanisms responsible for this down-regulation.

Thomas *et al.* (2005) found a concomitant increase in total catechin content up to 3 years followed by a decrease in the fourth year of the pruning cycle, irrespective of the cultivar. PAL, the precursor of catechin biosynthesis, also increased up to 3 years and showed a sharp decrease thereafter. On the other hand, the root carbohydrate reserve and the caffeine content of tea leaves increased steadily up to the fourth year.

Types of pruning cycle

Pruning cycles are mainly of three types: quality oriented, quantity oriented, and quality and quantity oriented. Quality oriented cycles are of shorter length *viz.,* LP-DS-UP, LP-DS-MS-LP and LP-DS-MS-LoS-LP. Quantity oriented popular cycles are longer cycles *viz.,* LP-LoS-LS-LS-LP, LP-LoS-LS-LS-DS-LoS-LP, LP-LoS-LS-LS-MS-LS-LP, LP-LoS-LS-DS-LoS-LS-LP, LP-LoS-LS-DS-LoS-LS-MS-LoS-LP etc. The combination of quality and quantity cycles includes LP-LoS-DS-LoS-LP, LP-LoS-LS-DS-LoS-LP etc. It has been observed in Himachal Pradesh that even within a 4 year cycle, interchanging of DS operation with LS (i.e. HP-UP-DS-UP cycle vs. HP-UP-UP-DS cycle) affected the cycle productivity under both time and level of rejuvenation pruning (Tables 7.1 & 7.2, Jain *et al.* 1992)

Length of pruning cycle

Though, pruning is determined based on the interaction of elevation and type of tea grown at a particular place, the length of the pruning cycle under normal conditions may be estimated by the following formula:

$$\text{Length of pruning cycle in years} = \frac{\text{Elevation (in thousand feet)}}{1000} \pm 1$$

Addition or deletion of 1 (one year) is specific to type of tea as described below:

Assamica type: +1, China type: -1 and China hybrid : +0

For example, for tea growing at 5,000 feet elevation, the pruning cycle should be equal to 5000 ÷1000 ± 1, i.e. = 6 or 4 years

Based on length of pruning cycle, the plantation area to be pruned is worked out by:

Area to be pruned (ha) = Total Area (ha) ÷ Length of pruning cycle

Crop distribution is more even with longer pruning cycles than with annually pruned tea (Rahman and Mitra 1973)

OTHER MANAGEMENT PRACTICES DURING PRUNING SEASON

Vacancy infilling

The most important work after pruning is to mark the vacancies and infill them with suitable fast growing cultivars. For quick coverage of the vacant patch, the infilling should be carried out at the rate of "2n+1" where n is the number of existing vacancies. For example if there is a vacancy of 3 bushes, 7 infills need to be planted. Tocklai studied the response of different methods of infilling following rejuvenation pruning at 40-45 cm i.e. no infilling, infilling with one plant per vacancy or two plants or three plants in staggered hedge planting and observed that in the seventh year, the increase in yield was 9, 47, 56 and 71 per cent over control where normal pruning was done without infilling (Sarkar 1981).

CSIR-IHBT standardized a method of transplanting century old china hybrid tea bush which may be used for vacancy infilling as well as interplanting in low density plantations (Fig 7.6, Jain *et al.* 1995). Every month, tea bushes were uprooted and transplanted with different root lengths (between 15 to 45 cm) and with or without 45 cm tall three breathers. Considering survival, growth and production, February and March were the most suitable months followed by April and May for transplanting. The root length of the bushes being transplanted should be more than 35 cm. Leaving breathers on the transplants *per se* has little influence on the survival or growth of the plants but it may protect against trampling etc. Advantage of the method is that the section with high vacancy may be uprooted for new plantation and the uprooted bushes may be used for infilling and consolidation of the other section under development or rejuvenation. Other advantage is that the rate of recovery of the vacant patch is faster with the transplanted bush than nursery plant. A 30-35 months old Assamica type nursery plant has been reported to have poor survival after stump planting (Rahman 1977).

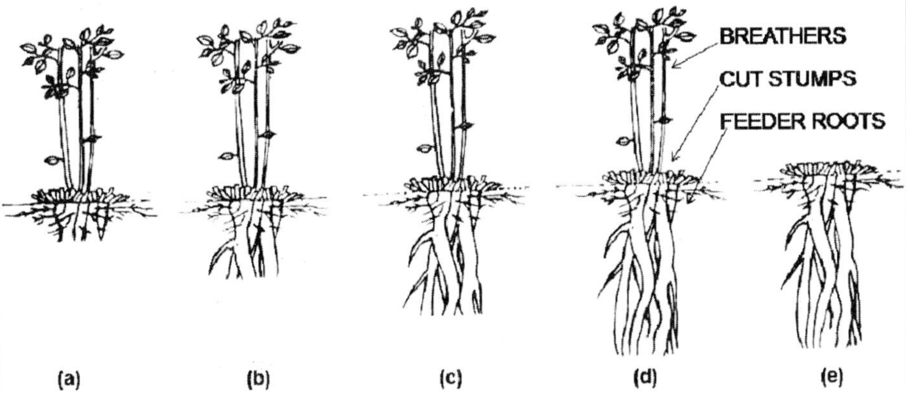

Uprooted old china hybrid tea showing (a) 15 cm root with breathers, (b) 25 cm root with breathers, (c) 35 cm root with breathers, (d) 45 cm with breathers, (e) 45 cm root without breathers.

Fig. 7.6. Method of transplanting century old china hybrid tea bush

Other operations

Other activities include removal of established perennial weeds, establishment of effective drainage system, shade regulation, land leveling for filling up the depressions etc.

Maintenance of soil fertility is of paramount significance. Cultivating green crops during pruned year is one option to achieve it. In the mid-country intermediate zone of Sri Lanka studies were conducted on post pruning cultivations of *Tripsacum laxum* and *Eragrostis curvula* grasses in between pruned tea rows for a period of 6 months or forking done both at the end and in the middle of the pruning cycle (Amarasekera *et al.* 1997). In cultivated plot, soil physical properties *viz.*, organic carbon content, total pore space and water retention characteristics improved and the mean cycle yield was also higher. Made tea yields were highest with *E. curvula*.

DEVELOPMENT OF UNPRODUCTIVE AND ABANDONED TEA PLANTATIONS

Rehabilitation of tea plantation by applying semi-hard pruning (35 cm) and mulching with the prunings was observed to better practice with regard to tea yield and quality, shoot regrowth, bush size, and economic effectiveness than either leaving the plantation unpruned and unplucked for 1-2 years, or hard pruning coupled with deep soil cultivation (Tabageri *et al.* 1991). According to Chakrabarty and Barpujari (1992) from Tocklai, for restoration of aged plantation, (i) remove as much of the unproductive bush frame as possible by pruning at a low height, (ii) thoroughly clean out the remaining part of the bush of all diseased and damaged parts, (iii) create conditions favourable for quick regrowth of shoots and roots, and (iv) maintain plant density by infilling. According to them, the plantation should be kept on rest at least for 2 months and fertilizers should be applied 5-6 months ahead of the pruning.

In Kenya, rejuvenation of 65 year old seedling tea plantation by normal cut-across pruning to 66 cm height and tipping at 10 cm above the pruning height with or without gap filling produced significantly greater yields (Mwakha 1994). Increase in productivity by 2-4 times with rejuvenation pruning, infilling and consolidation has also been reported by Tocklai (Jorhat 2006).

It has been concluded that though rejuvenation pruning and interplanting moribund tea with clonal plants has been effective in arresting the deterioration of old tea sections, raising their productivity and early payback period, yet it should not be used as an alternative to uprooting and replanting with high-yielding clones as the old plantation and infilling in unrehabilitated soil are unlikely to remain productive for long period (Chakrabarty and Barpujari 1992, Njuguna and Magambo 1993).

Development of dilapidated tea plantations in Himachal

Out of 2,312 hectares of the area under tea in Himachal Pradesh, 1094 hectares (47%) were lying in the abandoned / neglected state (Tea Board 1997). Tea gardens have been neglected by the growers owing to various reasons. The main reasons identified for not managing the garden/part of garden in the state are described below (Sud, 1998):

Size of holding: Most of the gardens in Himachal Pradesh are below 1 ha size and majority of such gardens is lying in abandoned / neglected state, because the owners of the gardens are not depended on tea.

Low returns: Many tea garden owners revealed that their garden/ part of garden are not taken care because of low productivity.

Labour shortage: Being labour intensive crop, the shortage of labour is another major factor in Himachal Pradesh. It appears that working in tea gardens in Himachal is rated as sub-standard job by the workers. The hard task in tea plantation in comparison to other agriculture/similar activities is another factor for non-availability of the working force.

Lack of interest: Continuous partitioning of properties, joint holding and being side business for majority of planters, are the other factors due to which the owners are not responding towards tea plantations.

Unawareness of potential: Owing to above factors, the planters are managing their gardens half-heartedly. They are not using the modern agrotechnology and following only the conventional practices. Neither they provide proper inputs, nor do they follow standard pruning / skiffing operations. Hence, they are getting the low returns. There is lot of gap between the existing potential of tea bushes and the harvested crop.

Other factors: Lack of technical guidance, finance, and unfenced field were main problems faced by the tea growers of Himachal Pradesh.

Development Plan

With the funding by Tea Board, the development plan of these plantations was implemented in two phases. In first phase (1997-2000) 150 acres of such plantation was targeted. The selected gardens were pruned for rejuvenation during winter months. These abandoned/neglected gardens were heavily infested with wild bushes, unwanted tree saplings, and other undesirable growth, and the frame of existing tea bushes was also knotty. Hence, in the beginning, all the wild growth was removed. Thereafter, the tea bushes were subjected to rejuvenation pruning at/near collar level in winter months. Light ground work was done for leveling the local depressions. Vacancies were also marked and infilling was done in the month of July-August, the rainy season.

During April-June, application of post-emergence herbicides was done for effective weed control measure. Fertilizers were administered during August-September, when the bushes had sufficient foliage. The bushes reached to tipping level by end June in majority of the gardens. After tipping, the plucking was done at the convenient interval. The targeted area was eventually successfully developed.

During second phase the target fixed was 100 ha area during the period 2000-2005, which was also accomplished successfully.

Sequence of operations for development of a dilapidated garden

1. Dilapidated tea garden is characterized with over growth of weeds, wild brushes etc. and unproductive, knotty and overgrown tea bushes (Figure 7.7). At the outset, removal of the wild growth in the plantation, killing the stumps of the perennial brushes with paraquat herbicide paste prepared with mobil oil or kerosine, applying on the fresh cuts or spraying glyphosate at 1% on the regrowth following cutting of the brushes at ground level (Singh *et al.* 1997).
2. Rejuvenation or deep pruning at collar level of the bush to remove old, dead and diseased bush frame during November/ December.
3. Cheeling (light hoeing) the ground for better aeration, water conservation and removal of the established weeds, and leveling the land by filling the depressions to avoid water logging and facilitate easy movement of the pluckers.
4. Marking the vacancies and subsequently infilling them @ 2n+1 (n = vacancy number) before onset of the monsoon.
5. Regular weed and other pest control measures (physical and chemical).
6. Application of balanced dose of fertilizers based on soil analysis after the pruned bushes attain sufficient foliage growth.
7. Tipping of the pruned bushes at 60 cm level from the ground in July-August to form plucking table in the first year.

8. Regular plucking at tipping level by plucking to *janam* in the rest of the season.
9. Routine garden management practices *viz.*, shade tree management, water management, pest control, foliar application of the nutrients etc.
10. Level-off-skiffing in winter.
11. Plucking at 65 cm level to *janam* during entire season in the second year.
12. Deep skiffing of the bushes done 5 cm below the tipping level in winter month (November/December).
13. Tipping, 10 cm above deep skiffing level in May in the third year.
14. Plucking to *janam* at the tipping level up to season end
15. Level-off-skiffing in winter.
16. Plucking at last year's plucking level to *janam* in the fourth year.
17. Light pruning at 40 cm from the ground in winter.
18. Tipping at 60 cm during June/July in the first year of second cycle.
19. Repetition of the cycle.

This package of agro technology could be repeated for development of dilapidated tea plantations of the Sikkim state (Sud *et al.* 2005).

Dilapidated/abandoned/ neglected tea garden

Garden pruned for rejuvenation after removal of wild growth

Post-pruning ground cheeling, depression leveling

Vacancy infilling, effective nutrition management, weed & pest control

Tipping, regular plucking and maintenance of garden

Fig. 7.7. Sequence of operations for development of dilapidated plantation

Pruning requirement of tea plantations of Himachal Pradesh

Tea plantations in Kangra valley are of chinary type, where tea bushes have multi-stems, small leaves and short internodes, from which orthodox black tea or green tea is manufactured. The plucked leaf from the garden is collected by the co-operative tea factories, which adjudge the price of leaf based on its fineness i.e. the relative proportion of 2+Bud, 1+Bud and 1+*banjhi* shoots.

With continuous unpruned condition of tea bush, the height of plucking table increases which poses problem in plucking. The pluckable shoot size becomes smaller and smaller, density of shoots increases, internodes become shorter, many shoots fail to grow, and incidence of *banjhi* shoots rises. All these factors ultimately result in low crop productivity and quality, high plucking cost and poor economics.

In order to overcome such situation, the bushes are required to be pruned at appropriate level, taking into consideration the condition of permanent frame, the thickness of primaries, the pruning cycle followed, the condition of top hamper etc. (Singh *et al.* 1999).

If the permanent frame is faulty (knotty or weak), it is recommended to prune at ground level so that new frame could be established, and this type of prune is called `Heavy Prune' or `Rejuvenation Prune'. This prune can be applied at 20 cm, if the frame is good. For better results, such pruning should be done in the month of November/ December. The recovery from such prune depends on pre-prune growth of bush, the starch reserve in the roots, physical characteristics of soil etc.

After rejuvenation prune, when the frame attains at least pencil thickness, the bushes should be cut back at 35 cm. Thereafter, the section should be put under a cycle of 4 or 5 years, following light prune 2-4 cm above the previous prune mark. After 3-4 light prunes, the frame becomes knotty, necessitating to go for medium prune, i.e. prune just below the cut back level (below 35 cm) and then the cycle should be repeated. Hence, it is evident that there is need to prune the tea bushes below 35 cm after 16-20 years. However, if regular pruning cycle is not followed or faulty pruning operation is adopted then the heavy prune may be required at early stage.

Main observations regarding pruning in century old chinary tea in Himachal Pradesh are as follows:

1. Chinary tea in Himachal Pradesh, being multi-stemmed, free from major primary root diseases, does not show any

mortality due to drastic prune like pruning at or below ground level. During, the planters, in general, were adopting the practice of '*denther*' i.e. pruning at collar level after every 5 to 6 years, keeping the bushes unpruned in between. Though the pruning system is faulty, yet it has not caused any mortality of tea bushes.

2. Before the establishment of the CSIR-IHBT (i.e. 1985), many tea gardens in Himachal Pradesh were in dilapidated state. The Institute recommended collar pruning (besides other management practices) as the first step to revive such gardens. Consequently, the state's production which was 6.77 lakh kg in 1985 rose to 17.11 lakh kg in 1998.
3. The tea bushes respond well to rejuvenation, provided other factors are favourable e.g. there is no waterlogging, the ground is *cheeled* for better aeration, the wild brushes etc. are removed and proper nutrition, particularly of potash is given.
4. Unlike Assam tea bush, it is very difficult to estimate the economic life span of chinary tea bush. The century old chinary tea plantations in Himachal Pradesh with better management practices have produced upto 3,000 KMTH.
5. In this state, the tea planters are not following the repleni-shment of older sections with new plantation. They prefer to go for rejuvenation prune, which has very low gestation period, with consistent rise in productivity.
6. About12 ha experimental tea garden in CSIR-IHBT, which was rejuvenated in the years 1984-86, the crop production (1673 KMTH) till 1998 was towards peak, the optimum yield potential of which is yet unknown (Table 7.3).

Table 7.3. Average performance of 23 CSIR Tea Technology Demonstration Plots

Year	Garden Average Yield (KMTH)	Demonstration Plot Yield (KMTH)
1984	504	504
1985	464	473
1986	554	1207
1987	626	1514
1988	672	1616
1989	742	1658
1990	1014	1762
1991	1272	1862

FUTURE AREA OF WORK

Higher production than consumption, competitive market, escalating cultivation costs and low profit margins are the global issues of concern. For cutting down the cost, mechanization of the operation is the need of the hour. Similarly, for value addition, organic farming movement is getting momentum. Hence standardization of pruning practices (including time and method) and pruning cycle under mechanized and organic farming situations is required. Observation on effect of mechanized cuts on plant growth and shoot regeneration pattern needs to be recorded.

During pruning heavy volume of pruning litter is removed from the bush. Being woody, its *in situ* decomposition takes long time. When pruning litter is removed from the field, heavy loss of mineral nutrients occur. Hence, for soil fertility management, efficient techniques for conversion of pruning litter to organic compost is another area of research. In addition, after pruning, knife clean out needs to be done for removal of the *banjhi* and unproductive left over shoots. This operation also accounts for high cost of production. An alternative cheaper method, such as spray of chemical which damage the unwanted growth without harming the tea bush need to be devised.

REFERENCES

Amarasekera AR, Abeysekera UP, Anandacoomaraswamy A (1997) The effect of post prune operations on soil physical properties and yield of tea. *Sri Lanka Journal of Tea Science* 65(1-2): 76-82.

Anyuka JCO, Mwakha E, Siele DK, Odhiambo HO (1992) On-farm demonstration trials on methods and timing of tea pruning in Kenya - a technical note. *Tea* 13(2): 84-90.

Baby UI, Premkumar R (1999) Blister blight disease of tea: how to achieve maximum control. *Planters' Chronicle* 95(7): 295-297.

Barbora AC, Banerjee MK (1988) Causes and prevention of knot formation in tea. *Two and a Bud* 35: 39-42.

Barman TS, Baruah U, Saikia J (1997) What to spray: urea/zinc/urea + zinc? *Two and a Bud* 44(2):23-24.

Barman TS, Saikia JK, Pathak SK (1998) Starch reserve and growth of tea. *Two and a Bud* 45(1): 14-18.

Barua DN (1989) Science and practice in tea culture. Tea Research Association, Calcutta, India, pp 341-368.

Barua SC (1999) Chemical growth regulation and its effect on yield in tea. *Journal of Plantation Crops* 27(3): 196-200.

Baruah S, Sarmah RC, Neog RC (1998) Training needs of small tea growers of Assam on tea culture technology. *Journal of the Agricultural Science Society of North East India* 11(1): 31-36.

Basu S, Dutta AC (1973) Kickers vs. no kickers. *Two and a Bud,* 20(1):39.

Bore JK, Isutsa DK, Itulya FM, Ng'-Etich WK (2003) Effects of pruning time and resting period on total non-structural carbohydrates, regrowth and yield of tea (*Camellia sinensis* L.). *Journal of Horticultural Science and Biotechnology* 78(2): 272-277.

Celebioglu G (1973) The effects of different crown shapes on the yield of tea plantations. *Yalova Bahce Kulturleri Arastirma ve Egitim Merkezi Dergisi* 6(3-4): 9-34.

Chakrabarty J, Barpujari N (1992) A few aspects of pruning. In: Chakravartee J (ed) Field Management in Tea, Tocklai Experimental Station, Jorhat, India, pp 66-77.

De Costa WAJM, Navaratne DMS, Anandacoomaraswamy A (2009) Physiological basis of yield variation of tea (*Camellia sinensis*) during different years of the pruning cycle in the central highlands of Sri Lanka. *Experimental Agriculture* 45(4):429-450.

DTRDC (2002) Pruning. http://www.dtrdc.org/pruning.htm.

Eden T (1976) Tea. 3rd edition, Longmans group Limited, London, pp 236.

Esartiya MV (1972) The effect of different pruning methods on root development in tea. *Subtropicheskie Kul'tury* 6: 61-62.

Gong ZhiHua, Huang YiHuan, Luo JunWu, Xiao WenJun, Tang HePing, Shen Cheng Wen, Gong ZH, Huang YH, Luo JW, Xiao WJ, Tang HP, Shen CW (2002) Studies on the effect of heavy pruning on the protective enzymes in senile tea plant. *Journal of Tea Science* 22(2): 152-155.

Gope B, Das SC (1992) Biology and effect of pruning and skiffing on the distribution of the scarlet mite, *Brevipalpus phoenicis* (Geijskes), on tea in North-East India. *Two and a Bud* 39(1): 23.

Hadfield W (1974). Shade in North East India Tea Plantations. I. The shade pattern. *J. Appl. Ecol.* 11:151-176.

Hazarika UN, Pradhan SK, Sarkar SK (1985) Effect of extended pruning cycles of varying durations on yield of tea in Darjeeling. *Two and a Bud* 32(1-2): 2-6.

Ho HsinFeng, Wang LianChuan, Lin ChengShih, Ho HF, Wang LC, Lin CS (1992) Effects of different pruning times and different plucking methods on tea growth, quality and economic evaluation in the tea district at a middle elevation. *Taiwan Tea Research Bulletin,* No. 11: 21-30.

Hudson JB, Durairaj J, Muraleedharan N (1988) Role of rejuvenation pruning in augmenting tea production in south India. Bulletin of UPASI Tea Scientific Department, No. 51, pp 22-25.

Hudson JB, Muraleedharan N (1998) Impact of cultural operations on blister blight control. *Planters' Chronicle* 93(8): 360-361.

Jain NK, Chakrabarty DN, Singh RD, Sud RK, Tamang MB (1995) Consolidation of vacancy by transplanting old china hybrid tea (*Camellia sinensis*) in hilly areas. *Indian Journal of Agricultural Sciences* 65(2): 93-97.

Jain NK, Chakrabarty DN, Tamang MB, Singh RD, Sud RK (1992) Studies on time and level of rejuvenation pruning in China hybrid tea in Himachal. *Journal of Plantation Crops*, **20**(S): 29-34.

Jain NK, Tamang MB (1990) Physiological basis of tea management practices. In: Proceedings of the International Congress of Plant Physiology. New Delhi. Feb. 15-20. Volume I, pp 469-479.

Jayamanne VS, Mahinda W, Wijayaratna CM (2002) Adoptability of new technology in the small-holdings tea sector in the low country of Sri Lanka. *Journal of Agriculture and Rural Development in the Tropics and Subtropics* 103(2): 125-131.

Jorhat (2006) Tocklai Tea Research Centre. R&D Activities in Tocklai. http://jorhat.nic.in/ttc.htm

Kathiravetpillai A, Abeysekera UP, Jayasekera APDA, Jayasinghe JASKV, Jayawickrema PK (1997) Effect of sprays of plant growth regulator hydrogen cyanamide (Dormex) on yield of tea. *Sri Lanka Journal of Tea Science* 65(1-2): 67-75.

Kilavuka CI (1980) An observation on the recovery of pruned tea under shade. *Tea* 1(1): 4-6.

Kulasegaram S, Kathiravetpillai A, Shahmagarajah V (1975) Approaches to higher tea productivity through proper bush management. Tea Research Institute of Sri Lanka, Talawakele, pp1-10.

Magambo MJS (1980) Crop physiology: Influence of reducing bush (down prune) on recovery from pruning and yield. *Tea* 1(2) :1-3.

Magambo MJS, Cannel MGR (1981) Dry matter production and partition in relation to yield of tea. *Experimental Agriculture* 17: 33-38.

Malenga NEA (1997) Long term effect of downpruning mature Indian hybrid tea on yield. Quarterly Newsletter Tea Research Foundation of Central Africa, No. 126, pp 18-21.

Manivel L, Kumar RR, Marimuthu S, Venkatesalu V, Rajkumar R (1998) Plant growth regulators for crop management in south Indian tea. *Journal of Plantation Crops* 26(1): 63-69.

Marimuthu S, Manivel L, Kumar RR (1996) Carbohydrate reserve in tea roots. *Planters' Chronicle* 91(1): 43-45.

Marimuthu S, Ranganath SR (1994) Effect of different pruning heights on chemical composition of green leaf and quality parameters of CTC teas. *Journal of Plantation Crops* 22(2): 115-118.

Mehta KG, Katoch DC, Misra SK (1972) The effect of the time of pruning on the yield of tea in Himachal Pradesh. *Journal of Research, India*, 9(1): 35-39.

Mitra H (1976) History of development of young tea pruning. *Two and a Bud* 23: 4-7.

Mkwaila B (1994) The impact of late pruning on mature tea. Quarterly Newsletter Tea Research Foundation of Central Africa, No. 113, pp 3-8.

Monkhouse AJW, Lennox RG (1976) The effect of pruning trash mulch on mature tea, 1969-1975.*Tea in East Africa* (161): 11-21.

Moorti TV, Negi YS, Pathaia M (1985) Economics of tea cultivation (a study of tea farms of tehsil Palampur, Dist. Kangra, H.P.) *ISAE* (mimeo), pp 62.

Mouli BC (1993) Blister blight control - new recommendations. *Blister blight control new recommendations*, UPASI, India, pp 8.

Muraleedhaan N (1995) Strategies for the management of shot-hole borer. *Planters' Chronicle*, Jan: 23-24.

Muraleedharan N, Mohan MS, Radhakrishnan B, Udayabhanu KG, Surendra MM (1994) Seasonal abundance and vertical distribution of purple mite, *Calacarus carinatus* (Green) (*Eriophyidae*: Acarina) infesting tea. *Entomon* 19(1-2): 53-56.

Mwakha E (1994) Yield and quality of old-seedling and young-clonal tea under different rehabilitation methods and fertilizers. *Tea* 15(1): 35-41.

Mwakha E (1997) Tea pruning for yield improvement. *Tea* 18(2): 144-148.

Nagarajah S, Pethiyagoda U (1965) The influence of lungs on carbohydrate reserves and growth of shoots. *Tea Quarterly* 36(3): 88.

Nakano T (1999) Modelling of the developmental process of lateral buds after skiffing in tea plants. *Japanese Journal of Crop Science* 68(3): 433-439.

Ng'etich WK (1996) Pruning in tea: a review focusing on experiments in east Africa. *Tea* 17(1): 41-46.

Njuguna CK, Magambo MJS (1990) Long term effect on yields of pruning clonal tea (*Camellia sinensis* L.) at different heights at high altitude in Kenya. *Tea* 11(1): 13-17.

Njuguna CK, Magambo MJS (1993) Effect of rejuvenation pruning and increasing plant density on yields of old seedling tea bushes interplanted with clonal tea plants. *Tea* 14(1): 13-20.

Nyabundi KW (2004) Effect of time of pruning and number of lungs on carbohydrate reserves of tea. Tea Research Foundation of Kenya Quarterly Bulletin No. 9(2), pp 2-3.

Odhiambo HO, Othieno CO (1992) Effect of height and month of pruning on yields of tea. *Tea* 13(1): 18-21.

Otieno W (1997) Epidemiology and management of Hypoxylon wood rot of tea. *Tea* 18(2): 175-183.

Owuor PO (1990) Advances in the improvement of tea quality through chemical and biochemical research in Kenya. *Tropical Science* 30(3): 307-319.

Pethiyagoda V (1964) Some observations on the dormancy of the tea bush. *Tea Quarterly* 35(2): 74-84.

Premkumar R, Baby UI, Muraleedharan N (1998) Forecasting of blister blight disease and its management - recent developments. In: Proceedings of the 35th UPASI conference, Coonoor, Tamil Nadu, India, 7 September 1998. Bulletin of UPASI Tea Scientific Department, No. 52, pp 34-41.

Radhakrishnan B (1997) Tea in Wayanad. *Planters' Chronicle* 92(10): 463-471.

Rahman F (1977) Stump planting of tea. *Two and a Bud* 24(1): 8-9.

Rahman F (1988) Physiology of the tea bush. *Two and a Bud* 35: 1-14.

Rahman F, Mitra H (1973) Factors affecting the choice of pruning cycle in mature tea. *Two and a Bud* 20(1): 11-13.

Rajalakshmi N (2003) Biochemical changes in tea infected by *Exobasidium vexans* causing blister blight. *Plant Disease Research, Ludhiana* 18(1): 1-4.

Ram CSV (1974) Pruning for rejuvenation. *Planters' Chronicle* 69(14): 279-282.

Ramaswamy Ravichandran, Ramaswamy Parthiban (1998a) Changes in enzyme activities (polyphenol oxidase and phenylalanine ammonia lyase) with type of tea leaf and during black tea manufacture and the effect of enzyme supplementation of dhool on black tea quality. *Food Chemistry* 62(3): 277-281.

Ramaswamy Ravichandran, Ramaswamy Parthiban (1998b) Occurrence and distribution of lipoxygenase in *Camellia sinensis* (L) O. Kuntze and their changes during CTC black tea manufacture under southern Indian conditions. *Journal of the Science of Food and Agriculture* 78(1): 67-72.

Rattan PS (1996) Hot and cold disorder. Quarterly Newsletter Tea Research Foundation of Central Africa, No. 121, pp 23-28.

Roberts GR, Silva ULLde, Wettasinghe DT (1976) The effect of cultural operations on the composition of the xylem sap of tea plants recovering from pruning. *Annals of Botany* 40(168): 825-831.

Satyanarayana N, Cox S, Cox S (1994) Factors influencing productivity of tea in drought. *Journal of Plantation Crops* 22(2): 87-92.

Selvendran RR, Selvendran S (1972) Changes in the polysaccharides of the tea plant during post-prune growth. *Phytochemistry* 11(11): 3167-3171.

Seneviratne D (1995) Shot-hole borer (*Xyleborus fornicatus*) and tea, a 100 years. *Journal of the National Institute of Plantation Management* 11(1-2): 34-48.

Singh RD, Singh B, Sud RK, Ravindranath SD (1999) हिमाचल के चाय बागानों का शोध एवं प्रसार कार्य द्वारा जीर्णोद्वार – एक विवरण *Journal of Council of Scientific and Industrial Research* (Hindi) 7(2): 73-77.

Singh RD, Singh B, Sud RK, Tamang MB, Chakrabarty DN (1997) Control of *Lantana* in non-cropped area and in tea plantations. *International Journal of Pest Management* 43(2): 145-147.

Sud RK, Prasad R, Singh B, Vats SK (2005) A survey of abandoned and neglected tea plantations in Sikkim for assessing the possibility of their rejuvenation and extension. *Journal of Council of Scientific and Industrial Research* (Hindi) 12(1): 25-29.

Sud, RK (1999) First Annual Progress Report of the project 'Improving tea production of 150-acres (dilapidated) in Himachal Pradesh' submitted to Tea Board of India: 2-3.

Sugha SK, Singh BM (1988) Effect of pruning cycle on severity of blister blight of tea. *Indian Journal of Mycology and Plant Pathology* 18(2): 199.

Sugha SK, Singh BM, Sharma KL (1990) Factors affecting develop-ment of blister blight of tea. *Indian Phytopathology* 43(2): 226.

Sugha SK, Singh BM, Sharma KL (1990) Factors affecting development of blister blight of tea. *Indian Phytopathology* 43(2): 226.

Tabageri LG, Mikeladze AD, Cheishvili ID (1991) Results of studies on measures for rehabilitating depressed tea plantations on podzolic soil. *Subtropicheskie Kul'tury* 4:136-141.

Tea Board (1997) Tea directory of H.P. Tea Board of India, Palampur, Himachal Pradesh, India, pp 1.

Tea Board (2002) Tea cultivation procedure: Pruning. http://www.teaindia.org/cultiv.html

Templer JC (1974) Pruning bearing tea. Tea Research Institute of East Africa Annual Report, pp 52-54.

Templer JC (1975) Pruning bearing tea. Tea Research Institute of East Africa Annual Report, pp 81-82.

Thomas J, Saravana M, Kumar RR, Pius PK (2005) Influence of age after pruning on the levels of flavanols and other bioconstituents in tea (*Camellia sinensis*). *Journal of the Science of Food and Agriculture* 85: 931-934.

Thomas PC (1979) Economics of rejuvenation pruning. In: Proceedings of the second joint area scientific symposium 1979. Bulletin, UPASI Tea Scientific Department, No.36, pp 64-68.

Tocklai (1972) Heavy pruning in Darjeeling. Serial, Tocklai Experimental Station. No. 193, pp 3.

Tubbs FR (1938) Report of the Plant Physiologist for 1937. Bulletin, Tea Research Institute, Ceylon, No. 18, pp 44-63.

Udayakumar S, Muraleedharan N (1998) Rejuvenation pruning and consolidation in Alageswar estate. *United Planters' Association of Southern India Joint Area Scientific Symposium-VII* (JASS-VII), Chikmagalur, Karnataka, India, 20-21 March 1998. Bulletin of UPASI Tea Scientific Department. 1998, No. 51, pp 26-30.

Ukers WH (1935) *All about tea* 1: 375-416. (The tea and coffee trade journal company, New York, 1985).

UPASI (2002). Pruning In: Hudson JB, Durairaj J, Muraleedharan N (eds) Guidelines on tea culture in south India. United Planters' Association of Southern India, Coonoor, Tamilnadu, Allied Publishers Pvt. Ltd.

Wadasinghe G, Senanayake YDA (1990) Effect of pruning and plucking on growth and development of tea (*Camellia sinensis* (L.) O. Kuntze). *Sri Lankan Journal of Agricultural Sciences* 27: 1-13.

Waheed A, Hamid FS, Naseer A, Noor A, Rauf A, Ahmad N, Alam N (1998) A comparative observation on agronomic characters by skiffing on tea (*Camellia sinensis* L.). *Sarhad Journal of Agriculture* 14(2): 93-96.

Wang DF (1991) Effect of pruning on carbon and nitrogen distribution in tea plants. *Plant Physiology Communications* No. 1, pp 26-29.

Weeraratna CS, Watson M, Wettasingha DT (1977) Effect of minera-lization of tea prunings on some soil characteristics. *Plant and Soil* 46(1): 93-99.

Whittle AM, Rattan PS (1995) Pesticide residues in tea. Quarterly Newsletter Tea Research Foundation of Central Africa, No. 120, pp 24-26.

Wijeratne MA, Premathunga P, Karunaratne WRMM (2002) Variation of root starch reserves of tea and its impact on recovery after pruning. *Journal of Plantation Crop*, 30(1): 33-37.

Xei XM, Yang XQ, Shen YM, Wang DF, Li LM (1993) Influences of different management procedures on the distribution of carbon and nitrogen in tea plants and on the quality of tea leaves. *Acta Agriculturae Nucleatae Sinica* 7(1): 29-36.

Yamashita M, Tanaka K, Sekiya N (1975) The effect of plucking times and the time and level of skiffing on the time of first plucking and yield in the following year. *Study of Tea* 49: 62-70.

Yilmaz G, Kandemir N, Kinalioglu K (2004) Effects of different pruning intervals on fresh shoot yield and some quality properties of tea (*Camellia sinensis* (L.) O. Kuntze) in Turkey. *Pakistan Journal of Biological Sciences* 7(7): 1208-1212.

8

TEA PLUCKING

R.K. Sud

CSIR-Institute of Himalayan Bioresource Technology,
Palampur (H.P.)-176 061, India. E-mail: rksud@ihbt.res.in

DEFINITION AND IMPORTANCE

Plucking in tea refers to harvesting of the apical shoots (the crop) growing above the pre-determined plucking level of the bush. Plucking is the most important operation in tea husbandry because:

- It is the crop harvesting operation.
- It removes the apical dominance, thus stimulates regeneration of new shoots for future crop.
- It checks the vertical growth of the bush and keeps it in vegetative phase.
- It determines the quantity and quality of the harvested crop.
- It influences the tea bush health.
- It is the costliest operation, accounting for 70-80% of total cost of green leaf production.
- It is the highest energy consuming operation, accounting for 63% of the total field energy consumption (Yadav and Srivastava 2002).

Until 1920s, the systems of pruning and plucking, as they exist today, were almost unknown even in China where the tea culture actually developed. Torgasheff (1926) described the then cultivation method as "primitive" and plucking method as "barbarous". The tender young tea shoots growing in flushes in naturally growing tea shrubs or trees were plucked as crop. Such plucking involved great care and difficulty particularly in high hills. From inaccessible heights, monkeys were reported to be trained to pluck tea in China (Maitland 1982).

With the passage of time, the concepts of pruning, tipping and maintenance foliage developed along with the experience in plucking, and better understanding of physiology of tea bush.

Fig. 8.1. Tea plucking in Kangra valley, Himachal Pradesh

BASICS OF PLUCKING

The bush of a cultivated tea has a permanent frame comprising of a number of branches called "pruning sticks". When the bushes are pruned, the cut ends of the pruning sticks lie in a horizontal plane. The new shoots that emerge from dormant buds on the sticks are called "primary shoots" or "primaries". These primaries are decapitated ("tipped") when they reach the pre-determined tipping level above sufficient maintenance foliage on the bush. After a primary shoot has been tipped, new shoots arise from axillary buds and the phenomenon is called "regeneration". Tipping stimulates the regeneration of dormant axillary buds on the primary shoots into actively growing lateral shoots which are then plucked as crop when they grow above tipping level. Repeated plucking of successive orders of lateral shoots, as crop results in the formation of a complex shoot system on each single primary shoot.

The fundamentals of tea plucking are mainly based on photosynthetic efficiency of tea bush and rate of shoot growth. During old times, the tea plants were plucked soon after the shoot growth without leaving behind any leaf, thus restricting their size to cabbage shaped bushes. Such bushes had a low yield potential. But with experience it was realized that retaining sufficient leaves on the bush improves crop production. In 1853, Williamson in Assam showed that plucking the harvestable shoots after leaving a few leaves on the stems below it before plucking away helped in sustaining productivity of the bush in the later season (Sarkar 1988). Thus, the concept of retaining maintenance foliage came into existence. A planter, Captain Harrison, also realized the importance of maintenance foliage in Kangra in 1860's who converted tea rows into continuous hedge and obtained higher production; similarly another planter, Captain Fitzgerland, raised the tea productivity three-folds by integrating plucking with high cultivation and better method of pruning (Gazetteer of the Kangra district 1905).

Photosynthetic efficiency of tea bush

The leaves, which are vital organ being site of photosynthesis for any plant, are harvested as crop in tea. Hence, in order to have a balance between the leaves as crop and leaves as maintenance foliage for photosynthesis, a plucking table is established, the mature leaves below the table serve as maintenance foliage and shoots growing above the table are plucked.

Physiological linkage between the maintenance foliage and growth of shoots (harvestable crop), was established in 1950's and was subsequently studied in depth. These studies explain the basis for different standards, systems and methods of plucking. It was reported that photosynthetic capacity develops gradually in young expanding leaf with age till it matures, even the third leaf on a growing shoot is photosynthetically less efficient than a mature leaf of same size (Barua 1960, Sivapalan 1977). The relative photo-synthetic rate of first leaf, second leaf and third leaf and fourth leaf (mother leaf) was reported to be 21, 28, 47, and 76 per cent, respectively than that of maintenance leaf (Rajkumar *et al.* 1998). Hence, most of the carbohydrates requirement of the bush and growing shoots for metabolism and growth are synthesized by the maintenance foliage. Radiotracer studies at Tocklai have also shown that the successive generations of shoots developing at the plucking table are the strongest sinks on a plucked tea bush and relative sink capacity of the bud, first leaf, second leaf and third leaf of a growing shoot is 100 per cent, 70 per cent, 40 per cent and 30 per cent, respectively (Tocklai 1979, 1984). Therefore, it is essential to retain sufficient foliage on the bush for the growth of the pluckable shoots. Any reduction in the

TIPPING

Definition and importance

Tipping refers to the process of decapitation of shoot tips of growing primaries arising from the dormant buds on the sticks in a pruned bush. It is done at pre determined height to establish a plucking table, to maintain the depth of maintenance foliage, to remove the apical dominance and stimulate regeneration of dormant axillary buds on primary shoots into pluckable lateral shoots, and to achieve good spread of the bushes.

Tipping level, established above pruning level, regulates the maintenance foliage depth. This level varies according to type of tea, climatic conditions, length of pruning cycle, standard of pruning and system of plucking to be followed (Figure 8.3).

Fig. 8.3. Tipping of China hybrid tea bush

Basis of tipping measures

On the basis of long term trial on annually pruned tea, Rahman and Barua (1980) reported that highest initial yield could be obtained from the low tipping measure of 10 cm, but sustained highest yield was possible with longer tipping measure of 20 cm. From the work done at Tocklai in the past (Barua and Weight 1959, Barua 1961), it has been concluded that from production point of view the ideal tipping height is the natural *banjhi* horizon (the mode height of dormant shoots on a pruned bush). This height in a normal Assam tea bush corresponds to 20 cm, equivalent to average height of 5 leaves of a primary shoot. However, the average height may vary depending on type of tea, shoot vigour, pruning level etc. For China hybrid tea, the average height of 5 leaves is lower than that of Assam tea. Similarly, the height will be more than 20 cm when primaries are originating from the thicker wood particularly in case of longer cycle and healthy bushes. Barbora and Banerjee (1988) opined that low tipping in the year of prune i.e. below 5-leaf measure, would result in the development of the forks and knots at the level of subsequent pruning. Formation of knots on the bush frame reduces its ability to produce optimum crop. Reduction of shoot size, die-back of plucking points and reoccurrence of *banjhis* are some of the common problems posed by knot formation at the pruning level.

The principal of 5 leaves tipping height is followed with harder system of plucking *viz., janam* plucking. With lenient system, *viz.,* mother leaf plucking, commonly followed in Sri Lanka, initial tipping is done over 2 leaves (fish leaf and first leaf), and after every alternate subsequent plucking 1 leaf is added to the plucking table. Objective in both the systems is to provide adequate maintenance foliage after a prune (Jain and Tamang 1988). Both the systems may have their own merits and demerits, but apparently are based on the rate of shoot growth.

The recommended tipping height for deep and medium skiff tea in north east India is the average height of 2 leaves and 1 leaf, respectively (Sarkar 1988, Chakrabartee and Barbora 1992) because these skiffed tea bushes bear partial maintenance foliage even after skiffing.

In south India, UPASI has recommended tipping height of 4 tiers, 3 tiers, 2 tiers and 1 tier of leaves for the pruning height of 45-55 cm, 55-60 cm, 60-75 cm and 75 and more, respectively for Assam tea, and 5 tiers, 4 tiers and 2 tiers of leaves for 35-45 cm, 45-55 cm and 55-75 cm pruning heights, respectively, for China hybrid tea. For rejuvenated tea, two stage tipping has been recommended, first at 55 cm and second at 65 cm (Hudson *et al.* 2002), because it resulted in better spread of the bushes and higher productivity in the rejuvenation cycle than conventional tipping and without adverse effects on the primary branch system (Satyanarayana *et al.* 1995).

Besides height, the stage of tipping is also important. Tipping-in-material, i.e. the amount of growth removed above tipping level, influences shoot growth in pruned bushes. The plucking table is normally formed early after tipping at 1+Bud stage, but if tipping is delayed to the 3+Bud stage, the first order laterals produce pluckable shoots earlier. Thus, the number of shoots plucked per primary, shoot weight and the thickness of the primaries increase significantly (Taparia *et al.* 1996, 1997). Shoot production capacity of different clones vary due to difference in tipping-in materials. Tipping of bushes should be started when their primaries attain green semi-hard wood, at least in 30 per cent of the bushes and tipped-in-material exceeds 3+B size (Hudson *et al.* 2002), so that the primaries get sufficient thickness to support regeneration of healthy shoots. If the recovery is slow and uneven after pruning, it is advisable to defer tipping until growth is well distributed all over the frame of the bush and eventually breaking back over grown shoots to the desired level (Hajra 2001). Simultaneously, if tipping operation gets delayed, it would be difficult to break the primary stems with fingers by the tippers due to their hardening.

Tipping measures for Kangra plantations

Sharma and Sharma (1992) recorded higher leaf yield with tipping measure of 15 cm above the pruning level at 37.5 cm. But for lighter pruned bushes (55 cm) they found that low tipping height of 5 cm gives better yield than higher tipping levels upto 25 cm (Sharma and Sharma 1998). Further, for rejuvenation pruning at 35 cm, they observed highest tea yield with a tipping / plucking margin of 22.5 cm above the pruning mark (Sharma and Sharma 1997).

Based on the experience with china hybrid tea, CSIR-IHBT has recommended the followings tipping measures for different types of prunes for Kangra tea plantations (IHBT 1998):

Rejuvenation prune (0-20 cm)	: Tip at 55 cm from ground level
Medium pruned or cut across (25-30 cm)	: Tip at 55 cm from ground level
Light pruned or cut across (35-40 cm)	: Tip at 60 cm from ground level
Deep skiffed (45-55 cm)	: Tip at 7.5 cm above skiffing mark

SHOOT GROWTH AND CROP DISTRIBUTION

Depending on growth, the shoots are of two types, (i) aperiodic shoots, the fast growing shoots originating from the pruned bush frame as primary shoots, which form the volume of maintenance foliage after decapitation, and (ii) periodic shoots, the shoots develop from the freely growing axillary buds, which exhibit definite rhythm of growth by passing through alternate phases of active growth and dormancy. The latter is referred to as periodicity of shoot growth. Tea shoot is said to be dormant or *banjhi* when terminal bud is less than half the length of the subtending leaf (Figure 8.4). The amount of growth produced by shoot between two states of dormancy is called "flush".

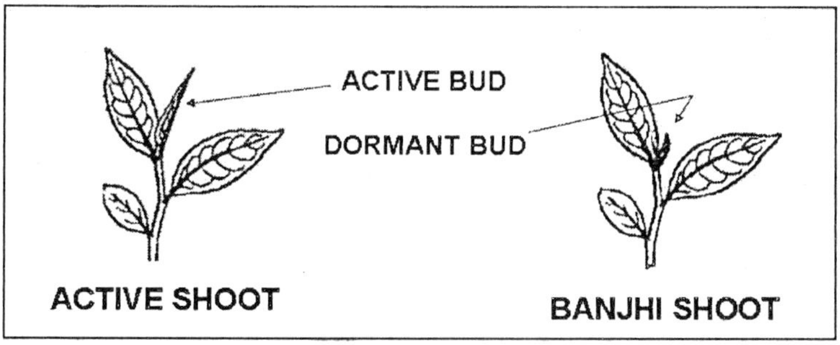

Fig. 8.4 Active and banjhi tea shoots

The ratio of number of *banjhi* shoots to active shoots is known as "dormancy index" (Wight and Barua 1955). In a tea bush, during growing season all the terminal buds are neither active nor dormant at a time, there being certain number of each type (Pethiyagoda 1964). Regular fluctuations in crop production are thought to be caused by regular changes in dormancy index (Tanton 1981 a&b). During crop season, active flush and cessation of flush can be judged by relative narrowing or broadening of dormancy index value in pluckable bush. Factors like nutritional constraint, moisture stress, hormonal imbalance, environmental conditions have been implicated for changes in the ratio. Restricted supply of nitrogen or inadequate availability of water and other nutrients will contribute for the buds going *banjhi* (de Haan 1941 & Bond 1945). This was further corroborated by anatomical studies where it was observed that the vascular tissues, especially the xylem of stem, immediately below the apices of *banjhi* shoots are ill developed (Bond 1945, Wight and Barua 1955, Barua and Das 1979). From economic and practical point of view, commercial tea cultivation is aimed at producing more active shoots or decreasing the *banjhi* proportion because higher level of dormant shoots are reported to adversely affect both productivity as well as quality of tea (Ranganathan *et al.* 1983).

Rate of shoot regeneration

"Shoot regeneration rate" denotes the time (days) taken by an axillary bud on a plucked shoot to grow into a pluckable standard shoot after the plucking. It is also called "shoot replacement cycle". It is reported to be influenced by the type of bush, climatic conditions and altitude (Mwakha 1985a, Janakie and Balasuriya 1996.). Time required for a dormant bud to grow into a pluckable size was 76-95 days in Kenya (Mwakha 1985b), 81 days in Sri Lanka, 42-49 days in Malawi and 36-42 days in Assam, India (Rahman 1988). Influence of plucking practices on regeneration rate is not known though it is an important phenomenon for formulating plucking programme (Clowes 1989b).

Rate of regeneration of shoots in Kangra is normally fast in April, July and August (28 to 32 days) than in other months (40 days or more). Bad plucking and stripping adversely affect the regeneration of shoots (Sud 1993).

Leaf period

The time interval between the unfolding of two successive appendages in a growing shoot is called "leaf period" or "phyllochron". Wight as early as 1932 suggested the consideration of leaf period for formulating plucking programme. Leaf period is not constant as it changes with climatic conditions, and is also affected by the types of plant in a given climate (Das 1984). Reports from various places reveal that leaf period is 9 days in Sri Lanka (Portsmouth and Rajiah 1957), 9-10 days in Java (Harler 1964), 3-7 days in north east India (Rahman 1977, Das 1984, Tamang 1985), 6-7 days in east Africa (Magambo 1975), and 6-8 days during the warm-wet season and 11-19 days during the cool-dry season in southern Tanzania (Burgess and Carr 1998). Leaf period was reported to be increased with increase in altitude, which was attributed to differences in mean air temperature (Janakie and Balasuriya 1996).

Rate of unfolding of leaves in plucked bushes in Kangra, Himachal Pradesh, was found to vary between 3-6 days during different months (CSIR Complex 1988). It was also observed that the unplucked shoots of regularly plucked (level of skiff or light skiff in the fourth year after prune) bushes produced upto 29 appendages including cataphylls. The leaf period was short at 4-5 days for the 1st and 2nd leaf, increased to 8-16 days upto 10th leaf, became regular at around 7 days upto 17th and thereafter increased to 9-17 days (CSIR Complex 1989).

Leaf period and crop distribution in Kangra plantations

It was observed that that the leaf period and crop distribution over the season has inverse relation (CSIR Complex 1988). The relationship has been shown in the Figure 8.5.

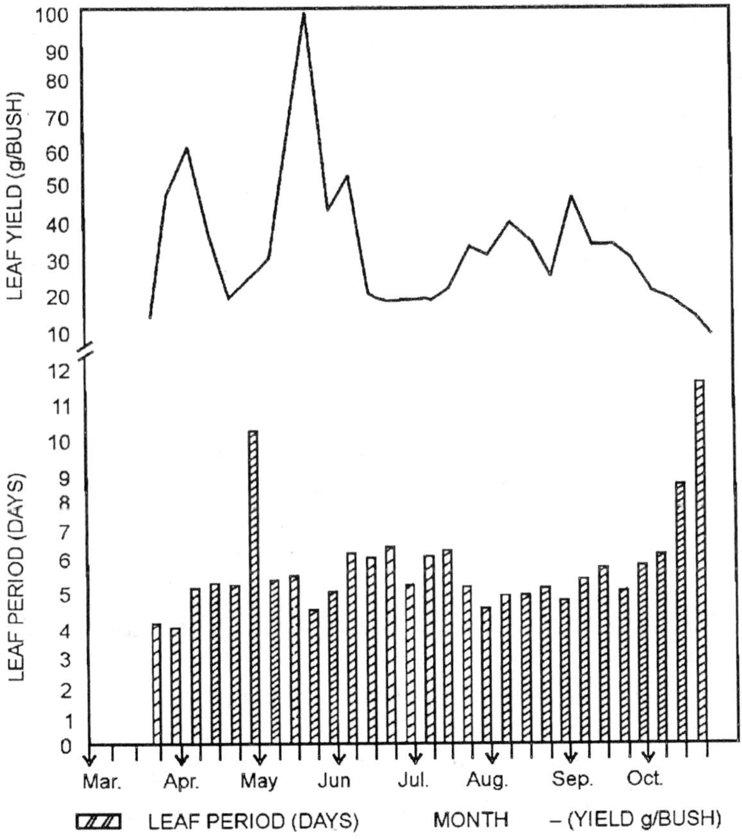

Fig. 8.5 Leaf period *vis-a-vis* tea leaf yield in different months

Growth of shoots in pruned and unpruned tea

In the pruned tea, the number of growing shoots is less but the shoots are bigger and heavier. These are not only easy to pluck but also have better quality both in terms of physical grade with minimum proportion of loose leaves in the crop and chemical constituents. With time, the shoot numbers increases but at the cost of individual shoot weight. In unpruned tea bush, shoots density is high and the shoots are light in unit weight and small in size. Many shoots fail to grow and turn *banjhi*. It could be because of (i) development of crows' feet on the stem or branches, (ii) interference with the movement of water and nutrients, and (iii) loss of vigour of apical meristem due to aging (Tamang 1981). Measurement of the growth of several thousand shoots by Tanton (1981b) indicated that an individual shoot grow exponentially to harvestable size but if left unplucked, they grow to the typical sigmoidal mode.

Pruning practice also affects the shoot weight. During pruned year, pluckable shoot is usually twice as heavy as similar standard shoot in the unpruned year (Mwakha 1988).

SYSTEMS OF PLUCKING

System of plucking determines which leaf to be plucked as crop and which leaf to be left behind on the bush as foundation for future crop. There are different systems of plucking in different tea growing regions. These systems are based on rate of shoot growth in the region and source-sink relationship in carbon assimilation. The common principal of these systems is to maintain the balance between production and bush health.

Depending on the severity of plucking, plucking systems are referred as "hard" and "lenient". A system that allows retention of new leaves at plucking table is "lenient" whereas a system where no new leaf is left behind on the table is called "hard" plucking. Growth, yield and quality of tea as well as the health of tea bush are influenced by the severity of plucking. Wettansinghe *et al.* (1981), Sharma (1983), Gvasaliya (1986), Kulasegaram (1988) and Sukasman (1989) observed that harder plucking system may lead to higher yield but adversely affect the bush health over a period of time. A hard plucking system encourages shoot number and has a negative effect on unit weight of shoots, but on the other hand, a lenient system has an opposite effect because many shoots become dormant below the plucking surface (Rahman 1988). Growing buds and 1+Bud shoots being the strongest sink, should be retained on the plucking table to reap the sink-induced accelerated photosynthesis of maintenance foliage (Manivel and Hussain 1986). So, formulation of plucking system should be based on source - sink relationship so as to harness higher production without any ill effect on bush health.

The common systems followed in different parts of world are "*janam* plucking", "fish leaf plucking" and "mother leaf plucking" or "single leaf plucking" (Figure 8.6). Detail of these systems is given below:

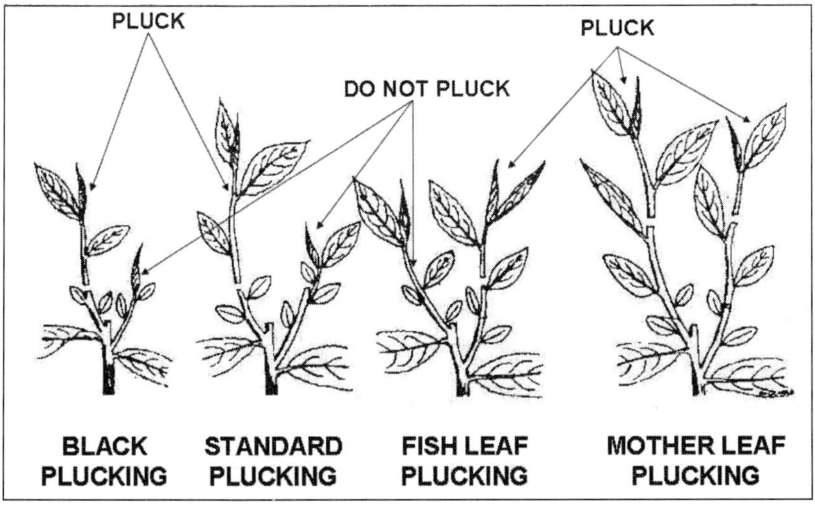

Fig. 8.6. Different systems of tea plucking

***Janam* plucking**

In north India where rate of shoot growth is fast, "*janam* plucking" is done. In this system tea shoots are plucked at plucking table to point of *janam* or scale leaf to avoid unnecessary creep in the plucking table. Depending on the size of the shoots left behind on the plucking table, the system is further classified as "standard plucking" and "black plucking". These two systems are practiced as per need to control the crop, quality, and insect-pests and diseases. Observations on assamica type of bushes show that when black plucking if carried out throughout the plucking season, the yield is reduced as compared to that under standard plucking (Tocklai 1983).

***Standard pluckin*:** In this system, all shoots at plucking table, excluding very fine shoots like small one-leaf-and-a-bud and unopened buds, are plucked. It is a general recommended practice carried out at 5-9 days interval. For Kangra valley and Darjeeling, standard or normal shoots consisting of large 1+B, 2+B and single *banjhi* should be plucked for maintaining the unique quality of tea of these regions.

***Black plucking*:** Plucking all shoots at plucking table leaving behind only unopened buds is termed as black plucking. It is somewhat hard plucking system, normally done at an interval of 8-11 days.

It is also recommended as a short term measure during first flush, when plucking goes beyond control owing to paucity of labour *vis-a-vis* fast shoot growth, during periods of severe pests infestations (Shanker *et al.* 2002), attack of blister blight disease, and tendency for higher dormancy state of the plucked bush towards the end of early and end flushes in unpruned tea (IHBT 1998).

Fish leaf plucking

All pluckable shoots, except immature one-leaf-and-a-bud and unopened buds growing at plucking table above the level of fish leaf (*golpat*), are plucked retaining the fish leaf on plucked stub at plucking table. This system is common in south India where rate of shoot growth is comparatively slow.

Mother leaf plucking

This is the most lenient system where shoots are plucked above the mother or first leaf. The system is followed in pruned tea where bushes are low tipped and depth of maintenance foliage is maintained by frequent stepping up of the table with mother leaf plucking. This system is appropriate combination of pruning and tipping measures with a view to maintain the depth of the maintenance foliage in steps. It is commonly adopted in Sri Lanka.

Studies on different systems of plucking in Kangra

Studies were conducted on different systems and methods of plucking for Kangra valley (Sud *et al.* 2003). The data reveal that different systems of plucking significantly influenced the production (Table 8.1). Compared with standard plucking, which is generally recommended in this region, production with black plucking was 14 to 32 per cent higher. Fish leaf plucking also increased the production by 12 to15 per cent but statistically it was non-significant. On the other hand, lowest production was recorded with mother leaf plucking particularly in the initial two years. The quality of tea shoots in terms of fineness, in general, was higher in standard plucking except in pruned year where it was higher in mother leaf plucking (Table 8.2). The labour requirement for plucking in the initial 2 years was more in black plucking but thereafter it was higher in standard plucking (Table 8.3).

Table 8.1. Effect of different systems and methods of plucking on tea productivity (Sud *et al.* 2003)

Plucking standard/ method	Productivity (made tea kg/ha)					
	UP4	LP	UP1	UP2	UP3	Mean
Standard	883	948	2131	1770	1871	1680
Black	1162	1076	2619	2230	2189	2029
Fish-leaf	-	1062	2403	2031	2152	1912
Mother leaf	-	499	1781	1785	1939	1501
SEm±	103	66	157	167	146	106
LSD	328	199	473	503	440	320
CV%	15	13	13	16	14	11

UP1-4: Unpruned tea (first year – fourth year), LP: Light pruned tea

Table 8.2. Effect of different systems and methods of plucking on tea quality shoots (Sud *et al.* 2003)

Plucking standard/ method	Quality shoots (% fine)				
	UP4	LP	UP1	UP3	Mean
Standard	69	72	67	73	71
Black	62	68	55	62	62
Fish-leaf	-	69	67	66	67
Mother leaf	-	83	64	68	72
SEm±	1	1	2	1	-
LSD	4	4	6	3	-
CV%	5	4	8	4	-

UP1-4: Unpruned tea (first year – fourth year), LP: Light pruned tea

Table 8.3. Effect of different systems and methods of plucking on plucking efficiency (Sud *et al.* 2003)

Plucking standard/ method	No. of mandays/ha				
	LP	UP1	UP2	UP3	Mean
Standard	609	1213	1401	1281	1126
Black	764	1294	1234	1054	1086
Fish-leaf	571	1070	1308	1140	1022
Mother leaf	405	1040	1411	1251	1027
SEm±	27	50	71	48	-
LSD	83	151	-	145	-
CV%	11	12	13	10	-

UP1-3: Unpruned tea (first year – third year), LP: Light pruned tea

In another study, Sud (1993) reported that black plucking needed 49 and 23% more mandays per interval in unpruned and pruned year, respectively, than standard plucking. This was due to the reason that black plucking entails removal of all small immature 1+Bud shoots whose plucking consumed more time particularly in unpruned year due to their higher density. But black plucked bushes took longer time to become ready for subsequent plucking owing to removal of small 1+Bud shoot thus reducing the plucking frequency. As a result, the overall mandays requirement for the season decreased to the level of that of standard plucking in pruned year and 22% higher than that of standard plucking during unpruned year

STANDARDS OF PLUCKING

Standard of plucking refers to the size of shoots plucked as crop (Figure 8.7). It determines the quantity and quality the crop. According to the standard, the plucking is of three types – fine plucking, standard plucking and coarse plucking. These standards are determined by the interval between the successive pluckings.

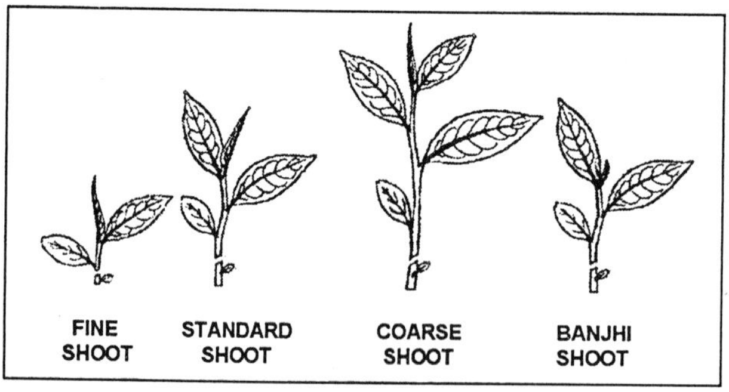

Fig. 8.7. Different types of tea shoot

Fine plucking

All one-leaf-and-a-bud, two leaves-and-a-bud and single soft *banjhi* shoots are plucked at plucking table. With this type of plucking total production of crop is severely affected. The quality of made tea would be very high and would fetch very high price. This plucking is done at shorter plucking intervals.

Standard plucking

All two-leaves-and-a-bud, large one-leaf-and-a-bud and single *banjhi* shoots are plucked. Shoots larger than two-leaves-and-a-bud are broken back. Plucking at weekly interval is suitable for this grade. Tea made from such shoots would be of good quality.

Coarse plucking

When grade of plucking is over two-leaves-and-a-bud it is coarse plucking. Here all two-leaves-and-a-bud and larger shoots, and double *banjhi* shoots are plucked without breaking back. Such grade of plucking is achieved when plucking interval is longer.

Yield, quality and economics are affected by the standard of plucking. In Sri Lanka, Botheju *et al.* (2000, 2001) compared normal estate plucking comprising 60-65 per cent tender shoots with plucking standards of only 2+Bud and 3+Bud shoots in terms of yield, quality and economics. They reported that the estate practice produced a significantly higher yield of black tea (4.5 kg/round) than plucking only 2+Bud (3.5 kg/round) and 3+Bud (3.7 kg/round). Net returns were also higher with the estate practice than the other standards despite better chemical quality and tasters' evaluations in the latter. Net returns were lower or more or less similar with 2+Bud plucking than that of 3+Bud planting. Mwakha (1991) in Kenya observed that the yield and plucker's productivity (kg green leaf/plucker) increased and the quality (total organoleptic scores) decreased with reducing standard from 1+Bud to 3+Bud. However, decrease in quality due to coarse standard was also clone specific (Nyirenda 1995). Asil (2008) in Iran observed that plucking based on the lengths of shoot growth was significantly better than those based on the standard conventional plucking.

Mahanta *et al.* (1990) evaluated the influence of three standards of plucking, fine (2+Bud), medium (3+Bud) and coarse (4+Bud) in Assam on cell-wall and soluble components in black tea and found that coarse pluckings had the highest cellulose, hemicellulose and lipid contents, and the lowest lignin content, while fine pluckings had the highest soluble solids and lignin contents.

Owuor and Obanda (1998) studied in detail the changes in black tea quality due to variations of plucking standard. They found that sum of Group I volatile flavour compounds (which impart inferior aroma) and C6 aldehydes and alcohols, in particular, increased with coarse plucking standard. The sum of Group II volatile flavour compounds (which impart a sweet flowery aroma) did not follow a particular pattern, but flavour index decreased with coarse plucking standards. The decrease in total theaflavin levels was due to a decrease in all the individual theaflavins. Although all the individual theaflavin levels decreased with coarse plucking standards, the decrease was greater in galloylated theaflavins, especially in theaflavin digallate, compared with simple theaflavin.

PLUCKING INTERVAL

Unlike many other crops, tea is harvested many times at a gap of some days. This time gap in days between two successive pluckings is called "plucking interval" or "plucking round". Crop yield, shoot composition and overall quality are largely affected by plucking interval and have been studied widely in different tea growing countries. Plucking interval is determined on the basis of rate of shoot growth, and standard and system of plucking. Leaf period and day degrees concepts have been developed for fixing the interval.

For harvesting maximum number of 2+Bud growing shoots in north east India, Wight (1932) suggested a plucking round equal to leaf period, since all growing shoots with one unfolded leaf (1+Bud) are usually left behind on the plucking table. But considering the impracticability of shorter round for plucking, he further suggested a generalized round of 7 days, which was, later, found suitable (Dutta 1961). But Rahman (1977) demonstrated that the suitable interval varies from 7 days to 14 days as rate of leaf unfolding varies during the crop season. Under south Indian conditions Sharma and Murty (1989) made an attempt to develop a system to predict the optimum interval between plucking rounds in different seasons to enable advanced planning and effective deployment of labour. They compared two systems, one based on the number of days required for axillary buds to reach the three-leaves-and-a-bud stage, accumulated day degrees and the growth hierarchy (number of shoot generation) existing on plucking surface, the other based on day degrees required for unfolding and expansion of the third leaf on shoot comprising two-leaves-and-a-bud. They found that by the use of first system, there was abrupt and marked variation in the interval requiring rapid and frequent changes in labour deployment, though second system was promising.

Shorter rounds result in more number of plucking points per unit area than extended ones (Ranganathan *et al.* 1983, Nathaniel *et al.* 1986, Sud 1993). Clowes (1989a) and Odhiambo (1989) observed that while shoot density per harvest showed increasing trend but cumulative shoots harvested showed reducing trend with increase in plucking interval. Increase in plucking interval from 6 to 22 days caused increase in fresh weight of shoot from 0.65 to 0.89 g/shoot (Odhiambo 1989). Mwakha (1991) observed that all shoot grades - 1+Bud, 2+Bud and 3+Bud were significantly affected by plucking interval, wherein mean shoot weight increased with increase in coarseness of plucking.

Due to nature of growth curve, there was 55 per cent increase in dry matter in a 3+Bud shoot than that in a 2+Bud shoot, which accumulated over a period of 3-5 days (Clowes 1989b). Under Kangra conditions the dry weight of a 2+Bud shoot doubled after 4 days of its formation (CSIR Complex 1990). Therefore, for harvesting heavier shoot without loosing fineness, the plucking interval may be fixed as follows:

Plucking interval (days) = (2 x Leaf period in days) -1

The management factors which affect the crop productivity do so through their effect on frame development and canopy depth of maintenance foliage both of which are essential prerequisites for sustained high production (Kulasegaram *et al.* 1975). Bush surface area and annual rise in plucking table have been reported to be affected by plucking practices (Barua 1969, Watson *et al.* 1982). A bush under continuous plucking manifests 'creep' of plucking table (Jain and Tamang 1988). In other words, there is continuous rise in plucking level of bush with plucking operation. Sarkar (1988) suggested that the creep should not exceed beyond 5.6 cm in pruned year and 4.0 cm in unpruned year otherwise it would result in loss of crop. Ranganathan *et al.* (1983) and Mwakha (1987) observed low creep due to shorter intervals as compared to longer ones. Similarly, consistent rise in plucking table with widening plucking interval have been reported by Odhiambo (1989). The percentage of active shoots was significantly reduced in a 7 days plucking round than shorter rounds (Nathaniel *et al.* 1986). Therefore, widening of plucking interval is expected to increase the dormancy index.

In Himachal Pradesh the effect of different plucking intervals, ranging between 3 to 17 days, was studied on yield, quality and economics (Sud 1993, Sud *et al.* 1996).

Effect of plucking interval on yield

The yield of tea bush depends on number of factors *viz.*, the number of plucking points per unit area of bush, the proportion of active to dormant shoots at a time, the frequency and duration of flushing (Herd and Squire 1976) and surface area of tea bush (Barua

1969). According to Rahman (1977), yield of tea is the product of weight of shoots and their number per unit area and these are generally complementary to each other.

It has been reported from Sri Lanka that though the average weight of shoot plucked on fortnightly round was distinctly higher than that of weekly round, the latter gave higher yield (Tubbs 1938 & 1949, Pethiyagoda 1967). Another report of Nathaniel *et al.* (1986) revealed that the shorter round of 4 days gave 16 per cent more crop over the standard round of 6 days. Similarly, Palmer-Jones (1977) in Central Africa, Dumur and Naidu (1985) in Mauritius and Mwakha (1987) in western Kenya observed the superiority of shorter rounds in respect of crop productivity. In another study at five regions in Kenya, yields declined with longer plucking intervals in three but increased in one location (Owuor *et al.* 2009). The increase in yield in the shorter round was mainly due to formation of more number of plucking points per unit area because the plucking act as a stimulus for further shoot regeneration by temporarily destroying the apical dominance (Portsmouth and Rajiah 1957). Besides, the higher mean weight of individual shoot plucked in longer rounds is not sufficient enough to compensate for reduction in total number of shoots, resulting in net loss of crop (Portsmouth 1957). However, the findings of experiments carried out in India contradict the above results. In north east India plucking was carried out in four types of clones at an interval of 3 to 14 days. The results indicated the differential behaviour of clones in respect of gross weight of harvest; however, it increased with increase in the length of plucking interval with some fluctuations in all clones (Tocklai 1971). Sarkar (1988) reported a linear relationship between plucking interval from 5 to 13 days and yield of tea both in pruned and unpruned sections. Report on long-term trial on plucking round from 1978-85 at Tocklai Experimental Station (Assam) showed that the longer round of 13 days brought the highest yield, while shorter round of 5 days the lowest, however the general trend in field production was in the order of 13> 9> 11> 7> 5 days interval (Barbora and Saikia 1989). The report further showed that the pruning/skiffing operations interact significantly with the length of plucking interval. Under south Indian conditions, Verghese (1977) observed longer interval of 15 and 25 days producing higher biomass of shoots. Similarly, Ranganathan *et al.* (1983), too reported highest total yield for three cycles with longer plucking interval. Grice (1984) in Central Africa had also found higher yield with longer intervals. Early work of Templer (1971) revealed that plucking twice a week as compared to once a week gave significantly higher yield but later (1977), he reported that longer plucking intervals were more productive due to greater shoot weight owing to extra time available for growing in longer plucking interval. Interestingly, report from Central Africa (Malawi) showed that increasing interval from 7 days to 10/11 days brought higher yield which remained constant when interval was increased to 14 days but declined significantly when the plucking interval was further broadened to 21 days and there was no significant difference either between 7 and 21 days round or 10/11 and 14 days plucking intervals (Clowes 1989a). Based on shoot growth patterns of different clones during the season, the optimum plucking interval in Kenya varied between 7-17 day, (Nyirenda and Mphangwe 2001).

In Himachal Pradesh, increase in plucking interval from 3 to 17 days resulted in consistent and significant increase in shoot size (Sud 1993, Sud *et al.* 1996). It was observed that by extending the plucking interval from 3 to 17 days, the total number of plucking frequencies reduced to one-sixth, but the shoots plucked were about 9 times heavier; thus the product of shoot weight and the total plucking frequencies showed a trend of increase with increase in plucking interval. The tea yield also showed the increasing trend with increase in plucking interval from 3 to 17 days during both unpruned and pruned years. The increase was to the tune of 50 per cent during unpruned year and 80 per cent during pruned year with longest interval of 17 days over shortest interval of 3 days. Total number of plucking points (recorded at the closure of season) was more due the shorter intervals, but the proportionate increase in the shoot weight at longer intervals due to extra time for growing was more, resulting in higher yield.

Effect of plucking interval on quality

Longer plucking round (or delay in plucking) reduces fineness of crop because of inclusion of coarse and overgrown shoots bigger than two leaves, and because of higher amount of dormant shoots, thus it marked loss of quality (Verghese 1977, Basu 1978, Sarkar 1988, Barbora and Saikia 1989, Owuor *et al.* 1997 & 2000). According to Basu (1989), with coarse plucking the percentage of bud, first leaf and second leaf comes down to 43 from 83 (in case of fine plucking), which brings down the chemical composition and, hence, quality of made tea. Another report from Malawi indicated that the proportion of 3+Bud size shoots increased while 2+Bud decreased as round length increased (Clowes 1989a). Amiri and Asil (2007) in Iran observed that black tea quality factors such as total theaflavins, thearubigins and total colour decreased with coarse plucking. Owuor *et al.* (2009) in Kenya concluded that tea quality improvs with short plucking rounds, thus one of ways to maximise tea quality is through shortening harvesting intervals with due emphasis to establish optimal harvesting rounds in the different tea growing regions to maximise yields and quality.

Bhatia and Ullah (1968) found a progressive decline in phenolic content from tender parts to the older parts of a shoot system. In a 3+Bud shoot, concentration of the polyphenols decreased from bud to third leaf, while stem of the shoot had the lowest concentration of the polyphenols (Bhatia 1961, Ullah and Jain 1980). Therefore, longer interval of plucking, which increases the proportion of coarser shoots in plucked crop, appears to lower down the polyphenols concentration compared to shorter intervals.

Reports of Tocklai Experimental Station, Assam, revealed that bud contains highest amount of caffeine which gradually goes on declining in lower leaves while stem of the shoot contains lowest content of it (Bhatia 1961, Tocklai 1978, Ullah and Jain 1980). The tender younger shoot, which is the consequence of shorter plucking intervals, was found to contain higher content of caffeine (Owuor and Chavanji 1986). Similarly, tender shoots also produce more theaflavins (TF) (Lelyveld 1986), hence shorter rounds ensure higher content of TF in made tea. In taster's evaluation also the relative values decreased in a linear fashion as the plucking interval increased (Barbora and Saikia 1989, Mitini-Nkhoma 1989).

In Himachal Pradesh also the fineness of plucked crop, averaged across the months, decreased continuously and significantly with increase in plucking interval from 3 to 17 days in both unpruned and pruned tea (Sud 1993, Sud *et al.*, 1996). The average range of variation in the fineness of crop due to different plucking intervals was between 77 to 49 per cent during unpruned year and fairly wider between 87 to 35 per cent during pruned year in 3 and 17 days plucking intervals, respectively. The decrease in fineness with increase in plucking interval might be due to two reasons. Firstly, with increase in plucking interval, the size of plucked shoot (in terms of number of unfolded leaves) exceeded above 2+Bud and the part of shoot below 2+Bud was considered coarse as have also been reported by Sarkar (1988). Secondly, with increase in plucking interval the proportion of *banjhi* shoots increased significantly. In a *banjhi* shoot the leaves below first leaf are regarded coarse and hence lower the fineness more sharply. The wider range of decrease in fineness during pruned year might be the result of faster growth as evident from rate of increment in dry matter accumulation and number of unfolded leaves per shoot recorded at different intervals.

Like fineness of the crop, there was significant decrease in the content of polyphenols and caffeine in the tea shoots with increase in plucking interval from 3 to 17 days during both unpruned and pruned years. Shoots plucked at longer intervals not only unfolded higher leaves but also contained longer stems thus resulted in decrease in polyphenol content.

Major changes which take place in tea leaf during manufacturing process are formation of coloured substances - TF and TR from colourless polyphenol. Both of these start forming with the initiation of fermentation during rolling of tea leaf. The TF concentration increases with progress of fermentation till a stage is reached when there is no further fermentation of TF. On prolonging the fermentation period, TF content ultimately shows a significant decline while TR goes on increasing (Roberts, 1958 & 1962, Deb and Ullah 1968). High levels of TF and adequate levels of TR could be related to high value of tea (Roberts 1958). The proportion of TF and TR ultimately determine the colour and brightness of made tea.

In a study conducted in Himachal Pradesh, the average values of all the made tea parameters *viz.*, TF, TR, total colour and brightness across the months during both years were highest in the tea made from the pluckings of 3 days interval, and there was significant decrease in the parameters with increase in plucking interval upto 17 days (Sud 1993). This was mainly due to consistent and significant decrease in content of polyphenols, which ultimately influences the parameters during manufacturing process. In tester's evaluation also, brightness, colour, flavour and total quality showed a trend of decrease with increase in plucking interval during both years, supporting the results obtained in biochemical analysis.

Effect of plucking interval on economics

Low cost, higher return and maximum profits are the basic objectives of all entrepreneurships, whether it is agricultural or industrial. In tea, plucking operation is a major input which absorbs around 60-85 per cent of labour cost and accounts for 25 per cent of cost of cultivation (Awasthi and Sarkar 1983, Mitini Nkhoma 1989, Rawlley and Beig 1990).

The plucker's productivity increased significantly with prolongation of plucking rounds from shorter (7 days) to longer (14 days) (Mwakha 1991). Shorter plucking round calls for over all higher labour force due to higher frequencies. Palmer-Jones (1985) estimated the plucking cost in 14 days round to be 64 per cent of that of 7 days round, but gross revenue was higher in latter. No doubt, high quality tea could be obtained from smaller shoots (Owuor *et al.* 1987) but there is doubt regarding economics of such tea. Sharma and Murty (1989) under south Indian conditions found higher net profitability with 3+Bud particularly with flexible intervals depending on growth of shoots instead of following 7 days flat round. According to them, the increase in profitability was largely due to reduced labour requirements.

The plucking efficiency in terms of quantity of leaf plucked in unit time was lower at shorter intervals and increased with the extension of plucking interval, in both pruned and unpruned tea. The plucking efficiency mainly depends on tenderness or succulency of shoot and weight of individual shoot plucked. In shorter intervals the shoots were tender and lighter in weight leading to lower efficiency. With increase in age of shoot from 3 to 17 days, its tenderness reduced and weight increased and consequently the plucking efficiency was enhanced (Sud, 1993). Over all, the number of shoots plucked at each interval has also influenced the efficiency.

Increase of plucking interval from 3 to 17 days was accompanied by marked decrease in plucking frequencies from 73 to 13 in unpruned and 60 to 11 in pruned years, respectively (Sud, 1993). Increase in plucking frequencies, at shorter intervals, called for more mandays requirement. The requirement was further escalated due to decrease in plucking efficiency at shorter intervals due to the reasons mentioned above. Owing to higher mandays requirement, the cost of plucking also showed consistent and significant increase with decrease in plucking interval. Contrary to this, the average mandays requirement per interval was lower at shorter intervals. This can be attributed to the higher succulency (tenderness) of the shoots at these intervals.

MANAGEMENT OF DIFFERENT FLUSHES IN KANGRA

Tea in Kangra valley flushes from March-end to October. Based on rate of shoot growth, the plucking period may be divided into three distinct flushes - the first or early flush (end March to end May), the main or monsoon flush (mid June to mid September), and the backend flush (mid September to season end). The distribution of crop in different flushes is as below:

Pruning Operation	Percent Crop Distribution		
	Early Flush	Main Flush	Backend Flush
Unpruned	40	50	10
Deep Skiffed	15	65	20
Light Pruned	8	70	22
Over all (60-80% unpruned)	35	50	15

Management of first flush

This flush has paramount importance as it is the quality flush of whole season. In this flush, the rate of shoot growth is rapid, and after unfolding 2-3 leaves the shoots turn *banjhi* affecting the crop quality. The flush comes in to full bloom (within 2-3 weeks), consequently the crop management becomes difficult. The following steps help management of the first flush.

1. Start standard plucking even when few bushes assume flush and follow up at an interval of 3-4 days.
2. From mid April to mid May, standard plucking at weekly interval should be adopted.
3. In the event of labour shortage and high occurrence of *banjhi* shoots, black plucking at 9-10 days interval can be followed.

Management of main or monsoon flush

Due to favourable weather conditions, the shoot growth is fast, particularly in pruned and deep skiffed sections. In unpruned sections, occurrence of *banjhi* shoots also increases.

1. Standard plucking should be followed at an interval of 6-8 days in unpruned, and 5-6 days in pruned/ deep skiffed sections.
2. In general, black plucking should be avoided. It may only be followed when situations as 3, above, exist.

Management of backend flush

In this flush, the shoot growth slows down with quick hardening of shoots with time. The percentage of *banjhi* shoots increases rapidly.

The active shoots loose lusture. The plucking table becomes uneven due to continuous plucking.

1. In unpruned sections, pluck black at 10-12 days interval to avoid excessive *banjhiness* on the plucking table.
2. In pruned sections, pluck standard at 7-9 days interval.
3. Avoid damaging the top layer of maintenance foliage.

YIELD VS QUALITY

Increase in yield associated with the declining plucking standard, or increasing plucking interval, was subjected to marked loss of quality (Basu 1978, Baruah *et al.* 1986 Sarkar 1988, Barbora and Saikia 1989), which means the yield may increase at the cost of quality or *vice versa.* Palmer-Jones (1974), who worked out the coefficient of regression together with the best estimate of expected future regression on theaflavin content at London Auction Price, indicated that the effect of plucking standard on quality is much less than its effect on yield. But later (1977) he considered unwise to transfer such research findings to commercial practice due to the reason that the commercial producers who employ coarse plucking may also have other practices which exacerbate the effect of coarse standard on quality. Clowes (1989 a&b) too reported considerable loss in yield following quality shoot plucking of 2+Bud size and advocated the selection of clones based on the quality of 3+Bud size shoot rather than 2+Bud particularly with regard to fiber content.

HARVESTING QUALITY CROP

In this era of competitive open world trade, it is necessary to produce quality. In case of tea, it is a general saying that quality of tea comes from the garden not from the factory. This is because the quality of made tea depends on the size of tea shoots plucked, proportion of active shoots in the harvested crop, chemical composition of the shoots, and post-harvest handling of the shoots during storage and transportation to the factory. The first step for quality production is harvesting quality shoots. Quality harvest is possible only when (i) tea bush bears actively growing shoots on the plucking table with lowest dormancy index (ii) shoots are harvested with care to have maximum fine shoot percentage and minimum loose leaves, and (iii) the harvested shoots are appropriately handled, safely stored and transported with minimum physical damage.

In Kangra valley, china-hybrid type tea is cultivated. The characteristics of this type tea plant are:

- Multi-stem bush, producing small size of shoots rich in aroma compounds or flavour precursors.
- Under continuous unpruned conditions, the shoot size becomes smaller and smaller with more incidence of *banjhi* (dormant) shoots and profuse flowering.

Problems with china hybrid tea are:

- On account of heavy flush during April-mid May and July-August (coinciding with wheat harvesting and paddy sowing, respectively) plucking management goes out of control.
- The size of two-leaves-and-a-bud is very small, internode is short and density of shoots is high, which makes unit shoot plucking difficult.
- Owing to small shoot size and shorter internode, occurrence of loose leaves with fine plucking increases, which causes decline in grade.
- High proportion of *banjhi* shoots, particularly in unpruned sections.

Hence in order to harvest quality tea shoots with higher percentage of fineness, it is necessary to adopt a set of agro-practices, which would induce production of active quality shoots and uniform flush distribution.

Shorter pruning cycle

A shorter pruning cycle will ensure production of higher proportion of active shoots, heavier shoots, low shoot density and minimum proportion of *banjhi* shoots. Though the total production of crop may suffer as compared to longer cycle but quality of shoots and fineness of plucking would be better.

A 4-year pruning cycle 'LP-LoS-DS-LoS' (i.e. every year prune 25%, deep-skiff 25% and level of skiff / unpruned 50% of the tea garden) has been found better for Kangra tea both in terms of quality and yield.

It is important to mention that for outstanding quality of the first flush crop, efforts should be made to get maximum crop during this flush by opting for summer pruning provided suitable weather conditions prevail.

Application of fertilizers

In order to ensure active shoot growth, recommended doses of nitrogen, phosphorus and potash fertilizers should be applied. The present recommendation of dose of nitrogen, phosphorus and potash fertilizers is 90-120, 90, and 90-120 kg/ha, respectively.

Foliar application of 2-3 rounds of Urea 2%, Zinc 2% and Magnesium 1% may be helpful in improving size of shoot and crop yield. Similarly, foliar spray of these nutrients along with potash 2% at weekly interval after plucking helps to combat drought conditions. It has been reported that application of nitrogen, phosphorus, potassium and magnesium increased amino acid and caffeine contents of tea.

Shade management

Shade trees are helpful in production of quality shoots in many ways:

- It improves succulence of the shoots, leaf area up to a limit and number of shoots.
- It improves the microclimate, making it cool and humid which is ideal for both the growth of tea shoots and pluckers in improving their productivity particularly in summer and rains.
- It creates suitable temperature for optimum photosynthesis. More photosynthates are directed towards pluckable shoots which favour shoot growth.
- It conserves the soil moisture by reducing evapo-transpiration losses.
- It adds organic matter and nutrients to the soil.

However, excessive shade is not good particularly in monsoon period as it can be congenial for blister blight disease. Under Kangra condition *Albizia chinensis* is suitable shade tree. *Albizia* shade trees should be maintained at a spacing of 12 meters both within and between the rows for harnessing its advantages.

Pest control

Attack of thrips and aphids is very common in tea during flush periods. Incidence of mealy bug in Kangra tea plantations is increasing every year. These insects suck the cell sap of the shoots and affect their growth, productivity and quality. Similarly, a foliage disease, blister blight, which appears during rainy period, inflicts heavy crop loss and quality deterioration, as the disease infects the young tea shoots harvested for tea manufacture. The infected shoots get distorted and tea made from such shoots is poor in appearance (leaf-style), infusion characteristics like briskness, brightness, and colour, and aroma corresponding to disease intensity. Hence it is necessary to control the disease. Similarly, weeds grow luxuriantly during April-May and July-August particularly in pruned and vacant sections, affecting the efficiency of the garden management, crop productivity and also quality when plucked inadvertently along with tea shoots. It is worth mentioning here that indiscriminate use of fungicide, pesticide and herbicides should be avoided. Resorting to integrated pest management should be the aim to avoid pesticides' residue problem.

Plucking management

Finally, it is plucking management, which governs the final quality of the harvested shoots. For quality tea shoots, it is essential that the plucking standard should be fine. Efforts should be made to harvest two-leaves-and-a-bud shoots. Coarse plucking should be avoided. A minimum of 60% fine crop should be targeted. Plucking interval should be kept between 5-8 days, depending on the shoot growth rate. Stripping of shoots and longer plucking interval should be avoided. In case of labour shortage, black plucking may be adopted to extend the plucking interval without loss of quality. In case the growing shoots are over the standard two-leaves-and-a-bud size, break back and discard the unwanted growth. It not only improves the quality of harvest but also protects the bush against diseases *viz.*, grey blight (Sanjay *et al.* 2003).

Handling plucked leaf

The shoots should not be hard pressed during plucking. The plucked crop should be stored in a cool shady place by spreading it on a neat and clean floor with maximum 15-20 cm thick layers. Undesirable shoots e.g. loose leaf, coarse leaf, hard *banjhis* etc. should be sorted out. The leaf should be given frequent turning to avoid heating up of the shoots during storage. The plucked leaf should be transported to factory as soon as possible. It should be loosely packed and should not be pressed during transportation.

Wilkie and Burton (1995) in Central Africa studied tea handling practices and damage occurring before leaves arrive at the factory and concluded that leaves should not be compressed at any stage during handling and should never be allowed to heat up to 40°C. Acceptable tea quality should result if leaves reach the factory within 5 hours of plucking and their temperature remains below 35°C at all times. For improving labour productivity and protecting the plucked leaf, particularly with share plucking, new plucking baskets are being designed (Wijeratene 2003).

FUTURE LINES OF WORK

The present need of tea industry is to produce quality tea. For quality tea production, tea bushes need to train to grow active, heavy and quality shoots. Hence location specific standardization of agro-practices *viz.*, pruning cycle, plant nutrition, water management, shade management and plucking strategies need to be done.

Availability of labour particularly during peak flush period is a serious problem in tea estates. Shifting to mechanical plucking could be a solution. But machine plucking has been reported to deteriorate the crop quality particularly when orthodox black tea is to be processed. Hence an effort to adopt integrated plucking programme i.e. the combination of mechanical and hand plucking to strike a balance between quality and cost of plucking is needed. Further, standardization of pruning cycle, fertilizer doses and plucking practices under integrated plucking programme may be another area of future research.

In mechanical plucking of tea, one of the main reasons for reduction in quality was observed the inclusion of cut leaves (both soft and hard, which do not get rolled properly during the process of manufacturing). Besides, the mechanical plucking also damaged many potential buds and immature 1+Bud shoots. This indicates that the efforts may be made to reduce the proportion of cut and loose leaves and prevent the harvesting of unopened buds, small 1+Bud shoots, and mature leaves. In achieving this, the following points may be considered:

- Raising the height of the plucking table by 1 to 2 cm at every successive / consecutive plucking interval.
- Selection of tea bush having horizontal leaves
- Modification of existing plucking machine or fabrication of new machines for selective plucking which can exclude the harvesting of undesirable shoot components.

Mechanical plucking reduces the cost of plucking drastically, but at the cost of quality. Investigations regarding standardization of manufacturing process to obtain quality tea from machine plucked crop may improve the economics of tea cultivation considerably.

In future researches pertaining to plucking studies, may also include observations on following growth parameters for better understanding of the cause and effect relationship:

- Regeneration of the lateral shoots
- Number of shoot generations produced
- Potential buds on plucking points
- Die back of plucking points
- Bush health including diameter of pruning sticks and weight of pruning litter.

REFERENCES

Amiri ME, Asil MH (2007) Determination of optimum harvestable length of shoots in tea (*Camellia sinensis* L.) based on the current shoot growth, rather than interval plucking. *Journal of Food Agriculture & Environment* 5(2): 122-124.

Asil MH (2008) Effects of plucking methods on yield and quality of black tea (*Camellia sinensis* L.). *Journal of Food Agriculture & Environment* 6(2): 337-341.

Awasthi RC, Sarkar AR (1983) A study on labour cost component in tea industry in Assam Valley. *Two and a Bud* 31(1-2): 14-18.

Barbora AC, Banerjee MK (1988) Causes and prevention of knot formation in tea. *Two and a Bud* 35: 39-42.

Barbora AC, Saikia MC (1989) Effect of plucking round on yield, crop distribution and standard of the harvested tea shoots *(Camellia sinensis* Lin.). *Two and a Bud* 36(1-2): 1-7.

Barbora AC, Sarma J, Barua DC, Barbora BC (1992) Effect of thermal time regulated plucking frequency on yield and quality of tea *(Camellia sinensis* L.). Paper presented in 31st Tocklai conference, Jan. 11-12.

Hajra Ghosh N (2001) Tea cultivation: comprehensive treatise. International Book Distributing Company, Lucknow, U.P., pp 225-267.

Harler CR (1964) The culture and marketing of tea. Oxford University Press, London.

Herd EM, Squire GR (1976) *Observations* on the winter dormancy in Malawi. *J. Hort. Sci* 51: 267-279.

Hudson JB, Durairaj J, Muraleedharan N (2002) Guidelines on tea culture in south India. United Planters' Association of Southern India, Coonoor, Tamilnadu, Allied Publishers Pvt. Ltd., pp 72-81.

IHBT (1998). Tea Plucking, Technical Folder 98/10, Institute of Himalayan Bioresource Technology, Palampur, India.

Jain NK, Tamang MB (1990) Physiological basis of tea management practices. In: Proceedings of the International Congress of Plant Physiology. New Delhi. Feb. 15-20. Volume I, pp 469-479.

Janakie B, Balasuriya, J (1996) Effect of altitude on shoot development of clonal tea with special reference to clonal selection and harvesting intervals. *Sri Lanka Journal of Tea Science* 64(1-2): 51-62.

Kulasegaram S (1988) Approaches to higher tea productivity through proper bush management. In Proceeds. Regional Tea (Scientific) Conference 1988, Tea Research Institute of Sri Lanka.

Kulasegaram S, Kathiravetpillai A, Shahmagarajah V (1975) Approaches to higher tea productivity through proper bush management. Tea Research Institute of Sri Lanka, Talawakele, pp1-10.

Kumar RR, Venkatesalu V, Manivel L (1993) Clonal variation in carbon dioxide assimilation in tea. *Photosynthetica* 29(4): 609-612.

Lelyveld LJVan (1986) Effect of field practices and factory processes on black tea quality. *J. Hort. Sci,* 61(4): 549-553.

Magambo MJS (1975) Some observations on growth of tea shoots. *Tea in East Africa* 15(2): 7.

Mahanta PK, Hazarika M, Baruah S (1990). Influence of plucking and processing on cellwall and soluble components in black tea. *Two and a Bud* 37(1): 17-19.

Maitland Derik (1982) 5000 years of tea - a pictorial companion. Gallery Books, 112 Madison Avenue, New York 10016, pp 87

Manivel L, Hussain S (1986) Relative sink capacity of developing tea shoots. *Two and a Bud* 33(1-2): 30-33.

Mitini-Nkhoma SP (1989) The effect of round length on the yield, pieces per kilogram and relative value. Tea Res. Found. Quart. News letter, No. 95, pp12-13.

Mwakha E (1985a) Tea shoots growth at different altitudes in Western Kenya. *Tea* 6(2): 19-24.

Mwakha E (1985b) Tea shoots growth in Kenya high lands. *Tea* 6(1): 5-13.

Mwakha E (1987) Plucking intensity in relation to hail damage in Western Kenya. *Tea* 8(2): 49-56.

Mwakha E (1988) Tea shoot growth in a pruning cycle in Kenya highlands. *Tea* 9(2): 71-75.

Mwakha E (1991) Clonal tea response to plucking standard and round length in the first year after recovery from plucking. *Tea* 12(2): 89-96.

Nathaniel RK, Jayasinghe HD, Dahanayake KD (1986) Effect of harvesting on the growth and yield of law grown clonal tea. *Sri. Lankan Journal of Tea Science* 55(1): 14-21.

Nyirenda HE (1995) The impact of the fourth leaf on the made tea quality of different clones. Quarterly Newsletter Tea Research Foundation of Central Africa, No. 118, pp 13-15.

Nyirenda HE, Mphangwe NIK (2001) Maximizing crop and total value of some TRF clones by plucking at optimum shoot age. Quarterly Newsletter Tea Research Foundation Central Africa, No. 143, pp 18-24.

Obaga SMO, Ng'etich WK (1989) Thermal-time and rate of shoot growth before and after pruning. *Tea* 10(2): 80-89.

Odhiambo HO (1989) Nitrogen rate and plucking frequency on tea: the effect of plucking frequency and nitrogenous fertilizer rates on yield and yield components of tea (*Camellia sinensis* (L.) O. Kuntze) in Kenya. *Tea* 10(2): 90-96.

Owuor PO, Chavanji AM (1986) Caffeine content of clonal tea: seasonal variations and effects of plucking standards under Kenyan conditions. Tea Res. Found. Kenya. *Food Chemistry* 20(3): 225-233.

Owuor PO, Kamau DM, Jondiko EO (2009) Responses of clonal tea to location of production and plucking intervals. *Food Chemistry* 115(1): 290-296.

Owuor PO, Ng'etich WK, Obanda M (2000) Quality response of clonal black tea to nitrogen fertiliser, plucking interval and plucking standard. *Journal of the Science of Food and Agriculture* 80(4): 439-446.

Owuor PO, Obanda MA, Othieno CO, Harita H, Tsushida T, Murai T (1987) Changes in the chemical composition and quality of black tea due to plucking standards. *Agric. Biol. Chem.* 51(12): 3383-3384.

Owuor PO, Obanda, M (1998) The changes in black tea quality due to variations of plucking standard and fermentation time. *Food Chemistry*, 61(4): 435-441.

Owuor PO, Othieno CO, Odhiambo HO, Ng'etich WK (1997) Effect of fertilizer levels and plucking intervals on yield and quality of clonal tea (*Camellia sinensis* (L.) Kuntze). *Tropical Agriculture* 74(3): 184-191.

Palmer Jones RW (1974) The economics of tea production and marketing in Malawi. Univ. Reading Dept. Agric. Econ. Dev. Study, pp 15.

Palmer Jones RW (1977) Effect of plucking policies on the yield of tea in Malawi. *Experimental Agriculture* 13: 43-49.

Palmer Jones RW (1985) Harvesting policies for tea in Malawi. *Experimental Agriculture* 21(4): 357-368.

Pethiyagoda V (1964) Some observations on the dormancy of the tea bush. *Tea Quarterly* 35(2): 74-84.

Pethiyagoda V (1967) Report of the Plant Physiology Division for 1966. Annual Report, Tea Research Institute of Ceylon, Part II, pp 48-58.

Portsmouth GB (1957) Factors affecting shoot production in tea (*Camellia sinensis*) when grown as a plantation crop. I. Some factors on the length of the plucking round on flush shoot production. *Tea Quarterly* 28: 3-7.

Portsmouth GB, Rajiah ES (1957) Factors affecting shoot production in tea (*Camellia sinensis*) when grown as a plantation crop. III. The time factor and new shoot production. *Tea Quarterly* 28: 21-29.

Rahman F (1977) Plucking of mature tea. *Two and a Bud* 24: 3-7.

Rahman F (1988) Physiology of the tea bush. *Two and a Bud* 35: 1-14.

Rahman F, Barua DN (1980) Plucking. In: Tea Encyclopedia, Sr. No. 201, filed under H2, Tea Research Association, pp 1-5.

Rajkumar R, Manivel L, Marimuthu S (1998) Longevity and factors influencing photosynthesis in tea leaves. *Photosynthetica* 35(1): 41-46.

Ranganathan V, Raman K, Nateson S (1983) Nutritional and physiological interactions with length of plucking rounds and banjhiness. Bulletin, United Planters' Association of Southern India, Tea Scientific Department, No. 38, pp 67-89.

Rawlley NK, Beig MV (1990) Comparison of plucking systems and leaf standards between N. east India and S. India. *The Planters' Chronicle*, October: 479-484.

Roberts EAH (1958) The chemistry of tea manufacture. *Journal of the Science of Food and Agriculture* 9: 381-390.

Roberts EAH (1962) Assessment of quality in teas by chemical analysis. *Two and a Bud* 9(3): 3-8.

Sanjay R, Baby UI, Sasidhar R, Premkumar R (2003) Impact of harvesting methods on grey blight incidence. Newsletter UPASI Tea Research Foundation No. 13(1), pp 3.

Sarkar SK (1988) A few aspects of plucking. In: Sarkar SK, Bezbaruah HP (ed) Field Management in Tea, Tocklai Experimental Station, Jorhat, Assam 785 008, pp 46-55.

Satyanarayana N, Spurgeon C, Sharma VS, Cox S (1995) Two-stage tipping in rejuvenated tea. *Journal of Plantation Crops* 23(2): 67-69.

Shanker A, Sood C, Kumar V,Ravindranath SD (2002) Insect and mite pests attacking tea plantation of Kangra Valley and their management. *Indian Journal of Entomology* 64(1): 53-57.

Sharma DK, Sharma KL (1998) Standardization of tipping/plucking height for a light pruned section of China hybrid tea (*Camellia sinensis* (L) O.

Kuntze) grown in Himachal Pradesh. Proceedings of the 12th Symposium on Plantation Crops, PLACROSYM XII, Kottayam, India, 27-29 November, 1996, pp 228-229.

Sharma KL, Sharma DK (1992) Standardization of tipping height in relation to pruning level of China hybrid tea (*Camellia sinensis* L.) grown in Himachal Pradesh. *Indian Journal of Agricultural Sciences* 62(11):758-761.

Sharma KL, Sharma DK (1997) Standardization of tipping height for a cut back section of China hybrid tea (*Camellia sinensis* L.) grown in Himachal Pradesh. *Himachal Journal of Agricultural Research* 23(1-2): 101-104.

Sharma VS (1983) Plucking styles. Bulletin, United Planters' Association of Southern India, Tea Scientific Department, No. 38: 33-48.

Sharma VS, Murty RSR (1989) Rationalization of plucking intervals in tea. Bulletin, United Planters' Association of Southern India, Tea Scientific Department, No. 43: 1-5.

Sivapalan K (1977) Tracer studies on some aspects of tea cultivation in Sri Lanka. Proc. Inter. Sym. "Improving crops and Animal Productivity by Nuclear and Allied Techniques". Colombo. ISNA, Nuclear Research Lab., IARI, New Delhi.

Squire GR (1979) Weather, physiology and seasonality of tea in Malawi. *Experimental Agriculture* 15: 321-330.

Sud RK (1993) Studies on interval and method of plucking on productivity and quality and quality of china hybrid tea. Ph.D. Thesis, Himachal Pradesh Krishi Vishvavidyalaya, Palampur, Himachal Pradesh.

Sud RK, Singh CM, Jain NK (1996) Economics of different plucking intervals in China hybrid tea (*Camellia sinensis*) in Himachal Pradesh. *Journal of Plantation Crops* 24(1): 29-33.

Sud RK, Singh RD, Singh B, Ravindernath SD (2003) Effect of methods and systems of plucking on productivity and quality of tea. *Bhartiya Vegyanic avm Audhyogic Anusandhan Patrika* 10(2): 146-150.

Sukasman (1989) Plucking and yield of quality tea. 3. Effect of plucking system on yield of shoot and pruning wood in tea. *Menaran Perkebunan* 57(3): 65-69.

Tamang MB (1981) A manual of tea culture for 3-months condensed coarse trainees. Tocklai Experimental Station, Tea Research Association, Jorhat, Assam, pp 35-44.

Tamang MB (1985) Studies on the nature of shoot production in clonal tea *(Camellia sinensis* (L.) O. Kuntze) as influenced by some important cultural practices. Ph.D. Thesis, University of Gauhati, Assam.

Tanton TW (1981a) The banjhi (dormancy) cycle in tea (*Camellia sinensis*). *Experimental Agriculture* 17(2): 149-156.

Tanton TW (1981b) Growth and yield of the tea bush. *Experimental Agriculture* 17(3): 323-331.

Tanton TW (1982a) Environmental factors affecting the yield of tea. I Effect of air temperature. *Experimental Agriculture* 18:47-52.

Tanton TW (1982b) Environmental factors affecting the yield of tea. II Effect of soil temperature, day light and dry air. *Experimental Agriculture* 18: 53-63.

Taparia M, Deka A, Sahewalla IP (1996) Effect of tipping standard on regeneration of pluckable shoots in pruned tea. *Journal of the Agricultural Science Society of North East India* 9(2): 195-199.

Taparia M, Deka A, Sahewalla IP (1997) Regeneration of pluckable shoots in pruned tea as affected by tipping-in-material. *Journal of Interacademicia* 1(1): 21-27.

Templer JC (1971) Annual Report for 1970. Tea Research Institute of East Africa, pp 19-20.

Templer JC (1977) Annual Report for 1976. Tea Research Institute of East Africa, pp 83-90.

Templer JC (1978) Tea plucking studies. Pamphlet No. 26/78. Tea Research Institute of East Africa, pp 20-39.

Tocklai (1971) Annual Scientific Report, 1970-71, Tocklai Experimental Station, Tea Research Association, Jorhat, Assam, pp 47-48.

Tocklai (1978) Annual Scientific Report, 1977-78, Tocklai Experimental Station, Tea Research Association, Jorhat, Assam, pp 49.

Tocklai (1979) Annual Report, 1978-79, Tocklai Experimental Station, Tea Research Association, Jorhat, Assam, pp 49.

Tocklai (1983) Annual Scientific Report, 1982-83, Tocklai Experimental Station, Tea Research Association, Jorhat, Assam, pp 21-22.

Tocklai (1984) Annual Report, 1983-84, Tocklai Experimental Station, Tea Research Association, Jorhat, Assam, pp 66-68.

Torgasheff BP (1926) Methods of cultivation. In "China as a tea producer". The Commercial Press Ltd., Shanghai, China.

Tubbs FR (1938) Report of the Plant Physiologist for 1937. Bullentin, Tea Research Institute, Ceylon, No. 18, pp 44-63.

Tubbs FR (1949) Plucking standard. *Tea Quarterly* 20: 85-92.

Ullah MR, Jain JC (1980) Seasonal variations in the chlorogenic acid content of tea. *Journal of the Science of Food and Agriculture* 31: 355-358.

Verghese MA (1977) Towards attaining 150 million kg tea production in 1985 - optimising plucking systems for maximum harvest. Proc. First Joint Area Scientific Seminar 1977. Bulletin, United Planters' Association of Southern India, Tea Scientific Department, No. 34: 35-48.

Watson M, Bandaranayake JIH, Wettasinghe DT (1982) The use of plucking machines for harvesting tea. *Tea Quarterly* 51(2): 58-66.

Wettansinghe DT, Nathanial RK, Karoon DD (1981) Effect of severity of plucking on the growth and yield of low-grown tea. *Tea Quarterly* 50: 123-130.

Wight W (1932) The structure and growth of the tea bush II. The tea leaf. *Quart. J.*, Indian Tea Assoc. Scientific Dept., Part IV: 206-217.

Wight W, Barua DN (1955) The nature of dormancy in the tea plant. *Journal of Experimental Botany* 6:1-5.

Wijeratene MA (2003) A new collapsible plucking basket. *TRI Update* 8(1): 2-4.

Wilkie AS, Burton, SJA (1995) Leaf handling study. Quarterly Newsletter Tea Research Foundation of Central Africa, No. 120, pp 26-39.

Yadav LS, Srivastava AP (2002) Critical operations and energy inputs of tea gardens in North East India. *Journal of Research, Birsa Agricultural University* 14(1): 105-108.

9

MECHANIZATION IN TEA

K.K. Singh

CSIR-Institute of Himalayan Bioresource Technology,
Palampur (H.P.) 176 061, India. E-mail: kksingh@ihbt.res.in

INTRODUCTION

In India Mechanization in general agriculture is not entirely of recent origin. Earlier ploughing and puddling were also done as farm operations in case of paddy. Now Indian and Japanese paddy transplanter are available for mechanized operation in paddy crop. For quite some time, the tea industry and the research institution had been considering the ways and means of mechanizing possible area of field operations in tea estates and factories. This issue has assumed considerable importance as tea production has been going up year after year, but it is becoming difficult to recruit labour and retain them due to rising cost of labour. This development is because of the change in attitude and mind set as the tea garden work has been considered as low standard job due to improved literacy and or higher wages in other works etc. Such difficulties in availability of manual labour are noticed in all other operations also.

Mechanization acquires all the more significance in tea culture in view of the high labour requirement for routine cultural operations *viz.* plucking, pruning, skiffing, plant protection measures etc.

Field operations, which can be mechanized, are as follows:

1. fertilizer application
2. chemical spraying

3. mechanical weed control
4. pruning and skiffing;
5. plucking.

The simple machines have been developed for field operations, have a limited capacity and are restricted to particular job. The power sprayer is designed as all purpose machines for plantation crops where as plucking, skiffing and pruning machines are is designed for tea only.

Engine Operated Sprayers

These machines are usually portable type mono block engine, mounted on the tank has been carried manually. This engine drives the sprayer pump and creates enough pressure for spraying of insecticide and pesticides. Some of the sprayer pump can be attached with the drive shaft of tractor/power tiller or any movable engine which creates pressure for spraying of chemicals. These pumps are also used in tea plantation.

Aeroplanes and helicopter Sprayer

A centrifugal pump is used to force the sprayer liquid through the nozzles. This pump works under a pressure of 3.0-8.5 kg/cm^2. The pump gets its drive from a wind driven propeller having four to six blades. In order to reduce the dead load plastic tanks are used to store the liquid. Depending upon the size of carrier the tank capacity ranges between 450-2200 litres.This system was used for application of fertilizers and chemical in unshaded tea. This operation is expensive although a large area can be covered in a short time. But the application efficiency is low in comparison to self-propelled ground machine. Aerial application of major nutrients has been done in Papua New Guinea and Malaysia (Sivaram, 1982). Similarly Zinc has been sprayed in Malawi (Malenga, 1979; Ellis, 1980). The result of aerial application of pesticide was not encouraging (Muraleedharan, 1986)

Tea Plucking Machine

Two types of tea plucking systems are followed:

1. Manual Operated Shearing Machine
2. Machine operated

Shear Harvesting

Shear harvesting machine is an important tool for maintaining a proper level of bush table. It has 25-30 cm shear blades with 700-800gm weight. This machine may be used from tipping level to production stage. Shear harvesting should be used only during high cropping months. The entire available crop can be harvested effectively during the rush period. Plucking average can be increased by 50-100 per cent and labour requirement can be reduced by 30 to 40 per cent as compared to hand plucking. A proper plucking table can be maintained throughout the pruning cycle (Figure 9.1).

Fig. 9.1. Shear harvesting of tea

As out of total cost of production, seventy per cent is the cost of leaf plucking and with increasing productivity of crops on one hand and decreasing labour availability on the other; the best option is to adopt motorized mechanical harvesting.

Motorized harvesting can be classified as follows

1. Hand held one-man operated machine.
2. Hand held two-men operated machine
3. Trailed type machine *i.e.* one man driven or two men driven.
4. Self propelled machine.

Hand held one-man operated machine:

Shear harvesting machine was developed for harvesting of tea shoots. A small motor drives the cutting action of the machine and a collection bag is attached behind the shearing blade where the harvested leaves would ve collected by a blast of air. It improves the productivity of manual plucking, particularly when flush is very high and the adequate numbers of workers are not available.

Fay (1950) made a trial with three men operated electrically driven portable motor through trailing cable and reported that there was neither significant change in yield nor a significant loss in quality of crop. Fuchinoue (1961) reported that there was a 30% reduction in yield and decline in quality and adverse effect on bush vigour. Whitham (1966) had reported that the shear harvester gave a lower yield and quality than the hand plucked bushes. The motorized harvested leaf was low quality containing immature and damaged leaves. The unopened buds are damaged which result into reducing the quantity and quality of tea leaf. Dale (1974) had compared the yield and quality of two varieties of tea plucked by hand-held battery-operated Ochial machine and hand-plucked on interval of 1, 3 and 6 week intervals for 3 months and observed that the portion of coarse leaf was very high with mechanical plucking, especially on 1 week interval. Separation of coarse leaf would reduce the crop advantage of large quantity plucked by machine. Shih *et al.* (1974) examined that the average yield of tea plucked by machine, shears or hand was 7858.3 kg, 7435.6 kg and 6318.5 kg per ha respectively. The best quality tea was hand plucked, but with careful pruning, and ideal cultural practices the quality in machine or shear harvested can also be improved. Scarborough (1975) had compared the yield and quality of mechanically plucked tea at 3-week intervals and recorded that the machine harvesting was disappointing as contained lot of coarse leaf. Tkeshelashvili and Khukhunaishvili (1975) found that there was 28 % good quality leaf in mechanical plucking. Rooster (1976) had compared battery-operated plucking machine with hand plucking and found that three times more tea shoots could be harvested by former machine as compared with hand plucking. Ndamugoba (1977) has compared mechanical plucking, light/hard plucking, hand plucking and recorded yield 571, 792 and 774 kg/ha, respectively. Reddy (1983) observed that 13% yield was increased from shearing China hybrid tea. It was also observed that a higher amount of coarser leaf was initially found and was reduced in future course whereas Sivaram (1982) reported that harvesting with tarpen trimmer was more expensive

Japanese Models : Single Man Operated Machine

Presently the common machine used in tea shoot harvesting was developed by Kawasaki Japan. The single man operated motorized harvester consists of a set of reciprocating shears carried on a suitable frame. The prime mover is the engine, which is mounted on the back and connected to the shear by means of flexible cable. The rotary motion is converted in to reciprocating motion. The length of the blade varies from 30 to 60 cm. The leaf cut by shear blade is carried by air blast in to a collecting bag. The Capacity of the engine is 27cc and the output of two-stroke engine is 1.75 HP at 6000 rpm. This engine can run by using fuel and oil mixture at a ratio of 25:1. Overall weight of the harvesting machine is about 9 kg and the cost of the machine is about Rs 50,000.

Two Men Operated Machine

This machine is operated by two labourers; each one walking on each side of the row of the bushes. In this type, the cutter bar may extend for half the width of the bush table or the full width of bush table. With a half width cutter bar the machine is carried along the same row twice in different direction. The machine will be lighter but the time required will be double. This machine was previously used in Republic of South Africa. Kwasaki PH 110 ES machine are used in Mauritius (Ellis 1984) with a cutter bar of 100 cm width. By a curved shape plucking machine the tea bushes are harvested in two passes by two men tea plucking machine. Jayaraman (1982) had experimented with curved blades tea plucking machine and harvested 113 kg of medium-sized leaves in 4 hours. Othieno and Anyuka (1982) had evaluated hand plucking, shear plucking and motorized machine plucking, at an interval of 14 and 28 days and found a narrow difference among three methods which were higher with a 14-day interval (1118 compared with 1021 kg/ha). They also reported that plucking was most rapid with machine, with little difference between the other two methods. Tea quality was best with hand plucking. Watson *et al.* (1983) had compared mechanical plucking by two small plucking machines at two-week interval and hand plucking at weekly interval. The plucking machines had a motor-driver hair-clipper type machine powered by a portable generator with a connecting cable and found the smaller machine was the more efficient, with a daily output of up to 215 kg of green leaf having 65 % leaf finenesses. Mwakha (1986) had evaluated motorized mechanical harvesting at 21, 42 or 63-day intervals, 1 or 2 cm above the previous level and compared with those of hand plucking at variable intervals. The highest yield (5927 kg/ha) was obtained from hand plucking, but this was not significantly different from that with mechanical harvesting at 63-day intervals with a 1 cm table rise at each harvest. Mwakha (1990) had further experimented mechanically plucking at 42, 56 or 70 days interval at the above mentioned heights (1.0 and 2.0 cms) and compared with hand plucking. He recorded double the yield and at least 35 times faster operations as compared to hand plucking. Gejima *et al.* (1993) had experimented with roll type plucking machine and found damaged tea leaves were 30-50% for the first crop of long tea shoots and 50-80% for the third crop of short tea shoots. The damaged leaves reduced the quality of plucked tea leaves. The results of a quality test indicated that the tea leaves plucked with the roll type plucking machine were of equal or better quality than those plucked with a shearing tool or commercial cutting type machine, because of the reduction of hard and woody parts of tea stems obtained with selective plucking (Figure 9.2).

Fig. 9.2. Demonstration of two men plucking machine

Presently two models (H120 and H140 from M/s Kawasaki Japan) straight blade and curved blade tea plucking machines are available in the market. Both the harvesting machines are similar in nature. The capacity of the engine is 40.6 and 41.5 cc. These motorized harvesters are run by petrol – 2T oil mixture (25:1) with fuel consumption of fuel 800 ml and 1000 ml per hour with 120cm, 140 cm, 17kg, 29 kg and 800 kg, 1000 kg, Rs 1,35,000 and Rs 1,55.000 is fuel consumption per hour, cutting span, weight of machine, green leaf harvesting per hour and the cost of H120 and H140 type two men tea plucking machine respectively. This machine can harvest about 0.5 ha area per day.

Picky Power Harvester

Picky harvester is a battery-operated machine with rechargeable battery of 24 watts and has a shear blade of 27.5X 25 cm. One set of battery can run for 2.3-3.00 hrs. The shear consists of two semi circular knives and rotation guide, which pushes the leaf in to the bag attached to the rear side of the unit. The weight of the harvester is approximately 9 kg inclusive of battery and the machine. The capacity of the machine is 100 bushes per hour and the cost of the machine is Rs 50, 000.

Indian Models Tal-Power Mobile Harvester

Single Man Operated Machine

A single man operated machine weighs about 10 kg and is equipped with 1.8 HP engine and generates 6200 RPM. The harvesting head consists of 45 cm blade. It harvests tea leaf under reciprocating action and throws leaf by an air blower. The cost of machine is about Rs 20,000.

Two Men Operated Machine

A two men operated machine weighs about 18 kg with Centrally Mounted Engine with a 100 cm reciprocating blade. This machine can cover about 0.4 ha and harvest 400 kg of green tea leaf. The cost of the machine is Rs. 35,000.

Two-Men Operated Machine developed by Institute of Himalayan Bioresource Technology Palampur (HP)

A two men operated machine has been developed in collaboration with Central Mechanical Engineering Research Institute, Durgapur (W.B.) This machine is mounted with a electronic start, fuel efficient air cooled 2 stroke 40.6 cc petrol engine capable of developing 2 HP at 7200 RPM Prime Mover. This motorized harvester is run by petrol – 2T oil mixture (25:1) as fuel @ 800ml per hour. It is equipped with a self-sharpening blade with low vibration and noise level. The length of the cutter bar is 1200 mm. The capacity of green leaf harvesting per hour is 800 kg. The average leaf composition recorded by mechanical plucking has been given in Table 9.1.

The comparative study different plucking system and their feasibility were recorded in Table 9.2 and 9.3.

Table 9.1. Percentage average composition of leaf in mechanical harvesting.

Bud	1 Leaf & Bud	2 Leaf & Bud	3 Leaf & Bud	4 Leaf & Bud	5 Leaf & Bud	Cut Leaf	Fine Leaf
3-5	8-10	28-30	20-25	12-16	5-9	8-12	42-52

Table 9.2. Comparative study of plucking system.

Parameter	One man plucking machine	Two men plucking machine Dome	Two men plucking machine Flat	Hand Shear	Manual Tea Leaf Plucking
Plucking cost (Rs./kg)	2.00-2.50	0.96-1.45	0.96-1.45	2.35-2.73	4-5.30
Green leaf plucked (kg/m. d.)	80-100	200-300	200-300	30-35	15-20
Efficiency by weight of leaf plucked (%)	500-533	1300-1500	1300-1500	175-200	Control
Efficiency by cost of leaf plucking (%)	200-212	365-400	365-400	170-194	Control
Time/ha Lean (hr)	20-25	4-5	4-5	100	Control
Flush	30-35	6-8	6-8	125-140	Control
Manpower engaged	2 + 1	3 + 1	3 + 1	1	1
Bush surface	Almost uniform	Uniform	Uniform	Non-uniform	Non-uniform

Agromatica DK Harvester

This is a pneumatically operated Italian machine. The harvesting system comprises of an engine coupled with a compressor with five harvesting heads. The harvesting heads have got reciprocating blades of 30 cm length and a leaf tray of 38cm X 38 cm X 18 cm. The handle of the harvester is 30 cm long and has got a cut-off trigger. Each harvesting head weighs 2.8 kg and the leaf tray can hold up to 1.5 kg of green leaf. The engine runs on diesel generating 10 HP motor with a single cylinder 406 CC. The harvesting machine can be started with recoiled system and lubricated with 20/40 oil. The compressor will generate a pressure of 13 kg/cm^2 at the head of the mains blower and 9-10 kg/cm^2 at the nozzles. The fuel consumption of the machine is 1.8 l/h. The tubes are 100m m each. This machine requires 5 workers for running of machine and two workers are required for collecting leaf

Fig. 9.3. Self–propelled machines

Mechanization of Pruning

Pruning is an important operation in tea carried out once in four to five year. Generally it is recommended to carry out once in four to five years. Due to shortage of labourers, pruning operation is delayed, resulting in poor recovery of the bushes. Mechanization of pruning operation will be very useful in completing the pruning in time. Kaaz Java cutter shoulder mounted double handled machine fitted with two cycle TL 43 engine on one side with a displacement/compressive volume of 42.7 CC. One man-operated machine has a shaft length of 1.54 m. The blade is a chip saw mounted on spinning disc with a diameter of 25 cm. The total weight of machine is approximately 7 kg and the maximum RPM is 8500. The machine is run by a mixture of gasoline and 2T oil at 25:1. The fuel efficiency of the machine 800 ml per hour at optimum speed and the price of the machine is about Rs 40,000. The capacity of the machine to prune by a single worker is 850 bushes.

Sharma and Venkata Ram (1980) tried two one man pruner operated by two stroke motors model KPE 200 and was found satisfactory and economic for skiffing at 85 cm and above. Model KFS 1200 was effective and economic for skiffing at 75 cm and above.

The number of bushes pruned, fuel consumption, and the benefit cost ratio of pruning and skiffing machine is tabulated below Table 9.4 to Table 9.9.

Table 9.4. Number of china hybrid tea bushes pruned per day

Particulars	One man pruning machine	Conventional method
LP	2000-2200	100-120
MP	1400-1600	80-90
DP/Rejuvenation pruning	1200-1300	40-50

Table 9.5. Fossil fuel consumed per ha of pruning

Particulars	One man pruning machine in litres
LP	9-10
MP	12-15
DP/ Rejuvenation pruning	15-18

Table 9.6. Pruning efficiency and B/C ratio of China hybrid tea with respect to conventional method

Particulars	Pruning machine with respect to bushes pruned	B:C ratio
LP	2000	3.5
MP	1875	3.24
DP/Rejuvenation pruning	2500	4.25

Table 9.7. Number of China hybrid tea bushes skiffed per day using different methods

Particulars	Machine		Conventional
	One man	Two men	(Manual)
LOS	6000	15000	800
LS	4000	10600	400
MS	2900	7100	200
DS	NS	5300	100

Table 9.8. Fossil fuel consumed per ha of skiffing

Particulars	Machine	
	One man skiffing Machine (in litres)	Two men skiffing Machine (in litres)
LOS	5	4
LS	8	6
MS	11	9
DS	NS	12

Table 9.9. A comparative study of skiffing efficiency with respect to number of bushes and B: C ratio of China hybrid tea with respect to conventional method

Particulars	Machine efficiency		Machine B: C ratio	
	One man	Two men	One man	Two men
LOS	650	837.5	1.06	1.16
LS	900	1225	1.25	1.37
MS	1350	1675	1.48	1.53
DS	NA	2250	N.A	1.67

Mechanical Up Rooting of Tea

Replanting is considered to be an important operation for long-term development programme for augmenting productivity. However, of late this has become a difficult task due to the non-availability of manual workers in time. To overcome this problem JCB and Proclaim have been used successfully.

The JCB and Proclaim can uproot 150 and 250 bushes per hour respectively. The operating cost of machine per hour is Rs 700 per hour, which is on par with the cost of manual uprooting.

Bio-composting

In present economic climate, low yielding fields of tea become an economic burden. Rejuvenation or total planting can be in some cases reverse the economics. So, chemical free tea may be produced from these gardens. A lot of biomass is produced in terms of pruning litters. The can be made after chopping and made powder by machine. It will increase the nitrogen and also the water holding capacity of soil, which may result into the higher yield and profitability

Advantages of Mechanical Uprooting over Manual Uprooting

1. The heavy-duty machines can saves time, and all the operations can be completed in dry weather.
2. Soil erosion measures can be adopted before the rainy seasons.
3. Shortage of manual labour can be over come.
4. Soil up to root zone can be loosening.

5. Perfect removal of roots can be obtained.

Limitations

It can be effectively used in area with mild irregular slopes.

Precautions

This operation should never be done during monsoon period as it could result in massive soil erosion. This operation can be supported with adequate soil erosion measures.

REFERENCES

Acuna, D and Swier, R. 1981. Self propelled tea harvesters and their operating costs. *Informe Tecnico Estacion Experimental Agropecuaria Misiones* **No 36,** INTA, Cerro Azul, Misiones, Argentina, 16pp.

Anon. 1958. Mechanical harvesting in tea. *World Crops* **10:** 226.

Anon. 1983. Tea harvester shows great promises. *World Crops*, **35: 1,** 13.

Chkhaidze G.I.; Patarava, B.D.; Bolkvadze, G.V. and Mgeladze, Ts, A.1971. Influence of mechanical and manual harvesting of leaves on general condition and physiology and biochemical changes in tea plants. *Subtropicheskie Kul'tury,* **5:** 57-63.

Chkhaidze, G.I. and Miminoshvili, P.B. 1964. The influence of mechanical harvesting of leaf and cuttings of the bushes on stem formation in tea plants. *Subtropicheskie Kul'tury,* 3: 3-16.

Dale, 1974.The Japanese [tea] plucking machine. Quarterly News Letter Tea Research Foundation of Central Africa **No. 34,** 3-6.

Dutta, S.K.1956. The effect of mechanical harvesting on the tea crop. *Two and a Bud,* 3:3, 4-6.

Ellis, R.T.1984. Zinc spraying. Quarterly Newsletter, *Tea Research Foundation (Central Africa)*: **Vol. 60,** 3-4

Fay, B.D. 1950. The mechanical plucking in tea. *Tea Quarterly,* **21,** 38-44.

Fuchinoue, Y; Fuchinoue, H. and Maruyama, T. 1961. Studies to establish the rational way to pluck the tea plantation cold locality. *Proceedings of Crop Science Society of Japan*, 229, 425-427.

Gejima, Y.; Okada, Y.; Nagata, M. and Ishikawa, K. 1993. Studies on the development of tea-plucking machine (Part 3). Trial construction of tea plucking machine. *Journal of the Japanese Society of Agricultural Machinery*. **55: 5,** 41-49.

Gokhale, N.G. 1959. Mechanization of tea in USSR. *Two and Bud* **6(1)** 9-11.

Hobman, F.R.1973. Tea growing. *Queens land Agricultural Journal,* **99:5,** 226-233.

10

PLANT NUTRITION

R.D. Singh

CSIR-Institute of Himalayan Bioresource Technology,
Palampur (H.P.)-176 061, India. E-mail: rdsingh@ihbt.res.in

Until end of the nineteenth century, manures were seldom used in tea plantations in south-east Asia. Inorganic fertilizers were not tried in tea until the first world war. Later on, the first inorganic fertilizer used in tea was ammonium sulphate (Barua 1989). Gradually, potassium, calcium, phosphorus, and magnesium became the important plant nutrients for tea plantations.

Fertilizers can be an expensive input in tea production, accounting for up to 15-25% of expenditure. Productivity is linked to fertilizer input (Verma and Muraleedharan 1998). Factors affecting fertilizer use efficiency, include climate (rainfall, temperature, relative humidity (RH), solar radiation and day length), soil (pH, organic matter, moisture, texture, parent material, clay mineral and topography or slope of the land), plant morphology (root growth, leaf orientation, root cation exchange capacity, branching, spread and growth habit) and cultural practices (competition from weeds and shade trees, pruning, shade management, plucking style and intervals, rate, method, time and number of splits of fertilizer application, use of proper nutrient ratios and bush health).

In south India, tea production has increased from 685 kg/ha in 1946 to 2451 kg/ha in 1991, despite only small increases in the tea acreage (Verma *et al.* 1993). It is estimated that 20% of the increase in crop yields was due to good fertilizer use and the other 80% was due to improved crop management practices. Fertilizer recommendations are aimed at suppressing the reproductive phase and encouraging vegetative growth.

DEFICIENCY OF IMPORTANT NUTRIENT ELEMENTS IN TEA

Removal of plant nutrients by the crop increased with increase in tea productivity. Also, the plant absorbs a nutrient in large quantities, much in excess of its requirement for growth, when its supply is plentiful. The efficiency of a nutrient may express shortage of another nutrient. Multi-nutrient deficiencies have also become a common feature in the tea garden (Barbora 1996). A nutrient deficiency can be overcome by application of the deficient element. According to Barua (1989), the deficiency symptoms of the major plant nutrients and important trace elements are as follows.

Nitrogen: Stunted plant growth; leaves pale yellow; defoliation of lower leaves followed by upper leaves.

Phosphorus: Mature leafs with smooth surface, which at times become slightly purple; leaves yellow at the tip.

Potassium: Foliage develops distinct marginal scorch; internodes grow longer with thin stem.

Calcium: Mature leaf develops downward curves, leaf tip starts showing marginal scorching; finally the leaves drop and plant gets stunted; lower surface of leaf shows translucent patches, which gradually convert into necrotic spots.

Magnesium: Mature leaf shows distinct inter-veinal chlorosis and defoliation starts from the lower leaves.

Sulphur: Foliage show distinct chlorosis; young leaf becomes pale-green to yellow, however veins remain green against a pale-yellow background.

Boron: Shoot apex dies; axillary buds abort and clusters of small shoots appear in the axils; leaf gets thick and dark-green; upper side of leaf stalk develop cork formation spreading to main veins and laterals.

Manganese: Foliage show inter-veinal chlorosis; irregular mottling, veins and surrounding areas remaining green; necrotic spots within the chlorotic patches close to leaf margin; severe deformation of young leaves and buds.

Zinc: Leaf gets smaller, darker and sickle shaped owing to unequal development of the lamina on either side of the mid-rib; internodes shorter.

USE OF FERTILIZERS

NITROGEN

Nitrogen and yield: A review of data from various countries shows that tea response to N is similar in all parts of the world including East Africa where results show a fall in N response as the application rate rises, due to the limited availability of other nutrients such as P (Wilson 1975). The acidifying effect of N fertilizers increases this problem and also leads to foliar Mn accumulations on some Mn-rich soils. Leaf analysis showed that in many cases increasing the N fertilizer rate would increase the crop yield.

In Valparai, Tamil Nadu, India it was found that for a particular yield target at the same level of organic carbon the calculated N dose for UPASI-9 is lower than that of SA-6, which indicates that UPASI-9 is more responsive to N application than SA-6 (Venkatesan *et al.* 2003).

In a trial with NPK (20:10:10), N was applied annually to seedling tea at 0-270 kg/ha and K_2O was applied at 0-135 kg/ha (Owuor *et al.* 1987). Leaf K contents declined with increment in rates of N application, even though the K_2O application rates also enhanced. It was concluded that high N application rates either inhibit K uptake by the tea bush or increase soil K leaching.

Despite the exclusive use of urea, which does not add any S to the soil, the S reserves in the soils were adequate to meet the S requirements of the crop (Wickremasinghe *et al.* 1986). Sulphur removal with the harvested crop was small (only 10-15 kg even with a harvest of 3000 kg made tea ha^{-1} $year^{-1}$).

In dilapidated seed-raised tea plantation, increasing N levels in the absence of Mg depressed the soil extractable Mg content (Wanyoko 1988). The soil extractable Mg content was positively correlated with the young shoot Mg content at the highest rate of N application. Similarly, the young shoot Mg content was positively correlated with the young shoot N content at the highest level of N application.

In a plantation with a low available P content yield was positively correlated with the available P content and that of easily hydrolysable N content. This indicates that the role of P is more important than that of N. In the plantation with a medium content of available P the effect of P on yield was not significant, and it further reduced in the plantation with a high level of available P (Nakaidze and Kvaratskheliya 1977).

In Assam, young and mature tea continuously receiving urea-N for 5 years showed significant decrease in soil available sulfate-S (Gohain and Dutta 1994). However, no change in soil available sulfate-S was observed following 5 years of continuous application of rock phosphate or muriate of potash. The ratio of total N: total S was significantly correlated with crop yield.

The annual rate of dry matter increase in 5-year-old plant increased with increasing N application rate (Tachibana *et al.* 1996). The increase in the rate of dry matter accumulation was more in case of tea younger than 7-years.

In Kericho, Kenya long-term application, over a 16-year period, of N on tea (clone AHP S15/10) could not influence tea yield and leaf N and S contents. However, leaf and soil P and K contents were significantly higher with NPK compared with NPKS (Kamau *et al.* 2003).

In a field experiment, conducted at Tocklai Estate, increasing application of N from 100 to 300 kg/ha resulted in increasing N content of top soil and plucked shoots. Application of N provided higher yield compared with the control for all years and 200 kg N/ha was found to be the optimum rate (Barua 1990). Similarly, Grice and Phiri (1986) have also reported that in mature tea, the optimum rate of N was 225-235 kg /ha. In Indian western Himalaya, Marwaha *et al.* (1977) have found that total yield increased linearly with increment in levels of N, giving 31.0, 41.5 and 70.3% yield increases at 90, 180 and 270 kg N/ha, respectively, compared with the control. Further, they have concluded that when the first flush yield was high (during March to June as a result of rainfall) the total yield for that year was also high. From the same geographical region, Sharma *et al.* (2002) reported 0.68 to 26.09% increase in yield of tea over untreated control owing to application of N with and without FYM and *Azotobacter*.

Salimi and Latifi (2008) recorded significantly higher yield with 100 kg/ha N under rainfed condition than that due to 180 and 360 kg/ha N under sprinkler irrigation in tea, in Iran. They also concluded that irrigation is more important than nitrogen in increasing tea yield. Owuor *et al.* (2008) reported 200 and 250 kg N/ha/year to be optimal dose for high yielding clones.

Sharangiya and Marshaniya (1988) have reported that application of N as ammonium nitrate at 100, 150, 200 and 300 kg N/ha increased mean leaf yields by 27.5, 30.5, 33.9 and 14.8 hkg/ha, respectively, compared with PK alone (111.0 hkg/ha). Ammonium sulphate at the same N rates increased yields by 27.4, 32.1, 23.9 and 23.4 hkg/ha, respectively. Urea at these 4 rates increased yields by 20.4, 37.7, 16.4 and 16.6 hkg/ha, respectively, and at 400 kg N/ha by 22.0 hkg/ha.

Application of N at 100, 150, 200, 300 or 400 kg/ha as increased green leaf yield significantly compared with the control (basal PK application), and was profitable (Sharangiya 1987). Application of N at the rate of 150 kg/ha was the best and provided maximum returns, and urea was the most profitable N source.

Similarly, Malenga (1987) reported average yield as high as 3691 kg/ha due to application of N at 250 kg/ha. Whereas, tea quality and value tended to decline as N level increased. The highest average yields amounted to 5259 kg/ha with N at 375 kg/ha.

Grice and Malenga (1985) have recorded that in case of seedling tea, reducing N level from 75 to 0 kg/ha caused reduction in yield from 2220 to 1925 kg/ha. Reducing the N level from 150 to 75 kg/ha lowered yields from 3099 to 2882 kg/ha, and a similar increase in N level raised yields to 3230 kg/ha. Increasing or decreasing N rates from 225 or 300 kg/ha had little immediate effect on yield.

Application of N at 300, 450 or 600 kg/ha provided the highest average yield (5031 kg/ha) due to the lowest N rate of 300 kg/ha (Rikhter and Lyashko 1979). The treatments had no appreciable effect on tea quality although it declined slightly with increasing N rates.

In a 125-year-old tea plantation in Palampur, Indian western Himalaya, N at 180 kg/ha had the most significant effect on 3-year mean tea yield, raising it from 563.4 (control) to 1273.7 kg/ha. Application of P_2O_5 at 180 kg/ha raised the yield from 861.0 to 1045.5 kg/ha, K was the least effective, even at the highest rate of 80 kg/ha, raising the yield from 926.9 to 982.3 kg/ha (Marwaha *et al.* 1977).

In a fertilizer experiment on tea in Peru over a period of 4 years the most profitable application per ha per year was of 180 kg N + 60 kg P_2O_5 + 90 kg K_2O, which raised the yield by 100% to 2600 kg (Cesare 1972).

Nitrogen sources: The common forms of nitrogen used in tea gardens are urea, ammonium sulphate and calcium ammonium nitrate. Studies in south India have revealed that ammonium sulphate gave better yields than calcium ammonium nitrate (CAN) or urea. CAN gave good results as winter applications, and urea was generally the least satisfactory N source (Ranganathan and Natesan 1987). Among different N sources, ammonium sulphate gave the highest yield (up to 5349 kg/ha, Watson and Wettasinghe 1982).

Similarly, Hasan and Chaudhury (1973) recorded consistently higher yield with ammonium sulphate, at all application rates. Whereas Guseinov and Kerimov (1975) recorded best results from ammonium nitrate regarding leaf yield and quality in comparison of urea and ammonium sulphate. They have also recorded that quality of tea plucked in July was better than that plucked in May or September.

As per Tuti *et al.* (1974) in West Java the substitution of urea for ammonium sulphate as a source of N had negligible effects on yields, soil properties and leaf nutrient levels. In support of this finding Hasan and Chaudhury (1973) have recorded that urea at higher rate of 120 kg N/ha was detrimental compared with the lower rate (80 and 100 kg N/ha).

In plants grown in sand culture, growth and leaf colour were better with NH_4-N. Plants receiving NO_3-N had higher K contents but lower P contents than those receiving NH_4-N. Mn, Fe and Al contents were higher in plants receiving NH_4-N (Ishigaki 1972).

PHOSPHORUS

Phosphorus requirement of tea is much less than that of N and K and the response of the plant to phosphatic fertilizer are also more erratic. In south east Asia, Sri Lanka was the first to observe positive response of tea plant to application of phosphatic fertilizers. Eden (1944) reported significant improvement in yield when mature tea was supplied with phosphate 34 kg/ha.

Phosphate fertilizer incorporated in the soil at planting time showed increment in the rate of growth of the plants, thus indicating that tea in acid soils had a demand for phosphate (Wilson *et al.* 1975). Response to P fertilizer in terms of yield in mature tea was not consistent. They have further concluded that phosphate provided better yield when there was an undisturbed mulch layer on the soil surface. Within the mulch layer many highly active roots were formed and thus phosphate was perhaps maintained in a more available form. Of several application methods, broadcasting between the rows followed by soil incorporation was the most effective, the optimum time of application being 40-45 days before plucking (Ikegaya 1974).

In tea cultivars TV-1 and T3E3, application of 50 kg P_2O_5/ha enhanced N uptake as well as the tea yield (Gogoi *et al.* 1994).

In a 7-year field trial on annual application of rock phosphate (RP) and triple super phosphate (TSP) up to 125 kg P_2O_5/ha to a clonal tea, TSP significantly increased soil pH. Compared with TSP, RP significantly increased the extractable P in the soil especially at high rates (Othieno *et al.* 1997). Soil extractable K and Ca were not significantly affected by the P sources. Mature leaf P was significantly higher in the TSP treatment. The mean annual yield was not affected by the fertilizer source, although the yield increased with the increment in the rate of P.

Placement of phosphatic fertilizers at 5-10 or 15-25 cm depth gave better results than broadcasting with or without dibbling-in (Ranganathan 1971). Rock phosphate was as effective as other forms of phosphatic fertilizers.

POTASSIUM

The tea plant in large quantities uses potassium. Wilson (1975) concluded that responses to K fertilizers leading to an improved return from N fertilization occurred where the soil pH is 4.9 or lower. Potassium concentrations below 1.75% in first leaf with bud and 1.57% in third leaf are a clear indication that lack of potassium is limiting yield. Yield responses follow the rehabilitation of the maintenance foliage in which photosynthetic activity creates the constituent materials for the crop. The level of K application should be related to crop yield and full benefit of K can be achieved provided other nutrients, particularly P, are not limiting (Wilson 1975).

In tea plantations in Georgia, K fertilizers (chloride and sulphate forms) provided 21-29% higher yield over untreated controls (Godziashvili 1984). The contents of polyphenols, theaflavins and thearubigins, important quality components of brewed black tea, were also increased by K application (Ruan *et al.* 1999).

In Coimbatore, Tamil Nadu, India the yield in all treatments with MOP was lower than the MOP containing sodium (Venkatesan and Kumaraguru 2002). Scorching of mature leaves was observed due to application of MOP containing 45,000 ppm sodium.

Nitrogen and P as a basal fertilizer along with graduated rates of two K fertilizers increased yields compared with NP alone (control) (Kakabadze and Kvachadze 1987). Application of K_2O at 200 kg/ha, as potassium metasilicate, produced leaf yield of 10147 kg/ha, and as potassium chloride it produced 9732 kg/ha. Systematic K application, especially as potassium metasilicate, increased soil humus and available P and K contents, increased pH and improved various other soil properties.

In a 5-year study on effect of K_2O at levels up to 120 kg/ha, with N at 80 kg/ha, on old and mature tea, response was little during the first 2 years, but subsequently yields enhanced by K_2O at 70 or 90 kg/ha (Ali *et al.* 1977).

Kangra local jat tea shoots contain high concentrations (2.51-2.71%) of K. Thus, the amount of K assimilated in the production of 1000 kg commercial tea and the quantity of K present in tea liquor were 26-27 kg and 48-53 mg/cup, respectively (Sud 1996). There was significant seasonal variation in green leaf K concentration, with the highest values occurring in April and May (first flush after dormancy). Also, there was variation in leaf K concentration among different Kangra local jats.

When applied as foliar treatment, sulphate of potash caused less leaf scorching on unshaded young as well as mature tea, the effect was distinct at higher concentrations (>2-10%, Rahman *et al.* 1978). Shaded mature tea was less scorched by the treatment.

In China, studies have revealed that introduction of balanced nutrient application, including K and Mg, greatly improved the nutritional status of soils and increased the yield of major tea types, e.g. green tea, black tea and oolong tea (Ruan and Hardter 2001). In addition to improved productivity, it is shown that quality, e.g. free amino acids, polyphenols, caffeine as well as theaflavin and thearubigin contents of black tea, was considerably increased. These results confirm that balanced nutrition including the nutrients K and Mg is an important measure to improve the productivity of tea.

Hardjono *et al.* (1974) have reported that in old-established tea plantation K (potassium sulphate) applied annually up to 200 kg/ha did not affect yield or leaf K content, however it increased the soil exchangeable K and the branch weights in the loppings. Although yield was not affected, K applications are recommended to maintain soil fertility and replenish the K removed by the crop.

CALCIUM

According to the study conducted by Wilson (1975), yield data, soil and leaf analysis from a series of fertilizer field experiments showed that the effect of excessive quantities of calcium on tea is to reduce the uptake of potassium and induce potassium deficiency with a consequent reduction in crop yield. The effect is cumulative and can follow repeat applications of small amounts of Ca such as are present in super phosphate. Application of potash fertilizer counteracts the effect of the Ca.

Studies on inter-relation of Ca and Mg (Wu 1994) showed that Ca accumulated in older tissues whereas Mg accumulated in younger tissues. The Ca/Mg ratio increased rapidly as tissues aged. The Ca/Mg ratio was generally >1 in leaves but <1 in the roots. The quality of made tea decreased with increasing Ca/Mg ratio in the leaves. The uptake of both elements from the soil depended not only on their availability but also on the Ca/Mg balance. Deficiency of either of these two elements or an imbalance in their ratio led to a decrease in growth, yield and tea quality.

MAGNESIUM

Gohain *et al.* (1998) have investigated effect of Mg on yield and quality of tea in Assam. Magnesium significantly influenced yield, with the highest yield (mean 2901 kg/ha) observed at a rate of 15 kg Mg/ha. A positive, but not significant, response was observed between Mg and quality of made tea, with quality parameters (theaflavin (TF) content, thearubigin (TR) content, brightness, total colour and TF/TR ratio) increasing with the rate of Mg up to 15 kg Mg/ha.

In Karnataka, assamica type seedling tea treated with magnesium sulfate 15 kg/ha as foliar spray in 5 split doses with zinc sulfate 10 kg/ha, 500 ppm Mn and 100 ppm B gave the highest overall yield (5683 kg made tea/ha) (Verma *et al.* 1998).

Application of Mg with K has been reported to provide synergistic effect on tea yield (Godziashvili 1984). Ruan *et al.* (1999) have also reported increase in free amino acids and caffeine in tea with the increment in the treatment including both nutrients *viz.*, K and Mg. Also, K fertilizer application reduced the incidence of Mg deficiency symptoms on old leaves when K was also deficient (Wilson 1975).

Flavour compounds in brewed oolong tea were considerably improved by K and Mg application Ruan *et al.* (1999). With increasing

K applied to soil, the Mg content increased in the bud, but decreased in the mature leaf. An increase in the soil water-soluble magnesium enhanced K uptake by the foliage (Sivasubramaniam and Jayman 1976). Magnesium deficiency symptoms can occur due to a lack of soil Mg, or due to an excess of K, despite the presence of adequate Mg, in the soil. Application of Mg has also been reported to enhance the effectiveness of N. The best results, with regard to yield, being obtained at 300 kg N/ha (Antiya 1985). It has been observed that Mg had no effect on crop yield when the soil pH was between 4.1 and 5.0 (Wilson 1975).

In China, foliar application of Mg fertilizer increased tea yield by 20.8%. It also increased contents of amino acids, caffeine and water extracts by 5.8, 9.2 and 2.3%, respectively, as compared to control (Zhu and Chen 2003).

SULPHUR

Gohain and Dutta (1994) recorded a significant decrease in soil available sulfate-S owing to 5 years of continuous urea-N application. However, no change in soil available sulfate-S was observed following 5 years of continuous application of rock phosphate or muriate of potash. The S status of 2 leaves and a bud shoot was not affected by differential treatments of NPK, but the ratio of total N: total S was significantly correlated with crop yield.

In north east India, Gohain *et al.* (2000) have observed that out of the 9 sources of S application (20 kg/ha), gypsum recorded the highest yield, followed by ammonium sulfate. Sulphur application did not significantly influence shoot contents of N, P and K, but did influence that of S. Sulphur application improved tea quality as determined by theaflavin and thearubigin contents, brightness and organoleptic traits. Some S treatments increased the content of caffeine in shoots. Whereas, in Tamil Nadu, Palani and Balasubramaniam (2001) have recoded that organic manure and sulfate of ammonia at 60 and 30 kg were superior among different sources of sulphur *viz.*, soil S, iron pyrite, gypsum, organic manure and sulfate of ammonia.

Ye *et al.* (1994) have concluded that application of S improved plant metabolism, increased photosynthetic rate, chlorophyll content and total N and P contents but inhibited K uptake. At certain levels of S application, nitrate reductase activity and free amino acid contents (particularly threonine and aspartic acid) in leaf increased. Polyphenol oxidase (catechol oxidase) activity and polyphenol metabolism decreased significantly with S application resulting in a reduced polyphenol/amino acid value, which was good for the quality of green tea.

At the UPASI Tea Experimental Farm, Verma *et al.* (1998) have observed higher yield with application of 60 kg/ha of sulfate of ammonia.

MICRONUTRIENTS

ZINC

After the discovery of zinc deficiency symptoms in the tea areas of Sri Lanka (Tolhurst 1962), deficiency of Zn was observed in other tea-growing areas and, within a span of two decades, zinc became an indispensable fertilizer for tea throughout the world. Aerial application of zinc oxide at 1.35 kg/ha in October and February enhanced yield by 12% (Dootson 1976). A single application of copper sulphate at 2.0 kg/ha in October increased yield by 5%, and by 14% when applied with zinc oxide (Dootson 1976). The response to Zn was not affected by N level (Dootson 1974).

There was an increase in germination, shoot growth, tea yield, chlorophyll, amino acid and soluble carbohydrate contents, and polyphenol oxidase and nitrate reductase activity with optimum Zn concentrations (Han *et al.* 1994). Excessive Zn application reversed these effects. Han *et al.* (1994) have recommended 0.2-0.5 % $ZnSO_4$ for foliar sprays and 100 kg/ha for soil application.

In Central Africa, in assamica type seedling tea, application of ZnO in 4 equal splits of 1.25 kg/ha each, in August, November, February and March, raised average yield from 2915 to 3366 kg/year; this was significantly better than application in 2 splits in a year (Malenga 1986).

Sobolevskii *et al.* (1984) reported that application of Zn or mixtures of Zn, Co, Cu and Mo with irrigation water increased leaf yield and raised the contents of tannins and extractive substances. When applied singly, Cu or Co could not increase yield significantly, while Mo caused reduction in yield (Sobolevskii *et al.* 1984).

Ghosh and Chakravarte (1980) reported increase in tea yield in Assam due to foliar-applied zinc sulphate at 12.5, 25.0 or 50.0 kg/ha each prior to skiffing. Basal application of zinc sulphate (4 % w/w) with the fertilizer mixture reduced premature leaf fall, flowering and the incidence of zinc deficiency in tea nurseries. It increased shoot dry matter production by 80% as compared to a maximum increase of only about 15% due to foliar spraying.

MANGANESE

The leaves containing high quantity of Mn made poor quality tea (Yoshida *et al.* 1981). Mn-rich leaves showed a higher polyphenol oxidase activity than normal ones (Yoshida *et al.* 1981). The older leaves contained more Mn than the younger ones and their Mn content was highest during June and July. Yoshida *et al.* (1981) suggested addition of gravel to improve the soil and shading of tea plants so that leaf Mn content can be reduced. The latter proved to be the more effective.

Prasad and Dey (1979) reported increase in Mn uptake with increase in rate of application of manganese sulphate from 0 to 40 kg/ha. They concluded that foliar application was more effective than soil application.

Sarishvili *et al.* (1976) recorded that Mn deficiency occurred in young leaves of tea. In cases of simultaneous Mn and Mg deficiencies, the characteristic difference was that the early symptoms of Mg deficiency occurred on the lower, older leaves. Leaf yellowing was caused by both deficiencies.

Gohain *et al.* (2000) reported significantly highest tea yield due to application of 10 kg Mn/ha/year. Manganese application positively affected uptake of leaf N and P.

COPPER

Foliar application of 0.1% $CuSO_4$ resulted in maximum activity of polyphenol oxidase (PPO, catechol oxidase) and nitrate reductase in isolated tea shoots (Han *et al.* 1993). Application of 2 g $CuSO_4$/pot increased yield, shoot growth, growth of small nutrient-absorbing roots and leaf polyphenol, caffeine and amino acids concentrations compared with control.

BORON

Boron uptake significantly decreased with increase in rate of Ca application. B uptake also increased with application of B as boric acid, up to 6 kg/ha, and was higher with foliar application. The highest spray concentration, however, was phytotoxic (Prasad and Dey 1979). Kalita and Mahanta (1993) recorded that B uptake was greatest in leaf, followed by stem and roots. Foliar application of boric acid up to 1 kg B/ha per year significantly increased yield by 12 %, however tea quality remained unaffected (Gohain *et al.* 2000).

ALUMINUM

The tea plant accumulates abnormally large amount of aluminum without any apparently adverse effect on the plant (Barua 1989)

SELENIUM

Huang *et al.* (2005) reported that color of extract of selenium (Se)-enriched green tea was more stable during storage as compared to that of normal tea. Besides it, the sweetness and aroma of the extract were significantly higher, while the bitterness of extract was significantly lower as compared with normal tea. These results indicate that Se application can enhance shelf-life of green tea.

TRACE ELEMENTS AS CHELATES

Dzadzua (1987) concluded that in young tea, application of chelate mixture of Fe + Mn + Zn + Mo + basal fertilizer combination increased leaf yield by 19%, compared with the control (basal fertilizer only). Leaf quality was improved in the year of application and in the 2 subsequent years, but in the 4th year both yield and quality declined.

Gohain and Barbora (2000) have observed that among different sources of Mg tested, magnesium chelate and magnesium sulphate recorded the highest yield (12 and 6% yield increase over control). All sources of Mg improved the quality of tea, with magnesium chelate having the highest positive effect.

FERTILIZER AND TEA QUALITY

Zyrin and Barabadze (1987) have reported adverse effect on leaf mineral nutrient content and ratio due to long-term (>10 years) mineral fertilization, which impaired leaf quality.

Utnelishvili (1974) recorded increased tea yield and leaf tannin, caffeine and extractives contents due to application of N up to 200 kg/ha (optimal). Higher N rates had an adverse effect on leaf quality. Similarly, Oniani (1976) reported basal application of N at 50 kg/ha along with PK (150:150 kg/ha) provided higher tea catechin content (up to 205 mg/g dry matter) as compared with 195 mg/g in tea receiving only PK. Nitrogen at 150 kg/ha had no adverse effect on tea chemical composition, but 300, 500 or 700 kg/ha of N reduced catechin content. High rates of N fertilizers have been reported to increase caffeine content and the compounds imparting inferior and superior flavour contents (Owuor *et al.* 1987). However, the flavour index, the ratio of compounds imparting superior flavour to those contributing to inferior flavour characteristics, decreased with increasing rates of N fertilizers. Although theaflavins generally increased while thearubigins decreased with N fertilizer rates, the relationships were quadratic. Aono *et al.* (1982) have recorded higher chlorophyll and amino acid contents in new leaf owing to foliar application of the N fertilizer urea (1 %). These effects were reflected in enhanced quality of made tea.

Ali *et al.* (1997) have reported that increasing N and B contents adversely affected quality. Sulphur showed a linear positive correlation with TF content and brightness of made tea. TR content and total colour was also marginally improved by S up to a certain level.

Devchoudhury and Bajaj (1988) reported reduction in theanine content while alanine, aspartic acid and glutamic acid contents increased with increment in K application from 0 to 45 kg/ha. An increase in K application from 45 to 180 kg/ha reduced the levels of theanine, glutamine, valine and total amino acids. An increase in K application also reduced the reducing and non-reducing sugar contents of the shoots, and the theaflavin and thearubigin contents of the made tea, but increased the fibre and protein contents of the shoot. The yield of fresh leaf and content of free amino acids and caffeine increased due to application of K and Mg (Ruan *et al.* 1997). The polyphenols content increased in leaves of K-treated plants but decreased in those of Mg-treated plants. The ratio of polyphenols: free amino acids decreased in all K and Mg treatments, thus improving tea quality. Potassium and Mg applications also increased the content of some important aromatic compounds in tea, including nerolidol. The quality of marketed oolong tea was closely correlated with Mg content.

Magnesium fertilizers also improved tea quality; the contents of amino acids, caffeine and water extracts increased by 5.8, 9.2 and 2.3%, respectively, compared with the control (Zhu and Chen 2003). In China, application of K and Mg fertilizers significantly enhanced yield and quality (contents of free amino acids, polyphenols, caffeine, theaflavin and thearubigin) of major tea types. i.e. green tea, black tea and Oolong tea. Resistance to drought and diseases were also improved with K application. Field experiments showed that potassium chloride had similar effects as potassium sulfate (Ruan and Wu 2003).

AGE OF PLANTATION AND FERTILIZER USE

According to a study in Assam (Rahman and Jain 1985), application of P_2O_5: K_2O at 0-180:0-180 kg/ha with N at 90 kg/ha up to 12 years and at 135 kg/ha thereafter revealed that P had an adverse effect on yield during the first 8 cropping years. During the next 12 years, P improved yield but only at up to 90 kg/ha. Potassium had a positive effect on yields in all years, and the highest yields were obtained with N: P: K at 135:90:180 kg/ha. The peak of productivity occurred in the 10th to 12th year and then declined; this decline was 51% in the 18th to 20th year on plots with nil P and K, and was only 9-14% on plots receiving P and K.

Kamau *et al.* (2008) reported that mean annual made tea yield was higher for the clonal tea compared to the seedling tea, and the yield increased with age within genotypes. The effect of genotype was greater than the effect of age on crop production. They also concluded that tea yield remained unaffected by age of plantation up to 80 years from planting. According to them, improved genotype and judicious N-application reduce probability of decline in production with age of plantation.

FERTILIZER USE AND ROOT GROWTH

Tsanava and Morchiladze (1974) have reported that increase in rate of application of N fertilizer from 100 to 300 kg/ha considerably promoted horizontal root spread towards the inter-rows, mainly of fine roots, which resulted in better growth of the bush. Similarly, application of K_2O at 100-200 kg/ha supported development of conducting and fibrous roots, though application of 300 kg/ha of K_2O caused detrimental effect on root growth (Morchiladze 1975).

MULCHING AND NUTRIENT UPTAKE

During formative stage tea plant take up fairly large amount of nutrients, which get locked-up in the plant and only a small part of it is taken up annually by the crop (Othieno 1979). When the plants are pruned, however, large percentages of the locked-up nutrients are returned to the soil. Some of the returned nutrient elements, particularly from leaves, can be utilized quickly and become available within a short time after pruning. A large uptake of K was observed suggesting that current K application rates are inadequate on K deficient soils. Mulching by plant material has been found to influence uptake of plant nutrients. Tea clone mulched with *Eragrostis curvula* or Napier grass (*Pennisetum purpureum*) showed higher leaf P content as compared with unmulched controls (Othieno 1980).

In tea plantations, fallen leaves and pruning litter left *in situ* contribute substantial quantities of nutrients to the soil. As per Dogo (1994), about 30 t/ha of organic material is returned to the soil annually as pruning, which comprises of 33% as easily decomposed foliage (with a C:N ratio of 15:1), and rest of it as woody matter (with 55:1 C:N ratio). In this study, overall nutrient content of pruning was 518, 142, 340, 223, 60 and 58 kg/ha of N, P_2O_5, K_2O, CaO, MgO and MnO respectively, most of which was in the foliage (Dogo 1994).

PRUNING VS FERTILIZER LEVELS

Weeraratna *et al.* (1977) have found that incorporation of pruning liter resulted in increased N and K availability and organic matter content of the surface (0-7.5 cm) and sub-surface (7.5-15.0 cm) soil. Incorporation also reduced leaching of applied N and K fertilizers. Addition of organic matter in the form of pruning litter during light prune and deep skiff provided highest N contents in top soil and plucked shoots (Barua 1990).

It has been observed that inadequate application of N and P before pruning adversely affected recovery of the bush (Ranganathan 1975). Application of P_2O_5 and K_2O at 70 and 140 kg/ha, respectively, in the presence of 110 kg N/ha was most profitable in the pruning year, when yields amounted to 1000 kg made tea/ha. However, the addition of 120 kg N, 100 kg P_2O_5 and 200 kg K_2O /ha in the pruning year and 260 kg N, 100 kg P_2O_5 and 200 kg K_2O in the following year gave 3100 kg made tea in the latter.

LEAF ANALYSIS FOR DIAGNOSIS

Analysis of leaf or any other plant part can be a useful tool for assessing the nutrient status of a tea bush. Nutrient composition can form basis for recognizing deficiencies and can serve as a preliminary source of information for fertilizer use. Mineral composition of tea plant may vary with elevation, plant type, age, leaf position, soil fertility, time and method of sampling, method of analyses etc. It is also influenced by genetic, environmental as well as cultural factors.

Chu (1976) analyzed content of N, P, K, Ca, Mg, Na, Mn, Cu and Zn in tea leaves in autumn. All but Cu and Na reached their highest concentrations in the leaves in autumn. Nitrogen, P and K were normally higher in young than old leaves, Mn and Ca were higher in old leaves, and Mg was similar at both ages; while Cu, Zn and Na contents showed no clear trends with age. The mineral contents of leaf were closely correlated with those in the soil. The mineral status of the tea crop could be diagnosed by determining levels of available nutrients in the surface soil or sub-soil.

Sud *et al.* (1990 and 1995) have observed that content of Ca, Mn, Zn, Cu and Fe in green tea leaves was related to the uptake capability of tea plants as affected by different weather parameters. High temperature and high atmospheric evaporative demand assisted Ca uptake, whereas high RH and high rainfall reduced it. Low temperature and high RH reduced Mn uptake. High weekly evaporation, weekly RH and accumulated rainfall depressed the uptake of Zn, Cu and Fe into green tea shoots. Similarly, Hang *et al.* (1992) have reported that the contents of Fe, Mn, Zn and Cu were positively correlated with leaf age. The Ca and Mg contents of young stems were higher than those of young leaves, while the relative contents of Fe, Mn, Zn and Cu in these organs varied through the season.

Akhmedov (1975) concluded that for determining the nutritional status, sampling should be done during beginning of growth, at the end of the first phase of intensive growth, at the end of the first summer resting period, at the end of the second phase of intensive growth, and at the end of the growing phase.

Studies conducted by Burchuladze *et al.* (1976) suggested that tea shoots on the well-illuminated parts of the bushes are more suitable for nutritional diagnosis than mature leaves. Moreover, they found best results when sampling was done between 8 and 11 a.m. during late April and early May, late June and early July and during late August and early September. According to Sivasubramaniam and Jayman (1976), chemical analysis of the mature leaf, rather than the third leaf or bud should be preferred as it is more reliable index of the need of the tea plant for Mg.

In Sri Lanka, Hettiarachchi *et al.* (1997) have observed that with maturity of tea leaf N, K and P content in the leaf decreased while content of Ca, Mn, Fe and Al showed increment. In this regard, no consistent trend was found for Mg and S. It indicated that the first mature leaf was not an entirely reliable guide for confirming visual observations, particularly at a time of widespread Mg deficiency.

IMPORTANCE OF SOIL TESTING

In precision agriculture, soil testing is an important diagnostic technique that should be adopted to test the soils prior to planting and subsequently at periodic intervals (Natesan *et al.* 2004). Interpretation of soil analysis data, taking into consideration the crop need, physiological stage of the crop, and anticipated yield would help to move towards the best practices of scientific agriculture.

Filippova (1975) suggested that for evaluating nutrient dynamics in tea crop, soil and leaf samples should be taken in May, July and September. A positive correlation was observed between NH_4 N and hydrolysable N in the soil and total N in the leaf. Only a very low soil P content was reflected in an appreciable decrease of leaf P content. High available K in soil was also reflected in high leaf K content.

According to Sivasubramaniam and Jayman (1976), water-soluble soil K was better correlated with K requirements of the plant than either exchangeable or total K.

Ranganathan and Natesan (1988) have reported decline in the soil fertility with increasing crop productivity, though the pH, tilth and physicochemical properties showed increment. Thus, intensive cultivation systems like tea plantation require information on nutrient dynamics, the annual nutrient supply; and the amount of nutrients needed to maintain a set production level.

UPASI has developed targeted yield equations of N and K based on soil test crop response study using correlation and regression analysis (Venkatesan *et al.* 2003). Statistical analysis of the cumulative made tea yield indicated that the optimum ratio of NK manuring would be 1:1 particularly, when the rate of N is between 300 and 450 kg/ha. The basic data on percent contribution of soil (% CS), percent contribution of fertilizer (% CF) and nitrogen requirement (NR) were calculated and transformed into workable adjustment equations. It was found that at the same fertility status, the calculated N dose for a given target yield may vary for different cultivars. It indicates that one cultivar may be more responsive to N application than another cultivar.

REFERENCES

Akhmedov GS (1975) Time for sampling plant material in foliar diagnosis of mineral nutrition of tea. *Subtropicheskie Kul' tury* 2: 59-60.

Ali MM, Chaudhury SH, Mannan MA (1977) Effect of potassic fertilizer on the yield of tea. *Tea Journal of Bangladesh* 13(1): 10-17.

Ali TS, Deka A, Barbora AC (1997) Influence of nitrogen, sulphur, boron on certain quality parameters of black tea. *Journal of Interacademicia* 1(3): 233-239.

Antiya SE (1985) Influence of magnesium on the effectiveness of nitrogen rates. *Subtropicheskie Kul' tury* 3: 69-74.

Aono H, Tanaka S, Yanase Y (1982) Effects of foliar application of fertilizer on the growth, quality of new shoots of tea plants. *Study of Tea* 63: 23-32.

Barbora AC (1996) Integrated nutrient management for tea in north-eastern India. *Fertiliser News* 41(12): 77-79, 81, 83.

Barua DN (1989) Science, Practice in Tea Culture. Tea Research Association, Calcutta, p 509.

Barua DN (1990) Effect of long term application of nitrogen fertilizer on yield of tea. *Two, a Bud* 37: 1, 1-3.

Burchuladze IT, Danava VP, Datuadze OV (1976) Chemical diagnosis, productivity, quality of tea in relation to soil properties, fertilization. *Analiz Rastenii kak Metod Diagnostiki ikh Pitania i Effektivn Makro i Mikroudobrenii*, pp 221-230.

Cesare Guerra O (1972) Fertilizer study in mature tea in Tingo Maria. *Revista Universidad Nacional Agraria Selva* 2: 1-5.

Chamuah GS (1988) The effect of nitrogen on root growth, nutrient uptake of young tea plants (*Camellia sinensis* L.) grown in sand culture. *Fertilizer Research* 16(1): 59-65.

Chu HM (1976) Studies on the relationship between micro-, macronutrients in soils, tea leaves. *Taiwan Agriculture Quarterly* 12(3): 106-147.

Devchoudhury MN, Bajaj KL (1988) Effect of potassium on nitrogen, carbohydrate contents of tea leaves (*Camellia sinensis* (L.) O. Kuntze), quality of made teas. *Journal of Food Science, Technology, India* 25(2): 105-107.

Dogo YW (1994) Prunings as organic matter inputs in tea agrosystem in Kenya. I. Biomass, nutrients. *Tea* 15(2): 99-104.

Dootson (1974) Zinc X nitrogen applications. Quarterly News Letter, Tea Research Foundation of Central Africa 34: 11-12.

Dootson J (1976) The response of tea to aerial zinc, copper sprays. *Quarterly Newsletter, Tea Research Foundation of Central Africa* 43: 5.

Dzadzua OL (1987) The role of microelement chelates in increasing tea yield, quality. *Subtropicheskie Kul' tury* 4: 43-46

Eden T (1944) Studies on the yield of tea - V: Further studies on manurial response. *Emperical Journal of Experimental Agriculture*. 12: 177- 179.

Filippova NA (1975) The relationship between soil, tea leaf NPK contents. *Nauchnye Trudy NII Gorn Sadovodstva i Tsvetovodstva* 22: 86-91.

Filippova NA, Bushin PM, Troyanskaya AI, Filippova AL (1984) Fertilizer application with irrigation water for tea. *Subtropicheskie Kul' tury* 1: 50-54.

Fu JieQi, Cho M, Hoshino K, Hirai H, Kato H (2008) Green tea cultivation at the northern limit of economical cultivation - rate of uptake of nitrogen fertilizer by young tea plants after application at different periods, the efficient application method. *Japanese Journal of Crop Science* 77(3): 306-314.

Ghosh TK, Chakravarte J (1980) Response to foliar application of zinc sulphate in tea. *Two, a Bud* 27(2): 55-57.

Godziashvili BA (1984) Effect of potassium fertilizer forms on tea plantations in Georgia. *Byulleten' Vsesoyuznogo Nauchno Issledovatel' skogo Instituta Udobrenii i Agropochvovedeniya* 70: 67-70.

Gogoi AK, Choudhury MND, Gogoi N, Chakravartee J (1994) Effect of phosphorus on nitrogen metabolism in tea plant. *Two, a Bud* 41(1): 42-47.

Gohain KK, Dutta Amrit (1994) Effect of application of urea-N, rock phosphate, muriate of potash on sulphur nutrition of tea. *Two, a Bud* 41(1): 25-27.

Gohain T, Barbora AC (2000) Effect of different sources of magnesium on yield, quality of tea (*Camellia sinensis* L. (O) Kuntze). *Crop Research* 19(2): 324-328.

Gohain T, Barbora AC, Deka A (1998) Effect of magnesium on yield, quality of tea *Camellia sinensis* (L.) O Kuntze. *Crop Research Hisar* 16(1): 33-36.

Gohain T, Barbora AC, Deka A (2000) Effect of boron on yield, quality of tea. *Journal of Plantation Crops* 28(1): 68-71.

Gohain T, Barbora AC, Deka A (2000) Effect of manganese on growth, yield, quality of tea [*Camellia sinensis* L. (O) Kuntze]. *Research on Crops* 1(1): 91-97.

Gohain T, Barbora AC, Deka A, Ahmed KZ (2000) Effect of different sources of sulphur on yield, quality of tea (*Camellia sinensis* L. (O) Kuntze). *Journal of Interacademicia* 4(3): 370-375.

Grice WJ, Malenga NEA (1985) The effects on yield of changing nitrogen levels on mature tea. *Quarterly Newsletter, Tea Research Foundation of Central Africa* 78: 12-13.

Grice WJ, Phiri GSN (1986) Nitrogen, potash response in mature local (China hybrid) tea over three biennial cycles. *Quarterly Newsletter, Tea Research Foundation of Central Africa* 84: 4-6.

Guseinov DM, Kerimov VN (1975) The effect of nitrogen fertilizers on the quality of green tea, the final product. *Subtropicheskie Kul' tury* 3: 51-54.

Han WY, Xu YW, Wu BH (1993) Effects of Cu, Zn on the development, physiological metabolism of tea plants. I. Effect of Cu on the growth of tea plants. *Journal of Tea Science* 13(2): 101-108.

Owuor PO, Othieno CO, Kamau DM, Wanyoko JK, Ng'etich WK (2008) Long term fertilizer use on high yielding clone S 15/10: tea yields. *International Journal of Tea Science* 7(1/2): (19-31.

Owuor PO, Wanyoko JK, Othieno CO (1987) Effects of nitrogenous fertilizers on leaf potassium contents of tea, *Camellia sinensis* (L) O. Kuntze. *Tea* 8(1): 4-13.

Palani N, Balasubramaniam K (2001) Alternative sources of sulphur in tea. *Newsletter UPASI Tea Research Foundation* 11(2): 6.

Prasad A, Dey SK (1979) Studies on the uptake of manganese, boron by young tea plants. *Two, a Bud* 26(1): 9-13.

Rahman F, Jain NK (1985) Long term response of light leaf Assam tea (Camellia sinensis L. (O) Kuntze) to phosphate, potash application in North East India. *Journal of Plantation Crops* 13(2): 104-115.

Rahman F, Sharma PB, Nandi NC (1978) Suitability of muriate of potash (KCl), sulphate of potash (K_2SO_4) for foliar spray in young, mature tea. *Two, a Bud* 25(2): 97-98.

Ranganathan V (1971) Phosphorús nutrition of tea. Proceedings of the Seventeenth Scientific Conference, United Planters' Association of Southern India, Tea Scientific Department, 31-44.

Ranganathan V (1975) Certain aspects of manuring tea. *Potash Review* 27, 8: 1-6.

Ranganathan V (1980) A new recommendation. Application of zinc sulphate to nursery [tea] plants. *Planters' Chronicle* 75(1): 35-36.

Ranganathan V, Ganesan M (1976) Frequency of potassium application. Bulletin, United Planters Association of Southern India, Tea Scientific Department 32: 72-84.

Ranganathan V, Natesan S (1988) Soil testing in plantation crops. *Planters'—Chronicle* 83(8): 270-275.

Ranganathan V, Ram CSV, Natesan S (1987) Superiority of AS, CAN as sources of nitrogen over urea to tea crop. *Planters' Chronicle* 81(11): 399-401, 403.

Rao TN, Sharma PK (2001) Sulphur nutrition of tea - global review. *International Journal of Tea Science* 1(1): 20-27.

Rikhter MA, Lyashko MU (1979) The effect of high nitrogen rates on the yield, quality of irrigated tea. *Subtropicheskie Kul' tury* 1: 36-43.

Ruan J, Hardter R (2001) Productivity, quality response of tea to balanced nutrient management - examples from China tea gardens. *Plant nutrition: food security, sustainability of agro ecosystems through basic, applied research Fourteenth International Plant Nutrition Colloquium*, Hannover, Germany, pp 324-325.

Ruan Jian Yun, Wu Xun, Hardter R (1997) Effects of potassium, magnesium on the yield, quality of oolong tea. *Journal of Tea Science* 17(1): 9-14.

Ruan Jian Yun, Xun Wu, Hardter R (1999) Effects of potassium, magnesium nutrition on the quality components of different types of tea. *Journal of the Science of Food, Agriculture* 79(1): 47-52.

Ruan JianYun, Wu Xun (2003) Productivity, quality response of tea to balanced nutrient management including K, Mg. *Journal of Tea Science* 23(S): 21-26.

Salimi KM, Latifi SMM (2008) Determination of the best level of irrigation, nitrogen fertilizer in tea field in Fouman, Iran. *Journal of Agricultural Sciences Guilan* 1(9): 37-56.

Sarishvili IF, Dzhibladze ShG, Sarishvili LI, Egorashvili NL (1976) Diagnosis of the manganese nutrition level in tea growing in the humid sub-tropics of the Georgian SSR. *Analiz Rastenii kak Metod Diagnostiki ikh Pitania I Effektivn, Makro Mikroudobrenii* 113-121.

Sharangiya ShU, Marshaniya II (1988) Effect of forms, rates of nitrogen fertilizers on the productivity of tea plantations on yellow podzolic soils with mechanical plucking of leaves. *Subtropicheskie Kul' tury* 3: 35-44.

Sharangiya USh (1987) Economic effectiveness of using nitrogen fertilizers on tea plantations fertilized over a long period under conditions of mechanized plucking. *Subtropicheskie Kul' tury* 6: 17-21.

Sharma KL, Sharma DK, Ganesh Sharma (2002) Long term response of integrated nitrogen nutrition with bioresources on the yield of China hybrid tea (*Camellia sinensis* L.) grown in North-West Himalayas. *Proceedings of the 15th Plantation Crops Symposium Placrosym XV, Mysore, India*, 10-13 December 2002, pp 386-391.

Sivasubramaniam S, Jayman TCZ (1976) The use of fertilizer for tea in Sri Lanka. 2. Foliar, soil analysis with particular reference to potassium. *Tea Quarterly* 46(1-2): 4-11.

Sobolevskii VS, Nikol'skaya AA, Ariel' RS, Volokitina NA (1984) Effect of microelements on tea productivity, quality. *Zemledelie* 11: 48-49.

Sud RG (1996) Potassium content in different parts of Kangra local jat tea (*Camellia sinensis*) plant, black tea infusion, its seasonal, varietal variations in green tea shoots at Palampur. *Indian Journal of Agricultural Sciences* 66(11): 645-650.

Sud RG, Badyal J, Bhan S, Misra SK (1990) Estimation, seasonal variations of macro, micro nutrients in tea of Himachal Pradesh. *Sri Lanka Journal of Tea Science* 59(1): 29-36.

Sud RG, Prasad Rajendra, Bhargava Manoj (1995) Effect of weather conditions on concentration of calcium, manganese, zinc, copper, iron in green tea (*Camellia sinensis* (L) O Kuntze) leaves of north-western India. *Journal of the Science of Food, Agriculture* 67(3): 341-346.

Tachibana N, Ikeda T, Ikeda K (1996) Changes in nitrogen uptake with aging, under heavy application of nitrogen in tea plants. *Japanese Journal of Crop Science* 65(1): 8-15.

Tolhurst JAH (1962) Zinc deficiency of tea in Ceylon. *Tea Quarterly* 33: 134-137.

Tsanava VP, Morchiladze GM (1974) The effect of nitrogen fertilizers on the development of the tea root system. *Subtropicheskie Kul' tury* 3: 3-5.

Tuti Warsito, Angkapradipta P, Hardjono A (1974) Urea, ammonium sulphate as sources of N in an NP experiment on tea in production. *Menara Perkebunan* 42(4): 187-190.

Utnelishvili MI (1974) The effect of nitrogen fertilizers on tea productivity, quality. *Subtropicheskie Kul' tury* 3: 28-30.

Venkatesan S, Kumaraguru R (2002) Tolerance level of sodium in MOP, zinc sulphate. *Newsletter UPASI Tea Research Foundation* 12(1): 4.

Venkatesan S, Verma DP, Ganapathy MNK (2003) Targeted yield equations of nitrogen for clonal teas under South Indian conditions. *Journal of the Indian Society of Soil Science* 51(2): 178-183.

Verma DP (1997) Potassium nutrition of tea. *Journal of Potassium Research* 13(1): 93-100.

Verma DP (1997) Sulphur nutrition in tea. *Planters' Chronicle* 92(7): 297-299, 301-303.

Verma DP, Balasubramanian K, Udaya Bhanu KG (1998) Sulphur nutrition of tea. *Bulletin of UPASI Tea Scientific Department* 51: 37-40.

Verma DP, Mulky MJ, Sharma VS (1993) Nutrient management of tea in South India. Tea, culture, processing, marketing, Coonoor, India, 23-24 Mar (1992: 55-69.

Verma DP, Muraleedharan N (1998) Fertiliser use efficiency for maximization of tea productivity. *Bulletin of UPASI Tea Scientific Department* 52: 10-26.

Verma DP, Palani N, Udaya Bhanu KG (1998) Magnesium nutrition of tea. *Bulletin of UPASI Tea Scientific Department* 51: 12-14.

Wanyoko JK (1988) Magnesium, nitrogen applications to moribund tea: preliminary results. *Acta Horticulturae* 218: 191-206.

Watson M, Wettasinghe DT (1982) Effect of type of nitrogenous fertilizer on the leaf nutrient composition of low-grown tea in Sri Lanka. *Tea Quarterly* 51(1): 27-30.

Weeraratna CS, Watson M, Wettasingha DT (1977) Effect of mineralization of tea prunings on some soil characteristics. *Plant, Soil* 46(1): 93.99.

Wickremasinghe KN, Ananthacumaraswamy S, Amarasekera AR (1986) Effect of urea, ammonium sulphate on sulphur nutrition of tea. *Sri Lanka Journal of Tea Science* 55(2): 77-83.

Wilson KC (1975) Studies on the mineral nutrition of tea. II. Nitrogen. *Plant, Soil* 42(3): 501-516.

Wilson KC (1975) Studies on the mineral nutrition of tea. IV. Potassium. *Plant, Soil* 43(2): 279-293.

Wilson KC (1975) Studies on the mineral nutrition of tea. V. Calcium. *Plant, Soil* 43(2): 295-307.

Wilson KC (1975) Studies on the mineral nutrition of tea. VI. Magnesium. *Plant, Soil* 43(2): 309-316.

Wilson KC, Hainsworth E, Green MJ, O' Shea PBT (1975) Studies on the mineral nutrition of tea. III. Phosphate. *Plant, Soil* 43(2): 259-278.

Wu Xun (1994) Calcium, magnesium nutrition for tea plants, their soil control. *Journal of Tea Science* 14(2): 115-121.

Ye Yong, Wu Xun, Yao GuoKun (1994) Sulphur (S) nutrition for tea plant, its effect on tea quality. *Journal of Tea Science* 14(2): 123-128.

Yoshida K, Imanishi M, Shirei O, Terada T, Yoshimura K, Kitagawa Y (1981) Improvement in tea quality after luxury manganese absorption. *Bulletin of the Nara Agricultural Experiment Station* 12: 75-93.

Zhu YongXin, Chen FuXing. (2003) Improvement of magnesium nutrition for tea gardens on hilly red soil in southern Hunan Province. *Journal of Tea Science* 23 (S): 34-37.

Zyrin NG, Barabadze LA (1987) Effect of changes in the content, ratio of nutrient elements in flushes on tea quality as a result of long-term fertilization. *Subtropicheskie Kul' tury* 1: 57-62.

11

WATER MANAGEMENT IN TEA

K.K. Singh

CSIR-Institute of Himalayan Bioresource Technology,
Palampur (H.P.)-176 061, India. E-mail: kksingh@ihbt.res.in

INTRODUCTION

Tea (*Camellia sinensis*) is an important beverage and the world drinks more of it than any other beverage. Major tea growing countries are India, Sri Lanka, China, Indonesia, Kenya, Malawi and South Africa. India produces about 27.9 per cent of global share and 18.3 per cent of world export. In India 44.5 percent of the country 's entire plantation is in Kerala followed by Assam (19.6 per cent), Karnataka (15.2 per cent), West Bengal (8.6 per cent), Tamilnadu (8.3 per cent), Himachal (1 per cent) and other area 3.7 per cent. Recently tea cultivation was introduced in Manipur, Nagaland, Arunachal Pradesh, Bihar and Orissa in non-traditional areas.

It is a perennial crop requiring high amount of water throughout the year but in most of the area in India the crop is dependent on rain water. Rainfall, which often occurs most evenly in rainy season, also leads to deficit in dry periods. Irrigation is highly essential not only to as insurance against drought but also for producing higher amount of crop and to reduce seasonal dormancy during deficit and cold dry period. Both over irrigation and under irrigation affect the soil moisture status and environment and the projected benefit might results into loss. So an optimum quantity of water may be applied at proper time. It has also estimated that more than fifty per cent of tea area suffers from either water logging or drought or both. The period soil moisture deficit is from November to April and May to June – July in Assam and Himachal Pradesh with the variation in climatic conditions. So analysis of long term meteorological data will shows the excess and deficit moisture condition of the region.

STUDY OF HYDROLOGICAL DATA

In planning and design of irrigation or drainage system, an estimation of irrigation requirement for the crop. In the most of the cases weather data relating to water requirement of crop on temporal basis are not available so the network of meteorological observation from evaporation and rainfall data are available for sufficient period can be used for estimating irrigation requirement.

ESTIMATION OF CROP EVAPOTRANSPIRATION

ET_{crop} is generally estimated from field water balance or from soil water depletion. It can be predicted for a specific climate using the meteorological parameters and crop factors. Various types and sizes of lysimeters are used for precise measurement of ET_{crop}. Weighing type lysimeters offer an accurate method for measuring the evapotranspiration (Harrold 1958; Harrold and Dreibelbies 1951; Pierce 1960; Pruilt and Angus 1960; Tanner 1967; VanBavel 1961; Van Bavel 1961; Van Bavel and Myers 1962). Because of high initial cost of the equipment and installation several design of hydraulic weighing system have also been used as alternatives (King *et al.* 1956, Ekern 1958 and Hanks and Shawcraft 1965; and Dylla and Cox 1973). However as weighing type lysimeters permits more accurate determination of ET for short duration of time (Pruitt and Angus 1960), simple low cost mechanical weighing type lysimeters may be preferred over hydraulic weighing type, as the later requires elaborate equipment to ensure accurate measurement and elimination of temperature effects.

The effect of climate on crop water requirement is given by the reference crop evapotranspiration. There are several climatologically approach for calculating the reference crop evapotranspiration *viz.* Blaney criddle, Thonthwaite, Radiation, Penman, Modified penman evapotranspiration etc.

Blaney -Criddle Method

Blaney–Criddle (1950) observed that the amount of water consumptively used by crop during their growing season may be closely correlated with the mean monthly temperature and day light hours. This has generally giving accurate estimate of seasonal consumptive use owing to inclusion of locally developed crop coefficient factor (K).

The relationship may be stated as follows:

$$U= KF = \sum k f =\sum u = \sum ktp/100$$

In which,

U = Seasonal consumptive use of water by the crop for given period, inches

u = monthly consumptive use, inches

k= empirical seasonal consumptive use coefficient for growing season

f = sum of monthly consumptive use factor (f) for the growing season

k= sum of the monthly consumptive use factor (f) for the growing season

k= emprical consumptive use crop coefficient for the month , u/f

$$f = t \text{ X } p/100$$

t= mean monthly temperature, °F

p= monthly day length hour expressed as per cent of day hours of the year

It has also recommended the following relationship for f factor (expressed in mm/day) in Blaney – Criddle formula

$$f = p\,(0.46\,t + 8.13) \text{ using t in °C}$$

or

$$f = 25.4\,p \text{ X } t/100 \text{ using t in °F}$$

in which t = the mean of daily maximum and minimum temperature in °C or °F over the month considered, and

p = the mean daily percentage of annual day time hours for a given month and latitude.

Thornthwaite Formula

Thornthwaite (1948) assumed that an exponential relationship betwwen mean monthly temperature and mean monthly consumptive use. It can be express by

$$E = 1.6(10t/I)^{a}$$

In which

e = unadjusted potential evapotranspiration, cm per month (month of 30 days each and 12 hours day time)

t = mean air temperature °C

I = annual or seasonal heat index, summation of monthly heat indices (I) when,

$$I = (t/5)^{1.514}$$

a = an empirical exponent computed by the equation

$$a = 6.75\, I^3 * 10^{-7} - 7.71\, I^2 * 10^{-5}\ 1.792I * 10^{-1} + 0.49239$$

The draw back of this formula as enumaerated by Chang (1968) are as follows:

Penman Formula

Penman (1948) proposed an equation for evapotranspiration from open water surface, based on a combination of energy balance and sink strength. It is follows:

$$E_0 = \Delta Q_n + \gamma\, E_a / (\Delta + \gamma)$$

In which,

E_0 = Evaporation from open water surface mm/day

Δ = slope of saturation vapour pressure *vs.* temperature curve (de_a/dT) at the mean air temperature Ta, mm Hg per °C

e_a = saturation vapour pressure of evaporating surface (e_s) in mm Hg at mean air temperature T_a

T_a = mean air temperature °K =273 + °C

Q_n = net radiation (mm of water) =

$$Q_A\,(1\text{-}r)(0.18+0.55)n/N) - \sigma\, T^4_a\,(0.55-0.92\sqrt{e_d})(0.10+0.90\ n/N)$$

r = reflection coefficient of evaporating surface, 0.06 for open water surface

Q_A = Angot's value of mean monthly extra terrestrial radiation, mm of water /day

n/N= ratio between actual and possible hours of bright sunshine

σ = Stefan-Boltzman constant

e_d = saturation vapour pressure of the atmosphere, mm in Hg at dew point

The Tea Research Association (Annual Scientific Report 1978) found that the quality of tea deterioration during rainy season. It is due to high concentration of chlorophyll and chlorogenic acid in the shoot. However, yield can be increased by 9 per cent if rainfall deficiency is replenished by irrigation during November-April.

The Tea Research Association (Annual Scientific Report 1986-87) conducted experiments on response of tea to different water treatments. The effect of treatment and crop response under sprinkler method of irrigation and yield of made tea was described in Table 3. It was observed that there was significant increase of 12.8 per cent to 29.4 percent in crop yield due to irrigation treatment.

WATER APPLICATION METHODS

In tea gardens where the slope is mild the boarder iirigation system can be applied where as sprinkler irrigation system is used for water application to crop in undulating and high slop area.

Sprinkler system

Sprinkler system is mainly two types on the basis of spraying irrigation water:

1. Rotating head system
2. Perforated pipe system

Sprinkler system is mainly five types on the basis of portability:

1. *Portable system*: In this system it is designed to move from field to field or to different pumping sites
2. *Semi portable system*: In this system all the unit of sprinkler irrigation unit are movable accept water source and pumping unit.
3. *Semi permanent system*: It has portable laterals, permanent main lines and sub main lines and stationary water source and pumping unit. The main and sub mains are buried and risers with nozzles are located at suitable distance.
4. *Solid set system*: In this system the lateral are position in field in early in the crop season and remain for season.
5. *Permanent system*: In this system consists of permanently laid mains, sub mains, laterals and stationary water system. It is used in perennial crops *viz.,* tea.

Sprinkler selections and spacing

The actual selection of sprinkler is based largely designed information furnished by the manufacturer of the equipment. The required discharge of individual sprinkler is a function of water application rate and the two way spacing of sprinklers. It may be determined by the following formula

$$Q= S_1 X s_m X\ I/360 \qquad (2.1)$$

Where q= required discghage of iindividual sprinkler, l/s

S_1= Spacing of sprinklers along the laterals, m

S_m= Spacing of laterals along the main, m

I= Optimum application rate, cm/h

Capacity of sprinkler system

The required capacity of a sprinkler system depends on the size of the area to be irrigated, the gross depth of water applied at each irrigation and net operating hours. The capacity of the system may be calculated by

$$Q = \frac{A \times d}{F \times H \times E} \qquad (2.2)$$

Where:

Q= discharge capacity of the pump l/s

A= area to be irrigated, ha

D= d= net depth of water application, cm

F= number of days allowed for a completion of one irrigation

H= number of actual operating hours per day

E= Water application efficiency., per cent

Hydraulic design of sprinkler system

The hydraulic design of sprinkler is aimed to obtain a uniform coverage area, desired rate of application, the break up of sprinkler drops necessary to minimize structural deterioration of soil surface, efficiency desired to reduce the energy requirement in operating the system and maximize the area.

Discharge of nozzle

The discharge of sprinkler nozzle may be computed by the following

$$q= C_a \sqrt{2gh} \qquad (2.3)$$

q= nozzle discharge, m^3/s

a= cross section area of nozzle, m^2

H= Pressure head at the nozzle

G= acceleration due to gravity, m/s^2

C coefficient of discharge

Water spread area of sprinkler

The area covered by rotating head sprinkler may be estimated by following formula suggested by Cavazza (Pillsburry,1968)

$$R= 1.35\sqrt{dh} \qquad (2.4)$$

R= radius of wetted area covered by sprinkler, m

D= diameter of nozzle, m

H= pressure head at nozzle, m

Rate of application

The average rate of application of single nozzle is estimated by

$$R_a= q/360 \text{ X } A \qquad (2.5)$$

Where= R_a= Water application rate cm/h

Q= rate of discharge of sprinkler, l/s

A =wetted area of sprinkler, m^2

Uniformity Coefficient (C_u)

It measure the degree of uniformity obtained for any size sprinkler operating under given condition. It is expressed by the equation developed by Christiansen (1942)

$$C_u \ 100(1-\sum X/mn) \qquad (2.6)$$

Where m= average value of all observations (application rate)mm

N= total number of observations points

X= numerical deviation from mean, mm

DRAINAGE IN TEA

Tea is a water loving plant but it cannot sustain excess water, which is known as free water, or gravitational water. It is required to be drained so as to create favourable soil condition for tea plant growth. The process of removing the excess water from the land surface is known as surface drainage. If the excess water saturates the pore spaces of soil, the process is known, as sub surface drainage .The excess water that is drained is gravitational water, which is not available to plants. Drainage does not remove capillary water and there is no danger of over draining the soil by planning the drains too close. The excess water on land and in the soil may originates from excess of precipitation, over irrigation, seepage or from it may come from the are from surface and subsurface. The problem of water logging is mainly in lower Assam and Brahputra valley where the land slope is marginal and the rainfall intensity is very high.

Surface drainage

Surface drainage mainly uses the potential energy that exists due to land elevation to provide a hydraulic gradient for movement of water. The system can be considered in three functional parts *viz.*, Collection system, disposal system and outlet. The water in individual drains collected through the collection system moved through the disposal system and drain off at the outlet.

Types of surface drainage systems

1. Random field drains
2. Parallel open drains system
3. Parallel open ditch system
4. Bedding system.

Random type drainage: This system of drainage of water used to be practiced if the water is accumulated at low spots in the field and the runoff water accumulate from adjoining higher area and it may remove by random field ditches or by surface drains.

Parallel field drains system is most effective method of surface drainage and well suited for drainage of irrigation and precipitation. The field drains discharges drainage water in lateral bordering to field and the lateral in turn to main.

The parallel open drainage system is applicable in the soils that require both surface and sub surface drainage

Bedding type drainage system for surface drainage is essential a land forming process

DESIGN DISCHARGE OF OPEN DITCH

The rate at which the open drains have to remove water from an area depends upon rainfall, size of drainage area, characteristics of drainage area and nature of crop

Drainage coefficient is the depth of water in cm to be removed in 24 hrs from the entire drainage area. It is express in cm/day. The intensity of rain fall and the duration are inversely proportion to time allowed for removal of water.

Cypress creek Formula

$$Q = CA^{p} \qquad (2.7)$$

Where Q = design discharge

C= coefficient depending upon the degree of protection

P= Coefficient

A = Area

The Boston Society formula

$$Q = CA^{1/2} \qquad (2.8)$$

Q= Peak discharge

C= Coefficient

A= Area

DESIGN OF OPEN DRAINS

The design of the section of a channel is done using Manning's formula. The section may not be possible to use the most efficient hydraulic section.

Types of Drainage sections

Rectangular type

Trapezoidal type

Parabolic type

SUB SURFACE DRAINAGE

It this system excess water present at the root zone crop is removed through drainage pies.

Design of Sub surface drainage system

It is design by the Hooghoudt Equation

$$Dy/dt= -K\ 2\ ЛIad\ /\ ЛIa^2\ y/s \quad (2.9)$$

Where a= width of cross section

D= depth of drain

Y= depth of water to be lowered

S = depth of impermeable layer

LAY OUT OF DRAINAGE SYSTEM

The different types of drainage lay out are random, herringbone, gridiron and interceptor type

In any drainage project a combination of surface and sub surface drainage method is to be adopted. Similarly a combination of open drains and tiles drains are used for effectively drainage of an area.

REFERENCES

Abdul Hafeez AT, Abdullah AA, Abu Goukh A A 1977. Mulching an approach for increasing water use efficiency under the arid condition of Sudan. *Proc. 1st Agri. Conference* 2:453-465.

Abraham, TP 1965. Optimal fertilizers dressing and Economics of manuring. Indian *J. Agri. Econ.* **20(2):** 1-20.

Annual Scientific Report 1971-72, 1975-76, 1978-79, 1983-84, 1984-85, 1985-86, 1986-87, 1987-88, 1990-91. Tocklai Experimental station (Tea Research Association) Jorhat (Assam) India.

Barua, D. N. and Barua S.C. 1964. Seasonal dormancy in tea. *Two and a Bud*, Tea Research Association, Calcutta, India. 41-45.

Carr, M.K.V. 1972. The climatic requirement tea plants: a review. *Experimental Agriculture* **8**:1-14.

Carr, M.K.V. 1974. Irrigating seedlings tea in southern Tanzania: effect on total yield distribution of yield and water use. *J. Agri. Sci.*, Columbia 83:363-378.

Carr, M.K.V. 1977. Response of seedling tea bushes and their clones to water stress. *Experimental Agri.* 13:317-324.

Choudhary, P.C. and Chatterjee, B.N. 1967. Moisture conservation under mulches in the eroded terraced soil of Ranchi. *J. Soil and Water Conservation*. India. 15: 51-57.

Dabral P.O. 1986. Optimal resource allocation and production function for irrigated paddy. An unpublished M.Tech Thesis, Deptt. of Agri. Engg, IIT Kharagpur. (WB) India.

Doorenbos, J and Pruitt, W.O. 1984. Crop water requirements. FAO, United Nations Rome.

Dutta, S.K. and Sharma, K.N. 1962. Cold weather yield from unpruned tea. *Two and a Bud*. Tea Research Association, Kolkatta, India: 76-80.

Gore, K.P. 1979. Production and water allocation models. An unpublished M.Tech Thesis Deptt. Agri Engg. IIT Kharagpur.

Kadirvellu, A.K. and Rajan, S.D. 1970. Irrigation cum trash mulch in sugarcane. *Chennai Agri. J.* 57:189-193.

Palmer-Jones, R.W. 1977. Irrigation system operating policies for mature tea in Malawi. *Water Resource Research* 13(1): 1-7.

Pierce, L.T. 1960. Practical method of determining evapotranspiration from temperature and rainfall. *Trans. ASAE* 3:77-81.

Prakash, C. and Rao, D.H. 1986. Frequency analysis of rainfall data for crop planning at Kota. *Indian J. Soil Cons.* 14(2): 23-25

Pruitt, W.O. and Angus. 1960. Large weighing type lysimeters for measuring evapotranspiration. *Trans. ASAE* 3:13-18

Tripathi, P.K. 1988. Maximization of tea production by judicious resource management. An unpublished Ph.D thesis Deptt. Agri. Engg. IIT Kharagpur India

12

DISEASE MANAGEMENT IN TEA

Arvind Gulati * and R.K. Sud

CSIR-Institute of Hialayan Bioresource Technology,
Palampur (H.P.)- 176 061, India. *E-mail: arvindgulati@ihbt.res.in

Like any other cultivated crop, tea is prone to attack by several diseases, causing heavy crop loss (Satyanarayana and Barua 1983, Satyanarayana 1987, Agnihothrudu 1995, Baby 2002, Joshi *et al.* 2009). The diseases attack various parts of the bush, and result in defoliation, debilitation and even mortality of the bush. Fungal organisms account for most of the diseases, apart from a few diseases caused by bacteria, viruses, and algae. In this chapter, common diseases of economic importance and their management are briefly described under three categories, in accordance to the part of bush infected by these diseases.

LEAF DISEASES

Blister Blight (*Exobasidium vexans* Massee)

Blister blight of tea is economically by far the most important disease in the major tea growing countries. In India, the disease is a serious problem and inflicts heavy crop losses throughout South India, Darjeeling in West Bengal, and Kangra valley in Himachal Pradesh.

Blister blight, the biotrophic fungal-pathogen, attacks the tender shoots, harvested for the manufacture of tea. Infected leaves develop blisters with a shiny concave upper surface and a white powdery convex lower surface. Blister coalescence induces leaf distortion and cur-

copper-nickel combination at shorter intervals during the intervening monsoon spell, was recommended for effective and economic control of the disease in South India. Weekly sprays of Tridemorph plus the binder Vinofan-In in conjunction with copper oxychloride imparted good disease control during the blister season in Kangra valley (Gulati and Satyanarayana 1988). Practising black plucking instead of standard plucking during the blister season helped in reducing the fungicide dosage in the disease control (Gulati and Satyanarayana 1991). Among several EBI fungicides tested, Tridemorph, Bitertanol, Cyproconazole, Fenpropimorph, Hexaconazole and Propiconazole provided effective disease control (Agnihothrudu and Chandra Mouli 1990, Dutta *et al.* 1992, Agnihothrudu 1995, Chandra Mouli and Premkumar 1995, Premkumar *et al.* 1996). Propiconazole and Hexaconazole showed high efficacy in combination with copper oxychloride in the disease control (Arulpragasam *et al.* 1987, Chandra Mouli and Premkumar 1997, Premkumar 1997).

UPASI investigated the effect of ergosterol biosynthesis inhibiting (EBI) fungicides on blister blight disease, biometric and physiological parameters of tea plant and quality of made tea (Baby *et al.* 2004b). EBIs expressed antisporulant action, reduced the size of spores and their viability, inhibited spore germination and provided satisfactory control of the disease. Continual use of EBIs increased shoot length and dry weight of shoots. They also increased epicuticular wax content, leaf thickness, size of vascular bundle, stomatal conductance, chlorophyll content and rate of photosynthesis, but reduced stomatal size. Made tea manufactured from EBIs-treated shoots showed an overall improvement in quality.

Combined use of morpholine and triazole fungicides protected the plants from the disease at half the dosage recommended in the respective ergosterol biosynthesis inhibiting schedules and they were also economical (Baby *et al.* 2004a). Among the combinations, tridemorph-hexaconazole and tridemorph-propiconazole were superior compared to tridemorph-bitertanol. The fungicide treatments showed an increase in yield and quality of made tea.

Chemical control in relation to disease forecasting: Attention was focused on disease forecasting to reduce the number of fungicide sprays in blister blight management (Kerr and Shanmuganathan 1966, Kerr and de Silva 1969). Based on the relationship between relative humidity and spore production, ejection and germination, fungicide spraying could be suspended when the average RH over 3-days remained less than 83 per cent (Huysmans

1952). Likewise, Homburg (1953) suggested discontinuing fungicide application when the daily RH averaged below 80 per cent over 5-days and resumption of the fungicide treatment when RH exceeded 80 per cent in a period of 3-days. In Indonesia and Sri Lanka, the sunshine data was also employed as a guide to forecast the disease build up to postpone fungicide sprays (Visser *et al.* 1959, Mulder and de Silva 1960). However, the forecasting systems could not be successfully adopted in South India due to the highly conducive weather conditions of south-western monsoon for the disease development to discontinue or postpone fungicide application.

Two computerized linear multiple regression models, based on climatic factors, atmospheric spore load and disease incidence have been developed to predict the incidence of blister blight in relation to fungicidal sprays (Premkumar and Muraleedharan 1998).

***Breeding for disease resistance*:** Identification and development of resistant tea cultivars against blister blight have been the long felt need of tea industry to increase crop productivity. Efforts were made to get clones/*jats* resistant to blister blight (Venkataramani 1950, Huysmans 1952, Van der Knaap 1952, Semangun 1971, Debnath and Paul 1994). China *jat* was reported relatively less susceptible than Assam *jat* (Majid, 1971). Susceptibility of UPASI released clones and garden selections to blister blight was tested at UPASI Tea Research Institute. Cultivars SA 6, SP/4/5 and B/6/24 exhibited least susceptibility. However, clones exhibiting 'resistance' in one locality showed susceptibility in other areas (Agnihothrudu and Chandra Mouli 1990). In a specific situation, Venkata Ram (1961) observed breakdown of resistance in a tea bush after 12 years. Singh and Prodhan (1985) also reported on field performance of Tocklai released 17 clonal and one seed cultivars at Darjeeling. Phoobsering 312 and TS 378 (Nanda Devi) were reported 'fairly resistant' to blister blight. Under controlled conditions, clones of different locational origin manifested varying degree of susceptibility; and some seed raised chinary nursery plants exhibited resistant disease reaction (CSIR Complex 1995, IHBT 1997). Hypersensitive disease reaction was obtained in UPASI 10 and SA 6 (Gupta *et al.* 1996). Leaf-pubescence, internode-length, twig-angle, and leaf-tannin content were postulated to contribute resistance against blister blight (Martosupono, 1991).

Higher amounts of epicuticular wax and increased thickness of cuticle and epidermal layer act as physical barriers to hyphal penetration. Resistant clones contain higher quantum of chitinase in intercellular spaces of palisade tissues. Both chitinase assay as well as western blotting studies confirmed that the constitutive level of chitinase expression was higher in the resistant clone when compared with the susceptible tea clone (Jeyaramraja *et al.* 2005).

Findings of Punyasiri *et al.* (2005) suggest that epicatechin p-lays role in the resistance mechanism. Levels of (-)-epicatechin in tea cultivars resistant to blister blight were significantly higher than those in susceptible cultivars, while the reverse was true for (-)-epigallocatechin gallate. They also observed that the content of the methylxanthines, caffeine and theobromine in the leaf increased in the initial translucent stage of the disease, probably as a defense response to the fungal attack. Epicatechin and epigallocatechin levels were less than in healthy tissues at this stage but increase in the corresponding gallate esters suggested that they were being converted into esters. Although epicatechin and epigallocatechin levels decreased from translucent to mature blister stages, the decrease was not significant. The decrease in levels of epicatechin, epigallocatechin, and their esters on infection and the formation of cyanidin and delphinidin on oxidative depolymerization of the blisters suggests that proanthocyanidins may play a role in the defense mechanism. The high resistance of a purple green leafed cultivar may be attributed to the additional catechin source provided by the high levels of anthocyanins.

With mass spectroscopy and phloroglucinolysis studies, Punyasiri *et al.* (2004) demonstrated that infection of leaves of the susceptible tea plant to blister blight resulted in a shift of the proanthocyanidin stereochemistry away from 2,3-trans (e.g. catechin and gallocatechin) and towards 2,3-cis (e.g. epicatechin and epigallocatechin), and infection also resulted in increased gallic acid esterification of the initiating subunits of proanthocyanidins.

Immunological techniques could be also be valuable in screening commercially cultivars against the pathogen. Using these techniques, Chakraborty and Sharma (2007) analyzed the antigenic patterns of host and pathogen and concluded that pathogenicity of *E. vexans* to different tea cultivars was related to the level of antigenic similarity between the host and the pathogen.

Although absolute resistance in tea clones/seed stocks against blister blight is not reported, the difference marked in disease response of various cultivars suggests existence of natural variability in tea. But the above studies appear inconclusive on two major accounts: (i) susceptibility of tea cultivars to blister blight was studied mainly in single location field tests, without taking into account existence of variation in the pathogen population. The pathogen population is a dynamic entity, liable to change with environment and selection pressure, and (ii) no attempts were made to work out the genetic basis of resistance in selected cultivars. The earlier work, therefore, needs critical reappraisal with respect to the performance of tea cultivars against various isolates/races of blister blight pathogen, and mechanism and genetic basis of resistance in selected clones for developing sound breeding programme(s).

Little work has been done to elucidate variability in the pathogen population(s) through pathological and PCR based techniques for characterization and identification of pathogen races. Differential disease response in some clones against pathogen populations collected from different areas suggested the existence of variability in the pathogen (Gupta *et al.* 1996). Lack of information on pathogen variability has impeded the development of suitable strategy identification and development of durable resistant sources.

Recently, the effect of two chemical elicitors (acibenzolar-S-methyl benzo-(1,2,3)-thiadiazole-7-carboxylic acid S-methyl ester (Boost 500SC)) and salicylic acid in inducing resistance in tea plants against blister blight was studied by Ajay and Baby (2010). They found that nursery plants treated with the elicitors had reduced severity of the blister blight. Further, there was a significant increase in the activities of defense enzymes like phenylalanine ammonia lyase, peroxidase and β-1,3-glucanase on elicitor treatments in tea leaves challenged with the pathogen. Acibenzolar-S-methyl (ASM) at 0.14% registered the lowest disease severity, whereas treatments with salicylic acid were ineffective. They further observed that under field conditions, application of ASM at 0.14% resulted in disease protection of 25%, but its application with the standard fungicide in alternate rounds nearly doubled the disease protection.

Cultural practices: Adopting certain culture practices helps in reducing the disease. Avoiding excessive shade by pollarding branches of shade trees is an important culture practice to reduce the disease severity (Murthy 1996). Hard plucking during peak blister season, limited to two to three months in a year, reduced the disease without affecting the bush health (Eden 1947).

The economic consequences of blister blight, in terms of yield loss, quality deterioration, and control measures constitute a major constraint on cost-benefit ratio of the tea industry. An economical disease control, less dependent on fungicides, appears directly dependent on the development and availability of planting material with proven performance against blister blight pathogen. The possibility for developing biological control measures based on the natural tea-phylloplane microorganisms should also be explored (Balasuriya and Kalaichelvan 2000, Acharya *et al.* 2001).

Black Rot (*Corticium theae* Bernard and *Corticium invisum* Petch)

Black rot is an important disease of the mature leaf and stem in the plains of north-east India and the mid elevations in South India. The disease is more common in the badly ventilated areas, left without cleaning out of the dead snags.

The fungus produces slightly raised irregular patches on the leaves, brown, yellowish to chocolate brown and grey on the upper surface and evenly brown or grey on the lower surface. The fungus produces basidiospores in fructifications on the lower surface of leaves, which appear dusty due to the spore production. The infected leaves turn black as they rot during the wet weather. The dead leaves though detached from the stems often remain suspended onto the bush, held together by mycelial chords. The fungus persists through the winter from season to season by the sclerotia embedded in the cracks and crevices in the stem bark. The infected bushes suffer loss of yield and vigour, and debilitation.

Chemical control: Effective management of the disease is achieved by spraying copper fungicides during the active infection phase, taking special care to wet the undersurface of leaves and drench the frame, followed by spraying Vitavax 75 WP and Hiltavax 75 WP in the late season (Dutta and Barthakur 1992). Spraying thiophanate-methyl and benomyl is also effective (Takaya 1976).

Cultural practices: The disease incidence is less in the pruned tea. Following pruning, the dead twigs should be cleaned out to remove any resting stages of the fungus. Ventilation should be improved and over dense shade thinned out.

Anthracnose (*Colletotrichum theae-sinensis* (Miyake) Yamamoto = *Gloeosporium theae-sinensis* Miyake)

Anthracnose is a serious disease in Japan. The fungus produces reddish brown lesions on the leaves. The diseased leaves usually fall off. Foliar spray of benomyl and thiophanate methyl was found to be effective in disease management (Hamaya, 1981).

Net Blister Blight (*Exobasidium reticulatum* Ito and Swada)

Net blister blight is another serious disease reported from Japan and Taiwan. Infected leaves develop on the under surface white reticulate lesions which are slightly protuberant along the veins. The infected leaves may fall off and petioles may die back during severe outbreaks of the disease, resulting in reduced yield in the first crop in the following year. Copper fungicides, chlorothalonil, and captafol provide effective disease control (Hamaya 1981).

Grey Blight (*Pestalotiopsis theae* (Sawada) Steyaert)

The genus *Pestalotiopsis* is another economical important fungal pathogen causing grey blight disease in tea plant. The disease inflicts an estimated crop loss of 17 per cent in South India (Joshi *et al.* 2009). This disease has gained importance recently due to the extensive usage of mechanical harvesting in tea plantations because the pathogen gains entry through wounds.

The disease can be effectively controlled using fungicides with different modes of action. Biocontrol agents can also be incorporated in the spraying schedule. Mancozeb at 0.3 per cent or carbendazim or thiophanate methyl 0.05 per cent are effective to manage the disease. Affected bushes should be thoroughly drenched with the fungicide suspension. Recent studies reveal that systemic fungicides thiophanate methyl and carbendazim at 0.05 per cent were more effective, followed by the contact fungicides mancozeb and copper oxychloride at 0.3 per cent. Application of biocontrol agents such as *Trichoderma, Gliocladium* and *Pseudomonas* along with a food base give satisfactory control (Sanjay *et al.*, 2008).

Recently, Japanese workers examined the possibility of induction of disease resistance in plants by plant activators (PAs, i.e. compounds that induce disease resistance in the plants) (Yoshida *et al.* 2010). The tested PAs (AGREVO EX, a yeast extract preparation, probenazole, prohydrojasmon, and tiadinil) suppressed lesion development of the gray blight and anthracnose in the leaves, and the induced resistance continued for a month after the treatment. All the tested PAs induced disease resistance in tea plants systemically. In field trials also, the PAs were effective to the diseases of anthracnose and gray blight with control efficacies to a varying degree. Further, the growth and chemical composition of tea shoots were not changed by the PA treatment.

Joshi *et al.* (2009) collected 42 *Pestalotiopsis* isolates from five different regions in South India. Among them, 22 isolates showed diverse morphological characters like colour, size and length of conidia and virulence. The results indicate that within south India the diversity of *Pestalotiopsis* was high both morphologically and genetically.

Based on parent-offspring genetic analysis, Japanese workers proposed that the resistance of tea plants to the disease was controlled by two independent dominant resistance genes Pl(1) and Pl(2), (Takeda 2002).

Bird's Eye Spot (*Cercospora theae* Petch)

Presence of the disease has been reported in all major tea growing countries (Muraleedharan and Chen 1997). The pathogen generally attacks mature foliage especially the mother leaves and maintenance foliage.

Recently, Balasubramanian and Ponmurugan (2012) conducted a detailed study for the management of this disease using contact, systemic and botanical fungicides and microbial based biocontrol agents. Their major findings include (i) systemic fungicides were superior to biocontrol agents in controlling the disease, whereas the green leaf yield significantly increased with biocontrol agents mainly because of improvement in the biometric, physiological and biochemical parameters of the bush, (ii) *Streptomyces sannanensis* offered the better yield followed by *Trichoderma harzianum*, (iii) vermicompost-based bioformulation showed better efficacy than talc-based formulation, (iv) among bacteria, fungi and actinomycetes groups of biocontrol agents, *S. sannanensis* was better than *Pseudomonas fluorescens* and *Trichoderma atroviride* belonging to bacterial and fungal antagonists, respectively, in terms of disease protection and increasing tea yield, and (v) the tea quality parameters, *viz.*, total liquor colour, thearubigins, theaflavins, highly polymerized substances and caffeine were significantly improved in biocontrol treated plants.

A study by Balasubramanian *et al.* (2011) on pathogen related enzymatic activity revealed that *C. theae* was capable of producing the range of pathogenic enzymes and exhibited its maximal activity at 10th day harvest. Further, the study showed that phyopathogenic isolates of *C. theae* are capable of producing various carbohydrases, pectinases and proteases and defensive enzymes along with phenolic and oxidative enzymes for defensive response.

Other Leaf Diseases

Brown blight incited by *Colletotrichum camelliae* Massee is very common on old and damaged leaves. This disease is of little economic consequence. Leaf scab (*Elsinoe theae* Bitancourt and Jenkins), sooty mould (*Meliola camelliae* (Cattaneo) Sacc., *Capnodium theae* Boedijn and *Capnodium footii* Berk. and Desmaz.), red rust (*Cephaleuros virescens* Kunze = *Cephaleuros parasticus* Karsten (alga), Phyllosticta leaf-spot (*Phyllostcta theae* Speschnew and *Phyllostcta erratica* Ellis and Everth.), and several other less important minor diseases have also been reported (Agnihothrudu 1995, Chandra Mouli 1999, APSnet 2003).

Bacterial diseases have also been reported in tea: bacterial shoot blight (*Pseudomonas syringae* pv. *theae* (Hori) Young, Dye and Wilkie) from Japan (Hamaya 1981); Bacterial canker (*Xanthomonas campesteris* pv. *theicola* Uehara, Arai, Nonka and Sano) from Japan; and Crown gall (*Agrobacterium tumefaciens* (Smith and Townsend) Conn. from Kenya (Sudoi and Langat 1992). *Pseudomonas teashirensis* and *Pseudomonas theae* have also been reported from Japan and Georgia, respectively.

Phloem necrosis, caused by Phloem necrosis virus (Camellia Virus 1), has been reported from Sri Lanka (Mulder 1960) and possibly in India from Darjeeling (Raychaudhuri 1967). Yellows of tea leaves, suspected to be a viral disease, have been recorded from Japan (Hamaya 1981). A graft transmitted yellow mosaic disease of sporadic incidence in pruned and unpruned tea plants, associated with polyhedral virus particles, have been recorded from Kangra valley (Zaidi *et al.* 1992).

STEM DISEASES

Of the stem diseases, wood rot, stem and branch canker, collar canker, thorny stem blight, and die back are ubiquitously reported stem diseases of importance.

Wood Rot (*Hypoxylon serpens* (Pers.: Fr.) J. Kick, *H. nummularium* (Bull.: Fr.), and *H. vestitum* Petch)

The fungus enters the branches and stems of mostly the old tea bushes through sun-scald injury. The affected portions become dry, brittle and defunct. The fungus produces irregular, slightly raised, whitish-grey to dark encrustations on the affected wood. Selective pruning of dead and drying branches and painting of the large cut

Black root rot, red root rot, brown root rot, charcoal stump rot, and Armillaria root rot are the common primary root diseases reported from the major tea growing countries. Among the secondary root diseases violet root rot and diplodia root disease are also common.

Charcoal Stump Rot (*Ustulina deusta* (Hoffm.: Fr.) Lind)

The disease is the most common of the root rot diseases. The disease is capable of causing sudden death of the entire bush. Development of white fan-shaped mycelial patches overlying the wood underneath the bark, wood traversed by typical double black lines, and effused, carbonaceous fructifications at the collar region are the characteristic symptoms of the disease.

Black Root Rot (*Rosellinia arcuata* Petch)

The disease generally causes sudden death of the bush. The distinguishing features of the disease are black, more or less woolly mycelial strands on the root surface, white star-shaped mycelial patches overlying the wood. Black lines are also seen in the wood.

Brown Root Rot (*Phellinus noxius* (Corner) G.H. Cunningham = *Fomes noxius* Corner)

The disease may cause gradual or sudden death of the infected bush. The disease is identified by the presence of tawny brown mycelium underneath the bark. Thin brown reticulate lines develop in the wood. In the advanced stage of disease, roots become very soft and spongy. Diseased roots are encrusted with firmly attached soil and stone particles intermixed with brown mycelium.

Red Root Rot *(Poria hypolateritia* (Berk.) Cooke

The death of the diseased bush may be gradual and partial or sudden and entire. The infected roots are invariably encrusted with soil particles. Red to black strands of mycelium grows on the root surface. White mycelial patches sometimes develop overlying the wood. Red lines or sheets traverse the wood. Roots become soft and spongy as the disease progresses to the advanced stage.

Armillaria Root Rot (*Armillaria mellea* (Vahl: Fr.) Kummer)

The infected roots show longitudinal split at the collar region, and the cracks on the bark are filled with white mycelial growth. The fungus develops ribbon-like compact rhizomorphs either black or brown in colour.

Violet Root Rot (*Shaerostilbe repens* Berk. and Broome)

Water logging and poor soil aeration, are the predisposing factors for the disease development. The root lenticels in the affected bushes are enlarged, root tips turn inky black, bark peels away, and prominent violet coloured rhizomorphs can be seen overlying the wood. Improving soil drainage and soil aeration helps in controlling the disease.

Diplodia Root Disease (*Lasiodiplodia theobromae* (Pat.) Griffon and Maubl. = *Botryodiplodia theobromae* Pat.)

Bluish discolouration of the wood in the sliced root is very characteristic. Root surface is covered with small coal black hairy cushions of the mycelium, giving a sooty appearance at the advanced stage of the disease. The disease is more common in sandy loam soils.

MANAGEMENT STRATEGIES FOR PRIMARY ROOT DISEASES

The primary root diseases cause total loss by ultimately killing the infected bushes. These diseases are difficult to diagnose due to the lack of any aerial symptoms till the infected bushes are moribund. The root rot pathogens are highly recalcitrant to treat.

Chemical control: Soil fumigations were undertaken with Vapam (sodium N-dithiocarbamate), Methamsodium (sodium N-methyldithiocarbamate), and Durofume CP 1:1 (methyl bromide and ethelene dibromide 1:1), methyl bromide, D.D. Trapex, WN12 to arrest the disease spread (Venkata Ram 1961, 1965, Shanmuganathan 1969, Satyanarayana 1973). These chemicals tested mainly against *Poria hypolateritia, Ustulina deusta, Fomes noxiuz, F. lamaoensis,* and *Rosellinia arcuata* were found to be more or less satisfactory in controlling the primary root diseases. However, these treatments are uneconomical in view of high cost structure (Satyanarayana and Barua 1975, Satyanarayana 1987) and subject to criticism due to their non-selective action and harmful effects on the beneficial soil microflora (Barthakur 1999). Soil drenching of the affected area with mancozeb or carbendazim effectively controlled the black root rot (Chandra Mouli 1993).

Field management measures: Recommendations made include uprooting the dead bush along with apparently healthy ring of bushes (suspects), removal of all wood debris from the soil, and rehabilitation of the area for 18-24 months under non-hosts, Guatemala grass or Mana grass, before undertaking replanting (Tea Encyclopaedia- Ser. 70/1 file under I.3, 1971).

When a large area is affected, as an immediate measure of isolation, 45 cm wide and 120 cm deep trench should be dug to include a mantle of healthy bushes around the diseased pocket.

Ring barking of shade trees to deplete their roots of the starch reserves for rapid colonization by other saprophytic microorganisms has also been suggested as a preventive measure in the control of root rots.

Biological control measures: In the recent years attention has been paid towards the development of biological control against the root diseases. *In vitro* studies on the biological interactions of *Trichoderma* spp. and *Gliocladium virens* against major root pathogens suggested potential for developing bioagents for controlling root diseases. Among the various *Trichoderma* spp. tested, *T. harzianum* was found to be the best antagonist to *Poria* and *Armillaria*, and *T. viride* to *Rosellinia*. *Gliocladium virens* exhibited antibiosis to all the pathogens (Baby and Chandra Mouli 1996). Onsando and Waudo (1994) also reported on the antagonistic potential of *T. koningii*, *T. longibrachiatum* and *T. harzianum*, based on the inhibitory effects of *Trichoderma* isolates on *in vitro* growth of *Armillaria*. However, *in vitro* antagonism of *Trichoderma* spp. to *Armillaria* was not reproducible *in vivo* (Otieno 1998). The outcome of the field-scale experiments for control of primary root rots with 'Trichoderma biocides' also remains to be seen though no mortality of plants was observed in treated soils up to 60 and 84 months in infilled and replanted areas, respectively (Barthakur 1999).

Early detection, accurate identification, and quantification of the diseases in crop system (asymptomatic/presymptomatic tea bushes, soil and other matrices) are critical to their successful management. In recent years, highly specific immunoassays using polyclonal antisera or mono-clonal antibodies have been used to identify plant pathogenic fungi. However, absolutely no work has been done on immunoassaying of tea root rot pathogens. Possibilities of using plant growth promoting (PGPR) bacteria and arbuscular mycorrhizal fungi should also be explored from the point view of developing integrated disease management.

REFERENCES

Acharya K, Katel ON, Pokhrel B, Sherpa AR, Acharya R (2001) Biocontrol of blister blight of tea by *Pseudomonas fluorescens*. *Journal of Hill Research* 14(2): 110-111.

Agnihothrudu V (1995) Some more important fungal diseases of tea. *Indian Phytopathology* 48(3): 229-253.

Agnihothrudu V, Chandra Mouli B (1990) Blister blight of tea and its control. *Planters' Chronicle* 69: 407-409.

Ajay D, Baby UI (2010) Induction of systemic resistance to *Exobasidium vexans* in tea through SAR elicitors. *Phytoparasitica* 38 (1): 53-60.

APSnet (2003) Common names of plant diseases. The American Phytopathological Society, *APSnet* Online: 1-5.

Arulpragasam PV, Addaickan S, Kulatunga SM (1987) Recent devel-opments in the chemical control of blister blight leaf disease of tea - effectiveness of EBI fungicides. *Sri Lanka Journal of Tea Science* 56: 22-34.

Baby UI, Ajay D, Premkumar R (2004a) Combined use of morpholine and triazole fungicides for the management of blister blight of tea. *Tropical Agriculture* 81(1): 49-53.

Baby UI, Balasubramanian S, Ajay D, Premkumar R (2004b) Effect of ergosterol biosynthesis the tea plant and inhibitors on blister blight disease, the tea plant and quality of made tea. *Crop Protection* 23(9): 795-800.

Baby UI, Chandra Mouli B (1996) Biological antagonism of *Trichoderma* and *Gliocladium* spp. against certain primary root pathogens of tea. *Journal of Plantation Crops 24* (Supplement): 249-255.

Baby UI, Ravichandran R, Ganesan V, Parthiban R, Sukumar S (1998) Effect of blister blight disease on the biochemical and quality constituents of green leaf and CTC tea. *Tropical Agriculture* 75(4): 452-456.

Balasubramanian Gnanamangai Mythili, Ponmurugan Ponnusamy (2012) Evaluation of various fungicides and microbial based biocontrol agents against bird's eye spot disease of tea plants. *Crop Protection* (32) :111–118.

Balasubramanian Mythili Gnanamangai, Ponnusamy Ponmurugan, Rajendran Yazhini, Sekar K Pragadeesh (2011) PR Enzyme activities of *Cercospora theae* causing bird's eye spot disease in tea plants (*Camellia sinensis* (L.) O. Kuntze). *Plant Pathology Journal* 10(1): 13-21

Balasuriya A, Kalaichelvan J (2000) Is there potential in natural tea-phylloplane microorganism in the control of blister blight leaf disease of tea (*Camellia sinensis*). *Planter* 76(892): 409-417.

Barthakur BK (1999) Recent trends in controlling primary root diseases of tea in north east India. *Two and a Bud* 46(1): 20-21.

Lehmann-Danzinger H (2000) Diseases and pests of tea: Overview and possibilities of integrated pest and disease management. *Tropenlandwirt* 101(1): 13-38.

Majid A (1971) Menghadapi masalah peremajaan perkebuman the di Indonesia. Sidang KTP III Kosimi budidaya Teh – Kina, Bandung, pp 7.

Martosupono M (1991) Some factors influenced the tea resistance to blister blight. In: Proceedings of the International Symposium on Tea Science, August 26-29, Shizuoka, Japan, pp. 651-654.

Mulder D (1960) Report of the plant pathologist for the year 1959. Reports of Tea Research Institute, Ceylon 1959: 66-67.

Mulder D, de Silva RL (1960) A forecasting system for blister blight control based on sunshine records. *Tea Quarterly* 31: 56-57.

Muraleedharan N, Chen ZM (1997). Pests and diseases of tea and their management. *J Plant Crops* 25: 15–43

Murthy SBK (1996) Shade in tea plantations. *Planters' Chronicle* 91(4): 591-595.

Onsando JM, Waudo SW (1994) Interaction between *Trichoderma* species and Armillaria root rot fungus of tea in Kenya. *International Journal of Pest Management* 40(1): 69-74.

Otieno W (1998) Armillaria-Trichoderma interactions and management of Armillaria root rot of tea (*Camellia sinensis* (L.) O. Kuntze). *Tea* 19(1): 11-16.

Premkumar R (1997) Tea: blister blight control - recommendations on the usage of propiconazole. *Planters' Chronicle* 92(4): 169.

Premkumar R, Baby UI, Mouli BC (1998) Systemic activity and field performance of triazole fungicides against blister blight pathogen of tea. In: Mathew NM, Kuruvilla Jacob C, Licy J (ed) Development in plantation crops research: Proceedings of the 12th Symposium on Plantation Crops, PLACROSYM XII, Kottayam, India, 1996, pp 279-283.

Premkumar R, Muraleedharan N (1998) Forecasting of blister blight disease and its management - recent developments. Proceedings of the 35th UPASI Conference, Coonoor, Tamil Nadu, India. Bulletin of UPASI Tea Scientific Department No. 52: 34-41.

Premkumar R, Ponmurugan P, Manian S (2008) Growth and photosynthetic and biochemical responses of tea cultivars to blister blight infection. *Photosynthetica* 46(1): 135-138.

Punyasiri PAN, Abeysinghe ISB, Kumar V (2005). Preformed and induced chemical resistance of tea leaf against *Exobasidium vexans* infection. *Journal of Chemical Ecology* 31(6):1315-1324.

Punyasiri PAN, Tanner GJ, Abeysinghe ISB, Kumar V, Campbell PM, Pradeepa NHL (2004) *Exobasidium vexans* infection of *Camellia sinensis* increased 2,3-cis isomerisation and gallate esterification of proanthocyanidins. *Phytochemistry* 65(22): 2987-2994.

Raychaudhuri SP (1967) Virus disease of plantation crops in India. *Indian Horticulture* 12: 9-13.

Reitsma J, van Emden JH (1949) De bladpokkenziekte van de Thee. *Arch. Thee Cult.* 16: 218-231.

Sanjay R, Ponmurugan P, Baby UI (2008) Evaluation of fungicides and biocontrol agents against grey blight disease of tea in the field. *Crop Protection* 27(3-5): 689-694.

Saravanakumar D, Vijayakumar Charles, Kumar N, Samiyappan R (2007) PGPR-induced defense responses in the tea plant against blister blight disease. *Crop Protection* 26(4): 556-565.

Sarmah KC (1960) Diseases of tea and associated crops in north-east India. Memorandum No. 26. Tocklai Experimental Station, pp 68.

Satyanarayana G (1973) Annual Report, Plant Protection Deptt. Tocklai, 1972-73.

Satyanarayana G (1987) Tea diseases of importance in India and their control. *Reviews in Tropical Plant Pathology* 4: 89-108.

Satyanarayana G, Barua GCS (1975) Trends in disease control with reference to tea in north-east India. *Pesticides Annual*: 109-111.

Satyanarayana G, Barua GCS (1983) *Journal of Plantation Crops* 11: 27-31.

Schweizer J (1950) De situatie inzake de blister blight ziekte van de thee op Sumatra's Oostkust. *Bergcultures* 19: 207-211.

Semangun H (1971) Penyakit - penyakit tanaman pertanian di Indonesia. Yayasan pembina fakultas Pertanian. *UGM Yogyakarta*: 465.

Shanmuganathan N (1969) Studies on the parasitism and control of tea root rot disease fungi in Ceylon. *Tea Quarterly* 40(1): 19-25.

Shanmuganathan N, Arulpragasam PV (1966) Epidemiology of tea blister blight (*Exobasidium vexans*) II – The diurnal and seasonal periodicity of spores in the air over a tea estate. *Transactions in British Mycological Society* 49: 219-226.

Shanmuganathan N, Saravanapavan TV (1974) Comparative effecti-veness of copper and organic fungicides against tea leaf blister blight (*Exobasidium vexans*). *Plant Disease Reporter* 58(10): 928-932.

Singh ID, Prodhan SK (1985) Characteristics of Tocklai released planting materials from Clonal Proving Station, Darjeeling. *Two and a Bud* 32: 49-51.

Sudoi V, Langat J K (1992) A report on identification of fungi bacteria associated with the tea plant in Kenya. *Tea* 13(2): 76-83.

Sugha SK, Thakur BR (1998) Fungitoxic management of blister blight of tea. *Himachal Journal of Agricultural Research* 24(1-2): 160-164.

Takaya S (1976) Fungicides for the control of black rot of tea. *Tea Research Journal* 44: 19-21.

Takeda Y (2002) Genetic analysis of tea gray blight resistance in tea plants. *Japan Agricultural Research Quarterly* 36(3): 143-150.

Tea Encyclopaedia - Ser.70/1 file under I.3, 1971. Control of primary root diseases of tea. Serial No. 70/1, File under I.3. Tea Encyclopaedia, Tocklai Experimental Station, Tea Research Association: 4 sheets

Van der Knaap WP (1952) Blister blight resistentie van theeheesters en theenclonen. *Arch Cult.* 1: 69-98.

Venkata Ram CS (1961) Application of nickel chloride to tea plants (*Camellia sinensis*) and control of blister blight. *Current Science* 30: 57-58.

Venkata Ram CS (1965) Report of Plant Pathologist, Annual Report of UPASI Scientific Department Tea Sciences 1964-65, pp 16-28.

Venkata Ram CS (1966) Report of Plant Pathologist, Annual Report of UPASI Scientific Department Tea Sciences 1965-66, pp 24-37.

Venkata Ram CS (1968) Report of Plant Pathologist, Annual Report of UPASI Scientific Department Tea Sciences 1964-65, pp 18-28.

Venkata Ram CS (1969) Systemic control of *Exobasidium vexans* on tea with 1, 4- Oxathiin derivatives. *Phytopathology* 59: 125-128.

Venkata Ram CS (1974) Calixin, a systemic fungicide effective against blister blight (*Exobasidium vexans*) on tea plants. *Pesticides* 8: 21-25.

Venkata Ram CS (1975) Systemic activity and field performance of pyracarbolid in the control of blister blight pathogen of tea. *Z Pfl Krankh Pfl Schutz.* 82: 65-76.

Venkata Ram CS, Chandra Mouli B (1976) Systemic fungicides for integrated blister blight control. UPASI Tea Scientific Department, Bulletin No. 33, pp 70-87.

Venkata Ram CS, Chandra Mouli B (1983) Blister blight of tea - principles of control. 4th International Congress of Plant Pathology, Melbourne, Australia, pp 89.

Venkataramani KS (1950) Annual Report. UPASI Scientific Department Tea Section for 1949-50, pp 23.

Visser T, Shanmuganathan N, Mulder T (1959) The possibility of timing blister blight spraying according to sunshine records. *Tea Quarterly* 30: 39-43.

Yoshida Katsuyuki, Ogino Akiko, Yamada Kengo, Sonoda Ryoichi (2010) Induction of disease resistance in tea (*Camellia sinensis* L.) by plant activators. *Japan Agricultural Research Quarterly* 44(4): 391-398

Zaidi SNH, Raja Ram, Zaidi AA, Gulati A, Tamang MB, Chakraborty DN (1992) A yellow mosaic disease of *Camellia sinensis*: purification of virus particles. *Indian Journal of Plant Pathology* 10(1&2): 69-70.

13

MANAGING WEEDS IN TEA

R.D. Singh

CSIR-Institute of Himalayan Bioresource Technology,
Palampur (H.P.)-176 061, India. E-mail: rdsingh@ihbt.res.in

Weed management in tea is the second most expensive input after plucking (Sinha 1985). In tea plantations, grasses generally predominate the weed flora followed by broad-leaf weeds (Singh *et al.* 1994). The major weeds from different tea growing areas of the world are as follows:

Grasses: *Imperata cylindrica, Paspalum conjugatum, Setaria palmifolia, Cynodon dactylon, Saccharum spontaneum, Panicum repens, Pennisetum clandestinum, Arundinella bengalensis, Axonopus compressus* and *Paspalum scrobiculatum.*

Broad leaf weeds: *Borreria hispida, Polygonum chinense, Borreria alata, Eupatorium odoratum, Mikania micrantha, Mikania cordata, Ageratum conyzoides, Ageratum haustonianum, Scoparia dulcis, Oxalis acetosella, Artemisia vulgaris* and *Commelina benghalensis.*

Other weeds: Sedges are not serious weeds in tea plantation (Singh *et al.* 1994). Ferns like *Pteridium aquilinum* and *Nephrodium* spp. have also been reported to infest tea plantation. In the tea gardens having high humidity and limited sunny days throughout the year, mosses tend to cover soil surface under the canopy and a large part of tea trunk and branches (Ronoprawiro 1976, 1981).

CRITICAL PERIOD

In India, studies in young tea have revealed that the critical period of weed competition is generally from April to September, Delay in weeding during this period adversely affect branching, growth and yield of young tea (TRA 1978, Sinha and Borthakur 1992). In Sri Lanka, the critical period for weed competition in young tea is between 8-16 weeks after planting and the threshold period of competition is about 12 weeks after planting (Prematilake *et al.* 1999).

REASONS FOR WEED INFESTATION IN TEA

Plant architecture: The chinary or china hybrid tea bushes bear many stems arising from collar level, which develop the soil pockets providing shelter to weeds. Such soil pockets are more evident in the gardens practicing faulty application of inorganic fertilizers (mostly ammonium sulphate and urea) in the centre of the bush frame and eventually thinning the number of primaries. Consequently, the weeds growing in these pockets escape various manual and chemical treatments for weed control. If such weeds are perennial grasses or brushes, the problem is further aggravated because they can neither be uprooted successfully along with their underground propagules nor dug out, which otherwise may also damage tea stem. Over all, application of herbicides over bush frame for managing weeds is a risky operation.

Wider plant spacing and vacancy: Vacant and wide space within tea bushes also support weed infestation, as increase in plant spacing, from 15 to 150 cm, was found to increase weed infestation (TRA 1978).

Pruning cycle: Pruning or deep skiffing needs to be done in every 3-4 years depending on the pruning cycle. However, it has been observed that highest weed infestation in tea plantation occurs in pruned or deeply skiffed sections where the ground gets exposed.

LOSSES CAUSED DUE TO WEEDS IN TEA

Tea plantations are highly labour intensive. Unlike other crops, the tea leaves are plucked routinely at 7-10 days plucking interval. The period of heavy crop coincides with the period of active weed growth and infestation, affecting the deployment of labour for plucking.

In nursery, the environmental conditions are congenial for plant growth, which facilitate rapid weed growth competing with tea plants.

It calls for extra labour for nursery success. Basu (1974, 1975) had reported that in tea nurseries weeds infested with *Meloidogyne incognita* and/or *M. javanica* may infect tea seedlings and there was possibility of spread of infestation to mature tea.

In 2-year old clonal tea, adverse effects of predominant weeds *viz.*, *Paspalum conjugatum*, *Eupatorium riparium* and *Ageratum haustonianum* on stem height, and length and total number of primary branches appear as early as 2 months from the infestation (Soedarsan *et al.* 1976). These effects were attributed to the dense root system and high regeneration rate of these three weed species. Yellowing and reduction of the tea leaf area was also observed in case of infestation with *P. conjugatum* and *A. haustonianum*. It was also reported that retarding effects of *Imperata cylindrica*, *Artemisia vulgaris* and *Panicum repens* appear over a longer period.

In south India, weeds adversely affected growth of tea plants, as evidenced by reduction in height, number of leaves, leaf area and dry weight. Growth inhibition was in the order of *Crassocephalum crepidioides* > *Bidens biternata* > *Cynodon dactylon* > *Paspalum conjugatum* > *Conyza ambigua* > *Panicum repens*. Among the three phenological stages studied (the peaks of the vegetative, flowering, and fruiting stages), weeds at their peak flowering stage caused the maximum reduction in the growth of tea plants (Ilango and Satyanarayana 1996).

In newly planted tea, weed control during April-September is essential for establishment of the plants. The weed competition during this period has been reported to cause nearly 50% reduction in the number of primary branches and about three and half times decrease in the yield in the second year (Rao and Singh 1977). Work at TRA (1978) had shown that weeds in young tea accumulated 13,090 kg dry matter/ha in the first year and 8,400 kg in the second year and removed as much as 252 kg and 151 kg nitrogen/ha, respectively. When weeds were controlled during the critical period of weed infestation, weed dry matter and nitrogen removal between October and March was only 1,890 kg and 34 kg/ha, respectively. Another study indicated that nitrogen uptake by young tea plants was higher in weed free plots under herbicide treatment followed by manually weeded plots (Hoshina *et al.* 1980).

Weeds also retard the efficiency of farm workers. Certain weeds like *Bidens pilosa* and *Rubus* spp. often reduce the plucking efficiency of workers. In heavily infested tea sections, shoots of weeds get inadvertently harvested along with tea shoots, which adversely affect the quality of made tea.

WEED MANAGEMENT

Chemical methods

The common herbicides used in tea plantations are paraquat, glyphosate, simazine, 2,4-D sodium, 2,4-D amine, diuron, dalapon and MSMA. Linuron, methazole, metribuzin, dichlormate, dinoseb, oxadiazon, butachlor and fluchloralin herbicides were also used in tea; however they were trivial in performance in tea plantations. Though, most of the herbicides approved are safe to tea, phytotoxicity on tea may occur due to the reasons (Sinha and Borthakur 1992) like application of herbicides at rates higher than the recommended doses, improper or non-targeted spraying, spray drift, leaching of pre-emergence herbicides by heavy rains, and age of the tea bush.

Pre-emergence herbicides

In tea nursery: Besides manual weeding in tea nursery, application of simazine, atrazine, fluchloralin, oxadiazon, methazole for chemical weed control had been recommended by TRA (1978). These herbicides were recommended to be applied @ 2 kg/ha in April, at 3 weeks before planting of clonal cuttings. Treatments may be repeated after hand weeding when weed cover exceeded 50 %. Mixtures of simazine or atrazine with oxadiazon or fluchloralin were found more effective.

In young tea: Chemical weed control in young tea is distinct from that in mature tea, as young tea plants are relatively more susceptible to herbicide treatments and the weed flora is more diverse and intense. Ghosh and Ramakrishnan (1981) observed that, in young tea, oxyfluorfen at 0.125 kg/ha applied pre-emergence in May followed by oxyfluorfen (0.06 kg) + either paraquat (0.24 kg) or 2,4-D (0.8 kg/ha) as post-emergence controlled most of the problem weeds throughout the season. Presence of moisture on soil surface improves the bioefficacy of pre-emergence herbicides, in general.

It is normally advised that young tea plants should be shielded from the herbicide spray since they are more likely to be adversely affected than older plants. Also, application of dalapon and diuron in tea younger than 3 years is not recommended (CSIR 1985, 1990, Sinha and Borthakur 1992).

In Sri Lanka, oxyfluorfen at 0.29 kg/ha + paraquat at 0.17 kg/ha or glyphosate at 0.99 kg/ ha + kaolin at 3.42 kg/ ha provided better weed control than hand weeding in young tea (Prematilake *et al.* 2004). Also, a combination of inter-row mulching and above herbicides followed by hand weeding at least every 6-8 weeks was found the most appropriate weed management system for young tea (Prematilake *et al.* 2004).

In mature tea: Oxyfluorfen at 0.25 kg/ha provided good control of broad-leaved weeds in mature tea without any phytotoxicity when it was applied to clean soil or to growing weeds. Oxyfluorfen was comparable to simazine or diuron, each at 2 kg/ha (Rao and Kotoky 1981).

In mature tea, pendimethalin 0.75, oxyfluorfen 0.44, simazine 1.25, or atrazine 1.25 kg/ha were more effective pre-emergence treatments for suppressing seed-borne weeds (CSIR 1992, 1993). Subsequent weed growth in either of the cases could be controlled with spot treatment of 2,4-D and/or paraquat (Singh *et al.* 1992).

In pruned tea: Autumn application of pre-emergence herbicides (simazine, diuron and oxyfluorfen), before pruning, was found effective for better weed management in pruned sections, and was as effective as spring application (Rao and Kotoky 1980, Sinha 1985, Sinha and Borthakur 1992).

Post-emergence herbicides

Glyphosate: Glyphosate was evaluated to be the most effective herbicide in tea plantations, as more than 90% of glyphosate is used in tea plantations (Rao 1983). It controls many perennial weeds including *Imperata cylindrica*, *Setaria palmifolia* and *Cynodon dactylon* (Rahman *et al.* 1975, TRA 1976), when applied at 1.5 to 6 kg/ ha (Rao *et al.* 1976, 1977, Awasthi and Rao 1979). Normally, glyphosate is used at 1.5-2.5 l/ha (HPKV 1989, AICRPWC 1990 a, b). Kabir *et al.* (1991) had also observed that glyphosate 0.92 or 1.23 kg/ha provided effective weed control in tea of Darjeeling area.

UPASI (1978) concluded that glyphosate is a promising herbicide against hardy perennial grasses and deep-rooted broad-leaved weeds, and is not toxic to tea bushes even when sprayed directly on the bushes at a rate of 1.68 kg a.e./ha (Sharma and Satyanarayana 1976). However, studies at CSIR-IHBT, Palampur had indicated that in case of seed-raised china hybrid tea plantations, use of glyphosate even at the rate of 1.03 kg/ha may cause phytotoxicity. The susceptible tea bushes had showed loss of crop for 1-2 pluckings.

Glyphosate was also found effective in controlling brush weeds like *Lantana camara* - a troublesome weed in tea plantations in Himachal Pradesh and Uttarakhand (Singh *et al.* 1997).

2,4-D sodium: About 60% of the 2,4-D marketed in India is used in tea gardens (Rao 1983). Most of the paraquat - resistant broad-leaved weeds could be controlled by 2,4-D at 1.12 kg/ha (Sharma 1975, 1977), 2,4-D sodium at 0.8 kg/ha was very effective on its own. Addition of paraquat 0.3 kg/ha made it more effective against grasses (TRA 1978, CSIR 1989).

Paraquat: Paraquat is a unique contact herbicide for tea plantations, being quick in action and rain fast, without adverse effect on beneficial soil microbes (CIC 1989). It causes desiccation of the green tissues of the plants. Repeated application of paraquat (0.3 kg/ha) at short interval (weekly or fortnightly, depending upon rate of regeneration in weed) may kill even the persistent weeds.

Rotation of herbicides

Sharma (1977) had observed that the prolonged and continuous use of some of the herbicides led to a situation where resistant weed species and their populations tended to increase. The shift in weed flora under different herbicide treatments should be monitored regularly for deciding suitable herbicide programmes (rotation) to particular weed problems (Sharma *et al.* 1981, Sinha 1985). Under such situations, rotation of herbicides is generally recommended for broad spectrum of weed control.

NEW HERBICIDES

Glufosinate-ammonium at 0.38 kg /ha provided better weed control than paraquat dichloride at 2 weeks after spraying, and about 75 and 50% weed control for 5 and 6 weeks, respectively (Ilango and Sreedhar 2001).

In an experiment conducted in Valparai, Tamil Nadu, India napropamide at 2.0 kg /ha exhibited good weed control only up to 60 days after the application of the herbicide. Also, napropamide at 4.0 kg /ha was at par with oxyfluorfen at 0.25 kg /ha only up to 60 days (Victor *et al.* 2001).

Carfentrazone-ethyl at 4 and 8 g /ha combined with glyphosate 620 or 720 g /ha exhibited 85% weed control, while increment in the dose of carfentrazone-ethyl to 12, 15 and 20 g /ha combined with glyphosate at 620 or 720 g/ha provided total weed control and was comparable in its efficacy to the currently recommended tank mixture of glyphosate + 2,4-D at 720 + 1120 g/ha (Ilango 2003). Similar results were also reported by (Rajkhowa *et al.* 2005) with weed control up to 45 days after herbicide application.

QUALITY OF WATER AND BIOEFFICACY OF HERBICIDE SOLUTION

A study at Bangladesh Tea Research Institute exhibited significant response of glyphosate when applied using pond and rain water (Islam *et al.* 2008). Bioefficacy of 2,4-D was better when tap, stream or pond water was used, while lowest was observed with rain water. Paraquat in combination of other herbicides was most effective with tap water. Efficacy of glyphosate, paraquat and 2,4-D was positively co-related with increment in Mg and Fe concentration. Significantly negative co-relation of paraquat was observed with the increase of Ca concentration.

PHYSICAL AND MECHANICAL METHODS

In tea nursery: Planting of single leaf cutting of tea by inserting through a black polyethylene mulch beneath a cover of hoop-supported polyethylene sheet provided satisfactory weed control at the nursery stage (Smale 1991). Tabagari and Kopaliani (1984) also recorded improved weed control both in vegetatively propagated as well as seed-borne nurseries by mulching with black plastic, tea factory residues or peat.

In young tea: In high rainfall areas, clean weeding may lead to soil erosion, thus under such conditions the aim should be to control noxious weeds but leave a non-competitive ground cover of soft weeds (Sutidjo and Lubis 1971).

Mulching with black low density polyethylene (LDPE) sheet (Pirtskhalaishvili and Kimutsadze 1972, Zarnadze 1972, Korzun 1981, Tabagari and Kopaliani 1984) was more effective for weed control in mature as well as young tea. Similarly, Singh *et al.* (1993) had concluded that in young china hybrid tea planted on slope, LDPE mulch totally suppressed weeds in the inter-row spaces. LDPE mulch also enhanced plant growth and yield of tea compared to no mulch.

WEED BIOLOGY

Information on morphology, reproduction, biology, geographical distribution and economic importance of serious weeds of tea plantations are available only regarding *Mikania micrantha* (Dutta 1977) and *Solidago canadensis* and *S. lanceolata* (Gogiya *et al.* 1984). Studies on seed biology of certain weeds showed that stirring the top soil every month increased the germination of *Borreria hispida* and *Cyperus pilosus* indicating the wisdom of refraining from cultivation (TRA 1978).

Parker (1972) had stated that *Mikania* spp. are vigorous creeping and climbing weeds of tea plantations in south Asia, and their chemical control with 2,4-D, 2,4,5-T or paraquat was not always safe or economic. Biological control of this weed may be possible but a thorough taxonomic revision and identification of the species involved is a prerequisite.

Watson (1986) had reported that for chemical control of *Pennisetum polystachion*, a problem grass in Sri Lankan tea plantations, most vulnerable growth stage was when the plants were taller than 45 cm and before flowering.

Such details in the context of many serious weed species of tea plantations may have to be generated and collected for better and integrated weed management.

BENEFITS OF WEEDS IN TEA CROP MANAGEMENT

There are some reports accounting the utility of weeds in tea plantations. Manipura (1971) and Sutidjo and Lubis (1971) had supported that in high rainfall area, leaving a non-competitive ground cover of soft weeds by selective weeding reduces the risk of soil erosion. Deori *et al.* (1997) had reported increase in root dry weight of tea (in pots) when treated with *Ageratum conyzoides* roots powder and rock phosphate.

In young china hybrid tea planted on slopes, mulch of grassy weeds effectively controlled weeds in the inter row spaces of tea and it was statistically comparable to LDPE mulch in terms of yield (Singh *et al.* 1993).

Sidhu (1985) had observed that for integrated control of the gryllid *Brachytrupes portentosus* on tea in north-east India, retaining of some weeds on ground during peak period of its activity in March-June was more effective as it formed an alternative food source for the pest.

ECONOMICS OF WEED MANAGEMENT

There are no ultimate statistics available regarding economics of weed control in tea, however certain reports indicate that depending on the intensity of infestation, weeds were estimated to cause yield losses between 10 and 25%. Weed control in tea by herbicides was found to be advantageous with a benefit cost ratio of 10 and even more (Sinha and Borthakur 1992). Ghosh and Ramakrishnan (1981) had concluded that the cost of weed management with herbicides was lower than that with the traditional garden practices (manual). Rattan (1996) had opined that mulches are effective but often pose practical and economic problems. On the other hand, mechanical control measures damage the roots of the tea.

SUGGESTED CULTURAL PRACTICES

For better weed management in tea, the following cultural practices are suggested. Infilling and inter-planting in tea gardens/ sections having wider spacing and/or high vacancy is very important step to improve the weed management and to increase production.

In case of chinary or china hybrid tea bushes, during knife cleaning out operation, immediately after pruning, the lower and thin shoots should be removed carefully, so that up to a height of 15-20 cm above ground measure no thin, non-effective and prostrate growing shoot should remain on the stubs. This will help to reach the weeds growing within the collar of bush, even in unpruned year at later stage.

REFERENCES

AICRPWC (1990a) Project Coordinator's Report - 1988, All India Coordinated Research Programme on Weed Control (ICAR/USDA), CRRI, Cuttack, India: 88.

AICRPWC (1990b) Project Coordinator's Report - 1989, All India Coordinated Research Programme on Weed Control (ICAR/USDA), NRCWS, Jabalpur (M.P.), India: 104.

Awasthi RC, Rao VS (1979) Economic evaluation of control of thatch grass (*Imperata*) and sour grass (*Paspalum*) weeds under experimental conditions in tea. *Two and a Bud* 26(1):15-17.

Basu SD (1974) Weed hosts of root-knot nematodes in tea estates around Jorhat. *Two and a Bud* 21(2): 50.

Basu SD (1975) Weed hosts of root-knot nematodes. *Two and a Bud* 22(2): 91.

Rao VS (1983) Principles of Weed Science. Oxford & IBH Publishing Co., New Delhi: 402-404.

Rao VS, Kotoky B (1980) Autumn application of pre-emergence herbicides in tea. *Two and a Bud* 27(2): 61-62.

Rao VS, Kotoky B (1981) Oxyfluorfen new pre-emergence herbicide for tea. *Two and a Bud* 28(2): 40-42.

Rao VS, Rahman F, Singh HS, Dutta AK, Saikia MC, Sharma SN, Phukan BC (1976) Effective weed control in tea by glyphosate. *Indian J. Weed Sci.* 8(1): 1-14.

Rao VS, Rahman F, Singh HS, Dutta AK, Saikia MC, Sharma SN, Phukan BC (1977) Effective weed control in tea by glyphosate. *Program and Abstracts of Papers, Weed Science Conference and Workshop in India, 1977*, Paper No. 96: 58.

Rao VS, Singh HS (1977) Effect of weed competition in young tea. *Proc. 28th Tocklai Biennial Conference, 1977*: 15-19.

Rattan PS (1996) Weed control in young tea. Quarterly Newsletter Tea Research Foundation of Central Africa. 124: 27-30.

Ronoprawiro S (1976) Control of mosses in tea. Proc. Fifth Asian-Pacific Weed Science Society Conference, Tokyo, Japan, 1975: 365-369.

Ronoprawiro S (1981) The possibility of using glyphosate to control mosses in tea. *Ilmu Pertanian* 3(1): 9-19.

Sandanam S, Jayasinghe HD (1977) Manual and chemical control of *Imperata cylindrica* on tea land in Sri Lanka. *PANS* 23(4): 421-426.

Sandanam S, Rajasingham CC (1982) Effects of mulching and cover crops on soil erosion and yield of young tea. *Tea Quarterly* 51(1): 21-26.

Sarkar SK, Chakravartee J, Basu SD (1983) Effect of soil stirring on yield of tea in heavy soil. *Two and a Bud* 30(1-2): 50-51.

Scarborough IP, Kayange CW (1977) Herbicides for young tea. *Quarterly Newsletter, Tea Research Foundation of Central Africa* 48:10-11.

Sharma VS (1975) Gramoxone-resistant weeds and their control. *Planters' Chronicle* 70(3): 59.

Sharma VS (1977) Chemical weed control in tea fields. *Planters' Chronicle* 72(2): 79-80.

Sharma VS, Satyanarayana N (1976) Recent developments in chemical weed control in tea fields. *UPASI Tea Science Department Bulletin 33*: 101-108.

Sharma VS, Satyanarayana N, Ramachandran K (1981) Changing pattern of weed flora in tea fields in south India. *Bull. Bot. Surv. India* 23(3&4): 121-125.

Sidhu HS (1985) Integrated control of cricket. *Two and a Bud* 32(1-2): 45-46.

Singh RD, Shanker Adarsh, Gulati Arvind (1992) Management strategies of weeds, insects and diseases in Kangra tea plantations. *Kangra Tea Festival Souvenir, June 5-7, 1992, Palampur (H.P.), India*: 13-14.

Singh RD, Singh Brajinder, Sud RK, Tamang MB, Chakrabarty DN (1997) Control of *Lantana* in non-cropped area and in tea plantation. *International Journal of Pest Management* 43(2): 145-147.

Singh RD, Sinha BK, Sud RK, Tamang MB, Chakrabarty DN (1994) Weed flora in tea plantations of Himachal Pradesh. *Journal of Economic & Taxonomic Botany* 18(2): 399-418.

Singh RD, Sohani SK, Singh Brajinder, Chakrabarty DN (1993) Influence of long term mulching on yield of young tea and weeds. *Proc. Int. Symp. Indian Soc. of Weed Sci., Nov. 18-20, 1993, Hisar (Haryana), India*, 3(S): 34-36.

Sinha MP (1985) A perspective of weed control in tea. *Two and a Bud* 32(1-2): 35-39.

Sinha MP, Borthakur B (1992) Important aspects of weed control in tea. *In*: Field Management in Tea (compilation of lectures), Tocklai Experimental Station, TRA: 134-139.

Smale PE (1991) New Zealand has its own green tea industry. *Horticulture in New Zealand* 2(2): 6-9.

Soedarsan A, Noormandias, Santika H (1976) Effects of some weed species on the growth of young tea. *Proc. Fifth Asian-Pacific Weed Science Society Conference, Tokyo, Japan, 1975*: 87-91.

Sutidjo K, Lubis JR (1971) Some results of experiments and practical use of herbicides in tea estates in North Sumatra. *Third Conference of the Asian-Pacific Weed Science Society* 7: 10.

Tabagari LG, Kopaliani R Sh (1984) Preliminary data on agrotechnical measures in the interrows of young tea plantations, cv. Kolkhida, on podzolic soil. *Subtropicheskie Kul'tury* 3: 77-79.

Tabagari LG, Mikeladze AD, Cheishvili, ID (1988) Results of preliminary studies on some agricultural measures for restoring neglected and unproductive tea plantations. *Subtropicheskie Kul'tury* 5: 67-75.

Tea Research Foundation of Central Africa (1975) Herbicides for young tea. *Quarterly Newsletter* 39: 10-12.

Thakur RS (1993) Personal communication with Mr. Rajinder Singh Thakur- a progressive tea planter of Khalet Tea Estate, Palampur, H.P., India.

TRA (1976) Annual Scientific Report 1974-75, TRA, Jorhat, Assam, India: 74.

TRA (1978) Annual Scientific Report 1977-78, TRA, Jorhat, Assam, India: 95.

UPASI (1978) Annual Report 1976, United Planters' Association of Southern India: 16-19.

serve as alternative hosts for pests as well as a refuge for natural enemies and hence are a major component of the tea ecosystem (Kawai 1997). Common insect and mite pests of tea are grouped in this chapter includes sucking, lepidopteran, coleopteran, soil and mite pests.

Some of the common pests of tea are as follows:

Sucking pests

1. Tea mosquito bug, *Helopeltis theivora, H. antonii, H. schoutedeni* (Hemiptera : Miridae)
2. Mealy bug, *Nipaecoccus viridis, N. vastator* (Hemiptera: Pseudococcidae)
3. Thrips, *Scirtothrips bispinosus, S. dorsalis, Heliothrips haemorrhoidalis, Mycterothrips setiventris* (Thysanoptera: Thripidae)
4. Aphid, *Toxoptera aurantii* (Homoptera : Aphididae)
5. Scale insects

   Soft scales (Homoptera:Coccidae)

   a. Green scale, *Coccus viridis*

   b. Brown scale, *Saissetia coffeae*

   c. *Ceroplastes* spp.

   Hard scales (Homoptera : Diaspididae)

   Aonidiella spp., *Aspidiotus destructor*
6. Leafhopper, *Empoasca flavescens, E. onukii, E. viti* (Homoptera : Cicadellidae)
7. Painted bug, *Bagrada cruciferarum* (Hemiptera: Pentatomidae)

Lepidopteran pests

1. Tea Tortricids/bud tier, *Homona coffearia, H. magnanima, Adoxophyes honmai* (Tortricidae)
2. Leaf roller, *Gracilaria theivora* (Gracillaridae)
3. Red borer, *Zeuzera coffeae* (Cossidae)
4. Black inch worm, *Hyposidra talaca* (Geometridae)

5. Twig caterpillar, *Ectropis bhurmitra* (Geometridae), Looper caterpillar, *Buzura suppressaria* (Geometridae)
6. Case worm, *Eumeta* spp., Bag worms, *Cryptothelea* spp.
7. Cluster caterpillars, *Euproctis pseudoconspersa* (Lymantriidae), *Andraca bipunctata* (Bombycidae)
8. Nettle caterpillar and saddle back caterpillar, *Aphendala recta, Cania bilinear, Chalcocelis albiguttatus* (Limacodidae)

Coleopteran pests

1. Shot hole borer, *Xyleborus fornicatus, Euvallacea fornicatus* (Scolytidae)
2. Cockchafer beetles/white grubs, *Holotrichia* spp., *Schizonycha* spp. (Scaraebidae)

Soil pests

1. Termites, *Microtermes* spp., *Odentotermes* spp., *Postelectrotermes militaris, Coptotermes* spp. (Isoptera : Termitidae)

Mite pests (Acarina)

1. Red spider mite, *Oligonychus coffeae* (Tetranychidae)
2. Scarlet mite, *Brevipalpus californicus, B. phoenicis* (Tenuipalpidae)
3. Pink mite, *Acaphylla theae* (Eriophyidae)
4. Purple mite, *Calacarus carinatus* (Eriophyidae)
5. Broad mite, *Polyphagotarsonemus latus* (Tarsonemidae)
6. Kanzawa spider mite, *Tetranychus kanzawai* (Tetranychidae)

Minor pests

1. Leaf miner, *Topicomyia theae* (Diptera: Agromyzidae)
2. Gross hopper, *Hieroglyphus banian* (Orthoptera : Acrididae)

This chapter describes the nature of damage, symptoms and management of economically important insect and mite pests of tea.

SUCKING PESTS

Tea mosquito bug (Hemiptera : Miridae)

About 41 species of mirids in the genus *Helopeltis* have been reported in Asia, Australia and Africa. Two species of *Helopeltis viz.*, *H. schoutedeni* and *H. theivora* have become the greatest enemies of tea planters in Africa and Asia causing 55% and 11-100% crop loss, respectively (Wilson and Clifford 1992, Sundararaju and Sundara Babu 1999, Hazarika *et al.* 2009). Both males and females look like mosquito. Young nymphs look a bit like ants, and range in color from copper or orange to green. Nymphs and adults are more visible in early and late hours of the day and take shelter under tea leaves especially in the lower frame during day time or when disturbed. The nymphs often drops from the plant when disturbed. Nymphs and adults suck the sap from tender stems, young leaves and buds forming circular feeding punctures. The feeding spots develop watermarked area, which turn circular and pale green and subsequently dark brown within hours of feeding. Damaged leaves curl upward and desiccate in severe infestations. Initiation of new shoots is prevented due to the death of the stem and may result in total loss of the crop; *H. schoutedeni* causes dieback and stem canker. Female lay eggs singly or in groups of 2 to 3 eggs, inside tender green stems. It lays 4 to 10 eggs per day, total 220 to 224 eggs in her lifetime. The egg is oval, white in color with two long hairs of unequal length at the smaller end of the egg. Incubation period varies from 4 to 20 days and the life cycle is completed within 12 to 40 days. The species has many overlapping generations. However, the population peaks from June to September, coinciding with the rainy season.

Management

a. Plucking every 10-15 days will be helpful for reducing the infestation as the eggs are removed during plucking. Try to pluck out all the badly damaged buds which will stimulate the growth of new shoots. Light or moderate pruning will limit the population in the seriously affected fields.

b. Weed and alternate host plants (wild plants having feeding stains) should be removed from and around the tea fields. Regulation of shade is important as the heavily shaded and moist areas is relatively more damaged.

c. Encourage natural enemies (spiders, dragonflies, praying mantis, predatory coccinellids) to reduce the pest population.

d. Spray thiamethoxam 25% WG @ 1 g/L or profenofos 50% EC @ 2 ml/L during severe infestation.

Mealy bug (Pseudococcidae : Hemiptera)

N. viridis is widespread throughout the tropics and subtropics, attacking numerous plant species and often causing considerable damage (Clausen 1978, Sharaf and Meyerdirk 1987). It is an economically important pest of tea. Besides other tea growing regions of the country, this pest was reported in Kangra valley of Himachal Pradesh (Shanker *et al.* 2002, Sharma and Kashyap 2002, Gupta and Shanker 2007). Mealy bugs are covered with white powdery substances and waxy filaments. Females are soft bodied sluggish oval, flattened and dark brown colour. Males undergo four and female three nymphal instars. Both nymphs and adults congregate on shoot, stem, branches and suck the sap which results in drying of branches. In severe cases, the mealy bugs secrete honeydew on the plant parts from which black sooty mould develops, which affects the photosynthesis and quality of tea leaves. In severe cases it leads to death of the tea plants. Juvenile mealybugs can crawl from an infected plant to another plant and small 'crawlers' are readily transported by wind, rain, birds, ants, clothing, vehicle and may settle in cracks and crevices, usually on new plants. Ants, attracted by the honeydew carry mealybugs from plant to plant helping in their spreading and invading into new areas and fields. Plant protection products are not very effective in controlling mealy bugs because of waxy covering on their body and their habit of hiding in crevices. Systemic insecticides are found effective and used to control during severe infestation.

Management

a. Destroy ant colonies during land preparation because their nests are located near the soil surface or drenching soil with chlorpyriphos 20 EC @ 4 ml/L or apply 5% malathion dust @ 25 kg/ha.

b. Remove weeds and alternate host plants like *Hibiscus* spp., okra, custard apple, guava, etc., and destroy.

c. Sanitize equipments before moving to uninfected portions in a crop.

d. Spraying with strong jet of water may help in removing the bugs.

e. Apply sticky bands like Track-trap or Bird Tangle Foot on arms or main stem to prevent crawlers from reaching the bunch.

f. Encourage predatory coccinellid beetles *viz.*, *Cheilomenes sexmaculata*, *Rodolia fumida*, *Scymnus coccivora* and *Nephus regularis* to reduce infestation.

g. Release Australian ladybird, *Cryptolaemus montrouzieri* and *Anagyrus pseudococci*, *Leptomastix dactylopii*, *Hypoaspis* spp. to suppress mealy bugs population.

h. Spray commercially available entomopathogenic fungus, *Lecanicillium lecanii* (5 g/L) to reduce mealy bug population at weekly intervels.

i. Spray chlorpyriphos 20 EC @ 2 ml/L or imidacloprid 200 SL @ 0.5 ml/L at 15 days intervals during severe infestation.

Thrips (Thysanoptera : Thripidae)

S. dorsalis is highly polyphagous minute insects. It is present in tropical Asia (from Pakistan to Japan), Oceania, Subsaharan Africa, Venezuela and Central America. Wings are narrow and fringed (PRA *Scirtothrips dorsalis*, 2009). Thrips prefer tender/mature leaves, buds and flowers. It undergoes through five developmental phases: egg, two active larval instars that feed, followed by two relatively inactive pupal instars and winged, feeding adults. The first and second larval stages are found on the green plant parts from which the second stage larvae seek out some sheltered place (leaf litter or crevices of bark) and then pass through two resting stages called pre-pupa and pupa, respectively. Winged adults, male and female, are found normally on the green plant parts, where they feed. Eggs are inserted into young and soft tissues of leaves, stems and fruit. Eggs hatch after 7-8 days. Both nymphs and adults lacerate and suck the sap on leaves, buds and flowers. The underside of the affected leaves become silvery with black spots. Feeding marks on the unopened buds and parallel brown lines on the leaves which affect the quality. The growing shoots are stunted with shortened internodes and defoliation often occurs.

Management

a. Yellow/blue sticky traps (10-15 traps/ha) to be installed at crop canopy level for monitoring and mass trapping of thrips.

b. Timely pruning of tea bushes to be followed to avoid thrips infestation.

c. Spray neem formulation containing azadirachtin 1% @ 3 ml/L or azadirachtin 5% @ 1 ml/L.

d. Spray deltamethrin 2.8% EC @ 0.3 ml/L or profenofos 50% EC @ 2 ml/ L in severe infestation.

Black citrus aphid/tea aphid (Hemiptera : Aphididae)

Tea aphids are minute, dark brown to black insects which infests tender shoot and foliage of the tea plant (Gupta and Shanker 2007, Devi *et al.* 2010). Aphids seen on tender shoots of tea immediately after pruning. Both nymphs and adults found in the colonies on tender buds, leaves, shoots and suck the sap leads to reduction of growth. Continuous feeding makes the leaf wrinkled forming a boat shaped structure. Severe infestation leads to delayed recovery of the plant leading to poor crop productivity (Muraleedharan 1991). In severe cases, aphid excretes honeydew on the leaves which results in the development of black sooty mould, which affects the photosynthesis of the plant. Presence of honeydew and ants on the plants indicate the aphid infestation.

Management

a. Collect and destroy the affected parts of plants to prevent the further spread.

b. Encourage predatory coccinellids beetles to reduce the population.

c. Spray any commercial botanicals insecticides containing azadirachtin 5% (1 ml/L)/ or azadirachtin 1% (3 ml/L) or neem seed kernel extract 4%.

d. Spray phosalone 35% EC @ 2 ml/L in severe infestation.

Scale insects (Hemiptera : Coccidae)

Scale infestation is usually localized. They tend to remain confined to the individual bush and start feeding by sucking the sap by inserting their mouth parts. Feeding may lead to severe defoliation, stunting and dieback on young tea. However, the effects may not be very serious in case of mature tea. Soft scales feed on green stem or on the underside of the leaves mainly along the midrib, and produce honeydew. Sooty mould grows on the honeydew which affects the photosynthesis of the plant. Yellow feeding spots appear on the upper side of the leaves, the damaged leaves eventually drop off. The ants feeding on honeydew are sometimes the only sign of their presence.

Management

Scales insects are some of the most difficult pests to control because their scales protect them from contact with chemicals and their mouth parts are inserted in the plant tissue.

a. Select planting material free from scale insects from mother plants.

b. Spray ethion 50% EC @1 ml/L in severe infestation.

LEPIDOPTERAN PESTS

Tea tortricids (Lepidoptera: Tortricidae)

Among the leaf feeders, 19 species of tortricids mostly of the genera *Homona* and *Adoxophyes* are economically important pests of tea in Japan, China, India, Sri Lanka, Taiwan, Turkey, Republic of Georgia, Azerbaijan and Bangladesh (Cranham 1966, Nabeta *et al.* 2005). *Adoxophyes honmai* is a major pest of tea in central and southern Japan, whereas, *Homona magnanima* is prevalent in southern Japan. *H. coffearia* occurs in India, Indonesia and Sri Lanka; it causes 50% yield reduction in Sri Lanka (Hazarika *et al.* 2009).

Female of *A. honmai* lays oval shaped egg masses on the under surface of young tea leaves. Eggs hatch in pale yellow coloured neonates which disperse individually to construct leaf webs for feeding on the growing leaves and shoots. They pupate inside the web after completing five to six instars. *H. magnanima* has a seasonal occurrence similar to that of *A. honmai*, with four generations per year in central Japan. No development occurs below 10.3°C and 417 degree days are required to complete development from egg to adult emergence (Nabeta *et al.* 2005).

Leaf roller (Lepidoptera : Gracillaridae)

This is a minor pest of tea. Caterpillars are dark-green in colour with shiny brownish black head. Caterpillars roll the leaves and feed, and can be seen inside the rolled and tied leaves. In severe cases, it causes complete defoliation and drying of the plant. Some buds cannot be plucked because they are ruined by leaf nests.

Red borer (Lepidoptera:Cossidae)

The pest is recognized by the presence large pieces of sawdust, on the ground under the bush. This is a minor pest. Moths have white wings with many black spots. Eggs are laid like beads on a thread. Mature larva is about 3.5 cm long, purplish or reddish brown in colour. Larval duration is 4-5 months and it pupates in a special chamber for a month. Caterpillars emerge in 10 days. Larvae bore into young stems and tunnel downwards, devouring the woody parts, especially the pith. Tunnels are extended to thicker branches. As the larva grows, the tunnel is also extended which may run even up to root. Holes are made at intervals to eject the excreta and wood particles.

Management of Lepidopteron insect pests

a. Collection and destruction of egg mass and late instar larvae.

b. For red borer, cut the affected branches and destroy.

c. Release egg parasitoid, *Trichogramma chilonis* for six releases @ 20,000 parasitoids/release (1 tricho card) at fort nightly interval depending upon the incidence of pest.

d. Spray neem based formulation containing azadirachtin 5% @ 1 ml/L or profenofos 50% EC @ 2 ml/L or deltamethrin 2.8% EC @ 0.3 ml/L for the control of different caterpillars.

e. Spray deltamethrin 2.8% EC @ 1 ml/L against leaf roller and @ 0.25 ml/L against loopers.

COLEOPTERAN PESTS

Shot hole borer (Coleoptera : Scolytidae)

Eight to nine species of scolytids attack tea, of which *Xyleborus fornicatus* is the most serious pest in Sri Lanka, causing up to 91 to 100% infestation in the mid country wet and dry zones. Female lays

Clausen CP (1978) In: Introduced parasites and predators of arthropod pests and weeds: a world review. United States Department of Agriculture, Washington, DC. 545 pp.

Cranham JE (1966) Tea pests and their control. *Annu Rev Entomol.* 11: 491-514.

Devi KD, Maisnam S, Varatharajan R (2010) Density, diversity and differential feeding potentials of aphidophagous insects in the tea ecosystem. *J Biopest.* 3(1): 58-61.

Gotoh T, Nagata T (2001) Development and reproduction of *Oligonychus coffeae* (Acarina : Tetranychidae) on tea. *Int J of Acarol.* 27: 293-98.

Gupta M, Shanker A (2007) Bioefficacy of imidacloprid and acetamiprid against *Nippaecoccus vastator* and *Toxoptera aurantii* in tea. *Appl Entomol.* 21(2): 75-78.

Hazarika LK, Bhuyan M, Hazarika BN (2009) Insect pests of tea and their management. *Annu Rev Entomol.* 54: 267-284.

Ho CC (2000) Spider-mite problems and control in Taiwan. *Exp Appl Acarol.* 24: 453–62.

Kawai A (1997) Prospect for integrated pest management in tea cultivation in Japan. *Jpn Agr Res Q*. 31:213-17.

Kumar A, Thakur M, Nigam SK, Nadda G, Shanker A (2004) Omite bioefficacy against scarlet mote, *Brevipalpus phoenicis* (Geijskes) in Kangra Tea. *Pestology* 28(2): 21-25.

Muraleedharan N (1991) Pest Management in Tea. UPASI Tea Research Institute, Valparai, Coimbaore, pp130.

Nabeta FH, Nakai M, Kunimi Y (2005) Effects of temperature and photoperiod on the development and reproduction of *Adoxophyes honmai* (Lepidoptera: Tortricidae). *Appl Entomol Zool.* 40: 231-38.

NCIPM, Delhi, http://www.ncipm.org.in/

PRA *Scirtothrips dorsalis,* May 2009, Plant Protection Service, the Netherlands.

Shanker A, Sood C, Kumar V, Ravindranath SD (2002) Insect and mite pests attacking tea plantation of Kangra Valley and their management. *Indian J Entomol.* 64(1): 53-57.

Sharaf NW, Meyerdirk DE (1987) A review on the biology, ecology and control of *Nipaecoccus viridis* (Homoptera: Pseudococcidae). *Miscellaneous Publications of the Entomological Society of America* 66: 1-18.

Sharma DC, Kashyap NP (2002) Impact of pesticidal sporay on seasonal availability of natural predators and parasitoids in the tea ecosystem. *J Biol Control.* 16(1): 31-35.

Sundararaju D, Sundara Babu PC (1999) *Helopeltis* spp. (Heteroptera: Miridae) and their management in plantation and horticultural crops of India. *J Plant Crops.* 27: 155-74.

Takafuji A, Ozawa A, Nemoto H, Gotoh T (2000) Spider mites of Japan: their biology and control. *Exp and Appl Acarol* 24: 319-35.

Tea Board of India, http://www.teaboard.gov.in

TNAU agritech portal, http://agritech.tnau.ac.in/

Tocklai, http://www.tocklai.net/Cultivation/pests.aspx

UPASI, http://www.upasitearesearch.org

Wilson KC, Clifford MN (1992) Tea: Cultivation to Consumption. London: Chapman & Hall, pp 769.

15

PESTICIDE RESIDUES IN TEA

Yogesh B. Pakade* and Adarsh Shanker

CSIR-Institute of Himalayan Bioresource Technology,
Palampur (H.P.)-176 061, India. *E-mail: yogesh@ihbt.res.in

INTRODUCTION

Tea crop is subjected to attack from a wide range of pests. About 1034 species of arthropods and 82 species of nematodes are active in tea fields (Chen and Chen 1989). Pest incidence in tea is far more complex and more damaging in the Asian region (India, Sri Lanka, Bangladesh, Indonesia and China). The damage caused by different pests on tea crop depends on the habitat locality and weather factors. Pest damage in tea often leads to a significant impact on production and productivity both in short and long term, the latter leading to almost total debilitation and thus eroding the capital assets itself. Despite crop loss, pest infestation also adversely affects the quality of processed tea. To minimize the pest infestation on tea crop and for reliable economic return, the use of chemical pesticides has been the most dominating practice in almost all tea growing countries for nearly three decades. Extensive and indiscriminate use of broad-spectrum, long-persistent pesticides causes widespread contamination of environment (soil, water and air) and human health. Pesticides residues in tea cause special concern as over one and a half billion cups of tea are consumed daily all over the world. Tea is said to be the second most consumed drink after water (Singh 2002). The tea crop has unique features; it is perennial and grown mostly as a monoculture. The surface area per unit weight of tea is large and the interval between pesticide application and harvest is short. The level of pesticide residues on tea at the same application rate will usually be higher than in other crops (Chen and Wan 1988). To prevent any health problem to consumers, various international authorities have regulated the use of pesticides by fixing maximum residues levels (MRLs) for commercial purposes (Jaggi *et al.* 2001). Legislations were enacted to regulate the residues in tea crop (Hoff and Zoonen 1999).

MRLs of a group of pesticide have been established and reviewed periodically by different tea producing countries and various International agencies (Table 15.1). MRLs are fixed based on the bioeffacacy field trial data for a particular pesticide to arrive at the highest residue levels expected under use according to Good Agricultural Practice (GAP). The lack of uniformity in MRLs and in discrepancy in legislation against the use of pesticides in tea crop is a serious problem in view of environment, health hazards and also in evolving guidelines for global tea trade for management of the pesticide residues.

Table 15.1. Maximum residue limits (MRL) of pesticides used in tea plantations prescribed by various agencies / countries

Pesticides	MRLs (mg/kg)										
	1	2	3	4	5	6	7	8	9	10	11
Bromopropylate	5.0	-	5.0*	5.0	3.0	1.5	-	5.0	-	-	-
Captan	-	-	15.0*	-	15.0	15.0	-	-	-	-	-
Cartap	2.0	-	-	-	20.0	0.1	-	-	20.0	20.0	-
Chlorbenzilate	-	-	1.5	-	-	1.5	-	-	-	-	-
Chlorfenvinphos	-	-	1.0	0.2	0.5	0.3	-	-	-	-	-
Chlopyrifos methyl	0.1	-	0.1*	0.1	0.2	0.1	-	0.1	-	-	2.0*
Cyhexatin	2.0	-	2.0*	2.0	-	1.5	-	2.0	-	-	-
Cypermethrin	-	-	-	-	-	-	3.0	20.0	0.5	20.0	10.0*
Deltamethrin-	-	-	-	-	-	-	2.0	10.0	5.0	10.0	5.0*
Diazinon	-	-	0.3	0.1	0.5	0.5	-	-	-	-	-
Dichlorvos	-	-	2.0	-	2.0	0.3	0.1	-	0.1	-	-
Dicofol	8.0	45.0	2.0*	5.0	2.0	2.0	-	5.0	0.1	-	20.0*
Dimethoate	-	-	1.0	0.005	1.0	0.5	0.1	-	0.20	-	0.5*
Dinabuton	-	24.0	1.0*	-	-	-	-	-	-	-	-
Endosulfan	30.0	10.0	30.0*	30.0	3.0	0.1	-	30.0	-	-	20.0*

Ethion	5.0	-	2.0*	-	0.1	1.0	-	7.0	2.0	-	5.0*
Fenazaquin	-	-	-	-	-	-	-	-	-	-	10.0*
Fenitrothion	0.5	-	0.5	0.5	0.5	0.2	0.3	0.5	0.1	-	-
Fenvalerate	-	-	-	-	-	-	2.0	-	-	-	-
Hucythrinate	-	-	-	-	-	-	-	20.0	-	-	-
Heptachlor	-	-	0.10*	-	-	-	-	-	-	-	-
Hexachlorobenzene	-	-	-	-	-	-	-	-	0.01*	-	-
Methidathion	0.1	-	2.0*	0.1	0.2	0.5	-	0.1	0.1	-	-
Malathion	-	3.0	-	8.0	0.5	1.0	-	0.1	-	-	3.0*
Monocrotophos	-	-	5.0	-	0.05	0.2	-	-	-	-	-
Permethrin	-	-	-	-	-	-	3.0	20.0	2.0	20.0	-
Parathion	-	-	0.5	0.1	0.5	0.5	-	-	-	-	-
Parathion methyl	0.2	-	1.0	0.2	0.2	0.5	-	0.2	-	0.2	1.0*
Phosalone	-	8.0	2.0	-	1.0	2.0	-	-	-	-	-
Phoxim	-	-	-	-	-	-	0.5	-	0.1	-	-
Piperoxyl butoxide	-	-	3.0*	-	3.0	3.0	-	-	-	-	-
Propargite	-	-	5.0*	-	2.0	1.5	-	10.0	0.1	-	5.0*
Quinalphos	-	-	0.20*	-	-	-	0.2	-	0.1	-	3.0*

1=CODEX (commission appointed by WHO and FAO to determine pesticide residue level), 2= Environmental Protection Agency, USA, 3= Germany (*Specific for tea), 4= The Netherlands, 5= Italy, 6= Switzerland, 7= Peoples Republic of China, 8 = FAO / WO, 9 = Council of European Communities ("Indicates lower limit of analytical determination), 10 =Japan, 11 = India

Although it is desirable that tea is kept completely free from pesticide, it is not practically feasible in view of the pest and disease problems which cause 10-20% of annual yield losses (Chen and Chen 1999) and the economy of tea production. Therefore, the use of pesticide in tea has been an essential input of any pest management system. Depending on their toxicity, pesticides may cause risks to both human health and the environment. Pesticides used for tea plantation must not be phytotoxic and should in no way affect the aroma of tea. Residues of applied pesticides in brew should be below the quantities that create consumer hazards. In order to minimize the pesticide residue, it is important to understand the fate of pesticides applied on tea crop.

FATE OF PESTICIDES

Tea is perennial crop and the tea bush is applied with pesticides periodically in a year for management of pests. The canopy of the tea plant constitutes a larger surface area than other crops. Most of the pesticides after spray retained on the tea leaf/shoot and constitute the initial deposit. The lipophyllic pesticides, e.g. chlorinated hydrocarbons, synthetic pyrethroids some of the organophosphates and carbamate deposited on the waxy layer of the leaf, where as many of the systemic organophosphate and carbamate compounds penetrate into leaf surface rather quickly. Consequently these chemicals are deposited on the young shoot and mature foliage. Most of systemic pesticides are translocated to the entire parts of the plant (Chen *et al.* 1987). Large percentage of pesticide applied on the tea plants is lost at field level. Degradation of most of the pesticides on plant surface follows first order kinetics. The factors involved in the degradation of pesticides on tea plant include growth dilution, photodegradation, evaporation, pH and elution due to rainfall.

Growth dilution: At any time of pesticide application, foliage and shoots of the plant remain at different stages of growth. The weight of these immature shoots increases over a period of 6-14 days, depending on the plucking interval. It is estimated that approximately the weight of all shoots increases 13 times in a week and 80 times in two weeks (Muraleedhran 1994). Therefore, the weight of a bud increases during the plucking intervals and this helps in brining down the residues of pesticides. The immature buds by the time attain the size of pluckable shoots (2/3 leaves and a bud), the residues of applied pesticides on them will decrease considerably (Agnihotrudu and Muraleedharan 1990). The younger the tea shoots, the faster the growth rate and the larger contribution of growth dilution to the pesticide degradation on tea shoots. Studies indicate (Chen and Wan 1988, Hajara *et al.* 2000) that 45% pesticide residues are diluting due to growth factor. In case of organochlorine and pyrethroid pesticides, growth dilution contributes around 40-45% (Xue and Chen 1992).

Photodegradation: All pesticides are susceptible to photodegradation to some degree. The intensity and spectrum of sunlight, length of exposure, and properties of the pesticide affect the rate of pesticides photodegradation. The energy source of photodegradation (Hogenboom *et al.* 1999, Sanz-Asencio *et al.* 1999) is mainly provided from the absorption of photons from solar radiation. Thus the intensity of light received by the canopy applied with pesticides is an important factor which determines the residues in tea. During dry period, the overexposure to sunlight might cause considerable degradation of pesticide residue in tea shoots. The mechanism for most photochemical reactions involves the hemolytic cleavage of covalent bonds. For example in

Table 15.3. Percent of pesticide residue left during orthodox black tea manufacture

Pesticide	Green tea leaves	Step 1 Withering	Step II Rolling	Step III Fermentation	Step IV Drying
Dimethoate	100	52.0 ± 3.2	48.0 ± 1.6	48.0 ± 3.3	30.5 ± 2.4
Quinalphos	100	42.5 ± 2.6	41.6 ± 2.2	42.8 ± 5.0	36.0 ± 3.0
Dicofol	100	66.0 ± 6.0	64.0 ± 4.5	64.9 ± 6.2	48.0 ± 2.8
Deltamethrin	100	84.0 ± 4.0	84.2 ± 4.2	83.2 ± 4.8	68.0 ± 4.0

Source: Sood *et al.*(2004)

FATE OF PESTICIDES RESIDUE DURING PROCESSING

It is evident from many studies that during processing and cooking of food items, pesticide residues are lost (Coulibaly and Smith 1994, Nagayama 1996, Cabras *et al.* 1997). Green tea leaves (two and a bud), after plucking, are subjected to manufacturing process that involves steps like withering, rolling, fermentation and drying (Figure 15.1). There are chances of loss of pesticide during processing of tea. Tea leaves are withered under gentle hot air stream resulting in a major weight loss and further subjected to mechanical rolling, fermentation and again rolling. Tea is finally dried at a temperature higher than 100°C, at this temperature; thermal decomposition of some of the pesticides is high. Further, the drying process produces a concentration factor of 4 in tea. Therefore, theoretically the pesticide residue in tea shoots are the harvest time could increase by the same factor in made tea. However, studies showed that the residue in made tea processed from the corresponding sample of green shoots did not correspond/match to the theoretically expected value and account for the concentration of residue lost during processing (Chen and Wan 1988, Jaggi *et al.* 2000). Sood *et al.* (2004) reported that considerable amount of residues of most commonly used pesticides present in the black and green tea leaves is lost during the processing of tea (Table 15.3 & 15.4). Degradation of pesticides in green leaf takes place at the time of manufacturing mainly through evaporation; condense distill-ation and thermal decomposition (Chen and Wan 1988, Jaggi *et al.* 2000, Kumar *et al.* 2003). The residue reduction could be highest during drying as it might facilitate pesticide distillation during water evaporation (Maguire 1991, Sood *et al.* 2004). Each pesticide could be distilled in different amount in accordance with their physico-chemical properties like vapour pressure. Chen and Wan (1988) reported that 30-40% of residue is lost during processing of drying. Dehydration during green tea processing reduced the residues further to 39.8% (Sood *et al.* 2004). It has been reported that higher the vapour pressure, the greater the degradation of pesticides during processing (Chen and Wan 1988, Cabras *et al.* 1988a, Cabras *et al.* 1998b). During the drying process the water contained in the tissue could entertain pesticide molecules while heat could cause evaporation and degradation.

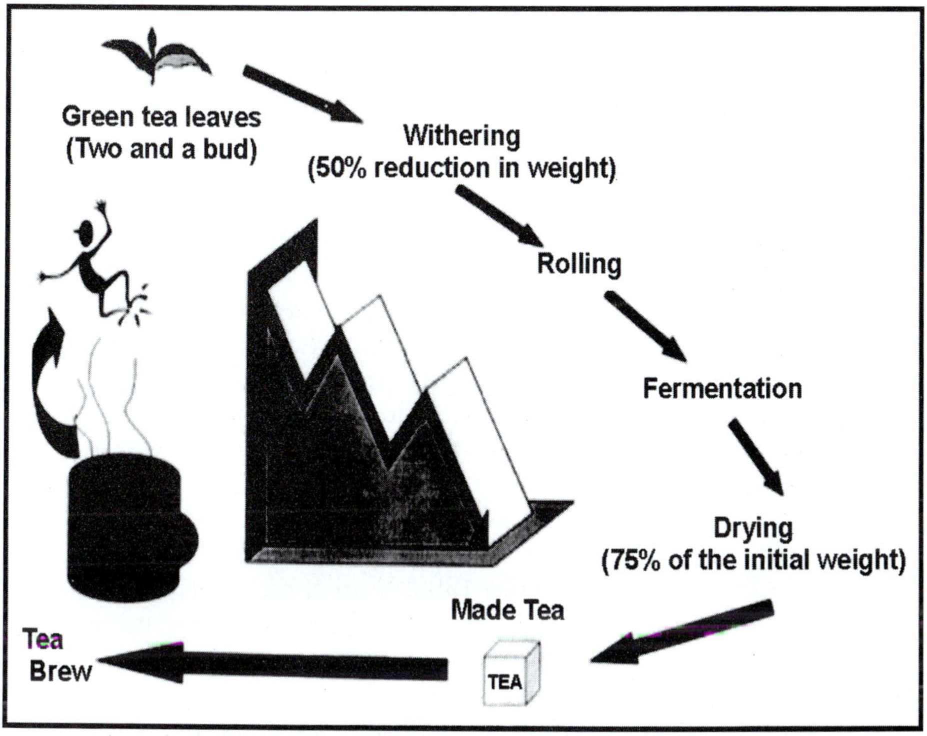

Fig. 15.1. Dissipation of pesticide during tea manufacturing process

Table 15.4. Percentage of pesticide residue left during green tea manufacture

Pesticide	Green tea leaves	Step 1 Withering	Step II Rolling	Step III Fermentation	Step IV Drying
Dimethoate	100	45.0 ± 2.2	39.8 ± 2.4	40.6 ± 1.8	23.4 ± 0.8
Quinalphos	100	46.0 ± 4.6	36.6 ± 1.4	36.2 ± 1.0	28.6 ± 1.5
Dicofol	100	50.0 ± 5.5	47.0 ± 3.8	48.2 ± 3.0	36.6 ± 2.0
Deltamethrin	100	76.0 ± 7.0	70.5 ± 4.4	68.8 ± 4.5	56.2 ± 3.2

Source: Sood *et al.* (2004)

FATE OF PESTICIDES DURING INFUSION

Processed tea, when infused with boiling water extract, smaller quantities of residues is transfer in the tea liquor. Generally, 3-25% of pesticides are transferred in brew which actually constitutes the dietary intake. The extent of pesticides transfer to the brew may be attributed to their water solubility (Table 15.5). Generally, those pesticides with high water solubility may potentially be transferred to the tea cup in significant amounts. The extraction rate of pesticide residues in water depends on their water solubility (Nagayama 1996, Wan *et al.* 1991) and the leaching ratio varies with pesticides but not with the residue concentration (Nagayama, 1996, Jaggi *et al.* 2001). Singh *et al.* (2001) reported the negligible transfer in infusion of water-insoluble pesticides such as organochlorine, synthetic pyrethroid and some water insoluble organophosphates. The organophosphates with relatively higher solubility in water were transfer to infusion in significant quantity. Though, water solubility plays an important role, the transfer of pesticides in brew may be explained by a relationship between water solubility and partition coefficient (K_{ow}) (Table 15.5) wherein pesticide with large K_{ow} entered the tissue strongly combining with it and not with the circulating water. Thus there is less transfer of pesticides in the water extracts on made tea. The pesticides in tea leaves are partitioned between infusion to residues, the extent depending on the water solubility (Zimnierii and Blasé 1982, Zongmao and Haibin 1988, Wan *et al.* 1991, Singh 2001). Transfer of pesticide in brew increased in water may increase in brewing time (Table 15.6) and also the concentration of pesticides in water may increase or decrease depending upon its fugacity and volatility proportions (Sunitio *et al.* 1988, Jaggi *et al.* 2001). The solubility is directly temperature dependent, as a result, the water insoluble pesticides like synthetic pyrethroid showed their transfer at a longer duration of brewing (Jaggi *et al.* 2001). Chen and Wan (1983) have observed that the rate of loss of organophosphates in between 30 to 70% due to their high vapour pressure. Further, the chemicals with high molecular weight are generally less soluble in hot water during the brewing time (2-3 minutes). Increased brewing time and continuous heating may aid in releasing the pesticides from the tissues.

Table 15.5. Relationships of physical properties with transfer of pesticides in tea brew

Pesticides	Solubility (mg/L)	K_{ow}	% transfer to infusion
Chloropyrifos	2.0	16255.5	3.14 ± 0.06
Ethion	Insoluble	N.A.	0.8 ± 0.02
Malathion	145.0	156.0	12.14 ± 0.8
Methyl Parathion	55.60	447.4	9.96 ± 1.0
Dimethoate	25,000	0.6	25.8 ± 2.23
Monocrotophos	10,000	1.6	19.78 ± 3.76
Phosphamidon	Completely miscible	-	33.0 ± 0.76
Quinalphos	22.0	1208	8.04 ± 0.36
Endosulfan	Insoluble	N.A.	1.80 ± 0.06
Dicofol	0.8	43954.0	0.10 ± 0.01
Deltamethrin	0.002	34506.0	-
Cypermethrin	1.0	34506.4	0.06 ± 0.0
Permethrin	0.2	34434.2	-

Source: Jaggi *et al.* (2001) N.A.: Not applicable

Table 15.6. Percent transfer of pesticides to tea brew at different brewing times

Pesticide	Percent transfer of pesticides to tea brew		
	2 min	5 min	10 min
Chloropyrifos	4.0± 0.75	6.7 ± 1.2	10.9 ±1.4
Ethion	0.7 ± 0.14	0.9 ± 0.8	1.16 ± 0.02
Malathion	10.2 ± 3.0	13.38 ± 1.0	13.8 ± 1.0
Methyl Parathion	10.9 ± 1.8	14.4 ± 1.2	18.26 ± 0.6
Dimethoate	26.2 ± 1.2	30.12 ± 2.0	32.5 ± 0.8
Monocrotophos	20.0 ± 2.5	26.12 ±1.6	26.0 ± 0.92
Phosphamidon	35.5 ± 4.2	38.0 ± 0.6	40.2 ±0.2
Quinalphos	10.0 ± 1.6	11.5 ± 0.2	14.8 ±1.6
Endosulfan	1.5 ± 0.32	0.5 ± 0.06	0.0
Dicofol	0.1 ± 0.02	1.28 ± 0.02	1.8 ± 0.08
Deltamethrin	0.0	0.24 ± 0.00	0.28 ± 0.0
Cypermethrin	0.1 ± 0.22	0.4 ± 0.02	0.32 ± 0.06
Permethrin	0.0	0.44 ± 0.00	0.48 ± 0.0

Source: Jaggi *et al.* (2001)

Health Organization (WHO) has established MRL for pesticides for variety of foods. Recent focus on pesticide residues on processed tea has lead issuance of Maximum Residue Limits (MRLs) on a group of pesticides by different producing and consuming countries and various International agencies (Table 15.1). Presently these MRLs are fixed for made tea (processed dry tea leaves) whereas, in view of the fact that made tea is not consumed as such and the consumable component is infusion (tea brew), it is therefore, pesticide transferred to tea brew (hot water extract) that should actually be considered for the fixation of realistic MRL.

CONCLUSIONS

It is indeed desirable that tea should be kept completely free from pesticide residues but this is impractical in view of the serious pest and disease problems and the economy of tea production. Many agencies/organizations aim at stricter controls on pesticide use. This will help to reduce these risks, but in the short term this could be at the expense of income earning opportunities for tea plantation workers and tea producing countries. The only alternative is to maintain the residues of pesticides much below the MRL. The consciousness about the pesticide residues in tea calls for immediate attention of the tea growers regarding restricted and judicious application of insecticides, herbicides, fungicides etc. in the plantation. The use of persistence broad spectrum pesticides should be discouraged and if required they should be used with utmost care. The use of environmental friendly (biopesticides) pesticides and newer proved formulations of existing pesticides with much lesser dosages ought to be incorporated in the entire pest management programme. If a new compound or formulation is included in the IPM system, the residue should be monitored continuously. Studies reveal that very small fraction of pesticides from made tea (maximum 40%, in case of water soluble compounds) get transferred to the brew. Thus MRLs should be based on the pesticides present in brew which is actually consumed rather than made tea. Generation of data on transfer of pesticides in tea brew and continuous monitoring of pesticide residues in tea samples are the imperative needs.

REFERENCES

Abou-Assaf N and Coats JR (1987) Degradation of [^{14}C] Isofenphos in soil in the laboratory under different soil pH's, Temperature's and moistures. *J. Environ Sci Health* B22 (3): 285-301

Agnihothrudu V, Muraleedharan N (1990) Pesticide residues in tea. *Planters Chronical* 85: 125-127.

Anonymous (1993) Anonymous Technical Bulletin of Fenazaquin, Department of Regulatory, Toxicology and Environmental Affairs Agricultural Products Research and Development. Dow Elanco European Area, Letcombe Laboratory, Letcombe Regis, Oxfordshire. OX 129 JT. United Kingdom.

Barrionuevo WR, Lancas FM (2001) Solid-phase extraction (SPE) and solid-phase microextraction of pyrethroids in water. *Quimica Nova* 24: 172-175.

Barrionuevo WR, Lancas FM (2002) Comparison of liquid–liquid extraction (LLE), solid-phase extraction (SPE) and solid-phase microextraction (SPME) for pyrethroid pesticides analysis from enriched river water. *Bull Environ Contam Toxicol.* 69: 123-128.

Cabras P, Angioni A, Garau VL, Melis M, Pirsi FM, Cabitza F and Cubeddu M (1997) Pesticides in distilled spirits of wine and its by products. *J Agric Food Chem.* 45: 2248-2251.

Cabras P, Angioni A, Garau VL, Pirisi FM, Brandolini V, Cabitza F, Cubeddu M (1998a) Pesticide residues in prune processing. *J Agri Food Chem.* 46: 3772-3774.

Cabras P, Angioni A, Garau VL, Pirisi FM, Espinoza J, Mendoza A, Cabitza F, Pala M, Brandolini V (1998b) Fate of azoxystrobin, fluazinam, kresoxim-methyl, mepanipyrim, and tetraconazole from vine to wine. *J Agri Food Chem.* 46: 3249-3251.

Chen and Chen (1989) An analysis of the world tea pest fauna. *J Tea Sci.* 9(1):13-22.

Chen Z and Chen X (1999) Plant protection in the sustainable development of the tea industry. *J Tea Sci.* 19:1-6.

Chen Z Y, Zhu Q Y, Tsang D, Huang Y (2001) Degradation of green tea catechins in tea drinks. *J Agric Food Chem.* 49: 477-482.

Chen ZM and Wan HB (1997) Degradation of pesticides on plant surfaces and its prediction - a case study on tea plant. *Environ Monito Assess.* 44:1303-1313.

Chen ZM, Wan HB (1983) Degradation of pesticide on plant surfaces and its prediction - a case study on tea plant. *Environ Monitor Assess.* 44: 303-313

Chen ZM, Wan HB (1988) Factors affecting residues of pesticides in tea. *Pesticide Sci.* 23: 109-118.

Chen ZM, Wan HB, Wang Y, Xue Y, Xia H (1987) Fate of pesticides in the ecosystem of tea garden. In Proc. International Symposium. Tea Quality-Human Health. Tea Research Institute, Chinese Academy of Agriculture Sciences, Hangzhou, China (November 4–9), pp 146-149.

Chen, Q. (1985) Tea-the Best Drinking for Health. *China Tea* 3: 28.

Coulibaly K and Smith JS (1994) Effect of pH and cooking temperature on the stability of organophosphate pesticides in beef muscle. *J Agric Food Chem.* 42: 2035–2039.

Deng Z, Tao B, Li X (1997) Effect of Tea on the Living Time of Musca Domesticas and the Anti-stress of KM Rats. *J Nanchang Univ.* 2: 69-72.

Desmarchelier JM, Ahern CM (1988) Insecticide-retentive carriers. 2-Fenitrothion-impregnated clays. *Austr J Exp Agric.* 28(2): 271-278.

Hajara B, Hazara G, Bisen JS (2000) Persistence and degradation of some insecticides in Darjeeling tea. *Journal of Plantation Crops* 28(2): 123-131.

Hatton CJ, Babcock JM, Schoonover JR, Dripps JE (1992) Fenazaquin, a new acaricide/insecticide. In: Proc. Int. Cong. Ent., pp 535.

Hoff VGR, Zoonen VP (1999) Trace analysis of pesticides by gas chromatography. *J Chromatograph A.* 843: 301–322.

Hogenboom AC, Niessen WM, Little D, Brinkman UA (1999) Accurate mass determinations for the confirmation and identification of organic micro-contaminants in surface water using on-line solid-phase extraction liquid chromatography electrospray orthogonal-acceleration time of flight mass spectrometry. *Rapid Commun Mass Spectrum* 13:125–133.

Huang Z, Zhang Y, Wang L, Ding L, Wang M, Yan H, Li Y, Zhu S (2009) Simultaneous determination of 103 pesticide residues in tea samples by LC-MS/MS. *J Sep Sci.* 32: 1294-1301.

Huang ZQ, Li YJ, Chen B, Yao SZ (2007) Simultaneous determination of 102 pesticide residues in Chinese teas by gas chromatography–mass spectrometry. *J Chromatogr. B* 853: 154-162.

Jaggi S, Sood C, Kumar V, Ravindranath SD, Shanker A (2000) Loss of quinalphos during tea processing. *Pestology* 24(12): 42–46.

Jaggi S, Sood, C, Kumar V, Ravindranath SD, Shanker A (2001) Leaching of pesticides in tea brew. *J. of Agri. Food Chem.* 49: 5479-5483.

Knuth ML, Heinis LJ (1992) Dissipation and persistence of chlorpyrifos within littoral enclosures. *J Agric Food Chem.* 40: 1257-1261.

Kumar S, Singla-Pareek SL, Reddy MK, Sopory SK (2003) Glutathione: Biosynthesis, homeostasis and its role in abiotic stresses. *J Plant Biol.* 30: 179-187.

Lambropoulou DA, Psillakis E, Albanis TA, Kalogerakis N (2004) Single-drop microextraction for the analysis of organophosphorous insecticides in water. *Anal. Chim. Acta.* 516: 205-211.

Macalady DL, Wolfe NL (1983) New perspectives on the hydrolytic degradation of the organo phosphorothioate insecticide chlorpyrifos. *J. Agri. Food Chem.* 31: 1139-1147.

Maguire JR (1991) Kinetics of pesticide volatilization from the surface of water. *J Agric Food Chem* 39: 1674-1678.

Murleedharan N (1994) Pesticide residues in tea: problems and perspectives. *Planters Chronicle* 9: 371-375.

Nagayama T (1996) Behaviour of residual organophosphorus pesticides in foodstuffs during leaching or cooking. *J. Agri. Food Chem.* 44: 2388-2393.

Nagayama T, Maki T, Kan K, Iida M, Tamura Y, Nishima T (1989) Residues of organophosphorus pesticides in commercial tea and their leaching into tea. *Nippon Noyaku Gakkaishi* 14: 39-45.

Ochiai N, Sasamoto K, Kanda J, Nakamura S (2006) Dynamic hollow fiber protected liquid phase microextraction and quantification using gas chromatography combined with electron capture detection of organochlorine pesticides in green tea leaves and ready-to-drink tea. *J. Chromatogr. A.* 1135: 6-11.

Royal Society of Chemistry (1994). The agrochemicals handbook (3rd ed.). Information system.Surrey, England: Unwin brothers Ltd.

Sanz-Asensio J, Plaza-Medina M, Martinez-Soria MT, Perez-Clavijo M (1999) Study of photodegradation of pesticide ethiofencarb in aqueous and non-aqueous medium by gas chromatography-mass spectrometry. *J. Chromatogr. A* 840: 235-247.

Singh D, Chhonkar PK, Pandey RN (2001) Soil plant water analysis: a methods manual. Division of Soil Science and Agriculture Chemistry, Indian Institute of Agricultural Research, New Delhi, India.

Singh N (2002) Studies on the biological activity of potato waste (peel) components for their possible applications. Ph.D. thesis, Mysore, University of Mysore, India, pp 347.

Sood C, Jaggi S, Kumar V, Ravindranath SD, Shanker A (2004) How manufacturing processes affect the level of pesticide residues in tea. *J Sci Food Agric.* 84: 2123-2127.

Sunito LR, Shiu WY, Mackay D, Seibes JN, Glofelty D (1988) Critical review of Henry's law constants for pesticides. *Rev Environ Contam Toxicol.* 103: 1-59.

The Tea Council, 2004, Available from http://www.teacouncil.co.uk.

Vidal JLM, Arrebola FJ, Liébanas, M.J. Gonzälez Rodríguez, Garrido Frenich A, Fernández Moreno JL (2006) Validation of a gas chromatography/ triple quadropole mass spectrometry based method for the quantification of pesticides in food commodities. *Rapid Comm. Mass Spectrometry* 20: 365–375

Wan H, Xia H, Chen Z (1991) Extraction of pesticide residue in tea by water during the infusion process. *Food Addit. Contam.* 8: 497-500.

Wong JW, Webster MG, Halverson CA, Hengel MJ (2003) Multiresidue Pesticide Analysis in Wines by Solid-Phase Extraction and Capillary Gas Chromatography–Mass Spectrometric Detection with Selective Ion Monitoring. *J Agric Food Chem.* 51: 1148-1161.

Xiong J, Hu B (2008) Comparison of hollow fiber liquid phase microextraction and dispersive liquid–liquid microextraction for the determination of organosulfur pesticides in environmental and beverage samples by gas chromatography with flame photometric detection. *J Chromatogr A*, 1193: 7-18.

Xue YZ, Chen ZM (1992) Advance in the investigation on the technology of chemical control Xinjian Sci. Technol, pp. 165-170.

Zeng JB, Chen JM, Lin ZQ, Chen WF (2008) *Anal Chim Acta* 619: 59-66.

Zimmerli B, Blaser O (1982). Preliminary study of the transfer of pesticides from tea leaves to the infusion. *Mit Geb Lebensmittelunters Hyg.* 73: 174-185.

Zongmao C, Haibin W (1988) Factors Affecting Residues of Pesticides in Tea. *Pestic Sci.* 23: 109-118.

16

ORGANIC TEA FARMING

R.D. Singh

CSIR-Institute of Himalayan Bioresource Technology,
Palampur (H.P.)-176 061, India. E-mail: rdsingh@ihbt.res.in

Agriculture practices have undergone various stages of transition and development from traditional farming to intensive agriculture for the `green revolution'. The ultimate goal has been boosting of farm production. In this process, irreparable loss has inadvertently been caused to the ecosystem and environment, yet agrochemicals are a necessary evil. Increasing demand and number of "Health or Reform Stores" in developed countries have also inspired the revival of the eco-friendly (natural/organic) farming systems in India and developing nations. Organic sector is still very small - less than 0.5% of the agricultural industries in most of the countries.

According to Lampkin and Padel (1994), organic farming is an approach to agriculture where the aim is to create integrated, humane, environmentally and economically sustainable agricultural production systems, which maximize reliance on farm-derived renewable resources and the management of ecological and biological processes and interactions, so as to provide acceptable labels of crop, livestock and human nutrition, protection from pests and diseases, and an appropriate return to the human and other resources employed. The fundamentals of organic/natural farming systems are maintenance of soil fertility by improving *in situ* biomass production, reliance on genetic diversity, nutrient recycling, maintenance of production, conservation of soil environment and ecosystem, limiting the use of fossil energy and of non-renewable sources, and to consider the farm as a close system having limited external inputs. Organic farm is over all a balanced system where desired economic realization is optimized within a sustainable socio-ecological milieu. Being an important plantation crop, organic tea plays a leading role in this endeavour.

The organic system of farming, in general, has two basic constituents - low energy input system and self sustaining system where external inputs are discouraged in view of the environmental and ecological degradation, more so in monoculture, like tea plantations.

Of late, in most of the arable crops productivity is showing stagnation or declining trend. The implication of yield stagnation or declining productivity is severe, since these trends have occurred despite rapid growth in the use of chemical fertilizers. Depletion of soil organic matter is the main cause of low productivity, which is considered one of the most serious threats to the sustainability of agriculture. Therefore, more organic matter should be used in the farmers' fields to sustain the soil fertility in an intensive farming system (Hossain 2001). On the other hand, large-scale application of manure and compost is constrained by their availability and quality at farm level, while the use of mineral fertilizers is limited by their escalating costs (Onduru *et al.* 2002).

In plantation crops, use of agrochemicals has been substantially higher than field crops. Consequently, threat from residual effects of pesticides applied in plantation crops is also high, especially so with made tea. Many diseases and insect pests attack tea grown in widespread monoculture. Due to absence of secondary tillage operation in tea plantations, weed infestation is also high. As the tea crop is generally harvested at weekly interval, too short a period for degradation of many agrochemicals, many tea importing countries have fixed the maximum residue limits of pesticides in made tea. This is likely to adversely affect tea exports among different countries, if pesticide residues are not monitored to meet the desired levels.

In addition, consistent use of pesticides, mostly soil applied persistent herbicides, pollutes ground water and may adversely affect the useful soil microflora like vesicular-arbascular mycorrhizal (VAM) fungi and plant growth promoting rhizobacteria (PGPR). Application of pesticides to soil also affects adversely the comparatively susceptible field crops, by way of water run off and seepage, particularly from plantations on hilly terrain. With the growing concern for pesticide residue free food and beverages, development of organic/ natural farming practices is the need of the time.

According to Sivapalan (1990 & 1993), traditional farming depends on relatively few external inputs, accumulate and recycle nutrients, effectively protect soils and their fertility and rely on genetic diversity. Thus, traditional farming principals can support generation of adaptive technologies that will reconcile environmental protection and economic benefits. Proper organic farming practices can provide better tasting tea, healthier plant growth, improved soil fertility and soil aeration, less soil erosion and greater water retention, rejuvenation of depleted soil, recycling of organic wastes generated by public and industry, thus cleaning up the environment, reduced environmental pollution, reduced pest and disease infestation, and increased demand, price, and saleability of tea (Dudeja 1997).

Hajra and Kabir (1993) suggest that use of organic mulch materials like *Crotalaria* sp., *Sesbania* sp., *Tripsacum laxum* (Guatemala grass) etc., botanicals, and release of natural pest enemies may be practiced for organic farming. They further suggest that the switch over should not be rapid but it should be planned properly. In this context it has been concluded that the transitional period between cessation of use of external inorganic inputs and switch over to organic farming is generally of 2-3 years (Dudeja 1991, 1997, Sivapalan 1993).

Though organic tea farming is being done in some tea estates in south India and Darjeeling hills and it is reported that India produces 1.3 m kg of certified organic tea and about 1.0 m kg additional tea is being produced by gardens under conversion, but there are no published records regarding technique/system of organic tea farming. A number of national laboratories in India are engaged in formulating a pest-management system; to minimize use of chemicals, involving application of neem based insecticide followed by contact or systemic pesticides and subsequently biocontrol measures (Chaudhuri 1993). Thus, various proposed cultural practices and components of an organic farming system have been or being tested in isolation or combination of more than one factor but there has been no concerted effort to develop a sound package of environmentally safe farming practices that could be adopted extensively by tea planters.

Tea is comparatively a new entrant in the field of organic products (Dudeja 1991). The first International Conference on Organic Tea and Coffee, held in 1991 at Frauenfeld (TG), Switzerland, proposed to have ecologically sustainable tea plantation where tea production and soil

Fresh tea yield was 78.1, 90.8, 94.5 and 67.7 kg/plot with mulches of fresh weeding, tea waste and cow dung, and no mulch, respectively.

In a study conducted to evaluate effect of different organic mulch materials on yield and quality of china hybrid tea, Sud *et al.* (2003) have observed that leaf plucked from the plot treated with 'farmyard manure (FYM) + humic acid and no mulch' showed significantly higher polyphenols and catechins than untreated control and was followed by the plot receiving 'FYM + humic acid + *Imperata* mulch'. Similarly, the made tea from the plot under 'FYM + humic acid and no mulch' contained significantly higher theaflavins, thearubigins, and caffeine than untreated control. However, the flavour profile of the made tea was not influenced by the treatments.

TECHNICAL EFFICIENCY

While analyzing the technical efficiency of organic tea small holdings in the mid-country wet zone of Sri Lanka, Jayasinghe and Toyoda (2004) have found that there is great potential to increase production by 55% through efficient use of the present technology. These results identify a significant relationship between compost application and labour on organic tea production. A significantly positive relationship was also found between technical efficiency and training of farmers. Large numbers of vegetative propagated tea plants in the farm enhance the efficiency. The characteristic crop diversification observed in the organic tea farming system has significantly reduced the efficiency.

A review of intensified systems of land use on unirrigated farms in developing countries for formulation of a basis to propose complementary sets of conditions necessary for farming systems to be sustainable in the long run revealed that studies on soil have generally been weak regarding soil organic matter and biological activity in the rooting zone of crops. All the cases illustrate the importance of leguminous species in the farming system, but in general these species have not been adequately studied.

SCOPE

Organic farming is a highly knowledge intensive, labour oriented, and more diverse and complex system integrating several organic recycling processes and enterprises. Thus, it requires high degree of motivation, firm commitment and innovative ability. In development of the organic farming the main hurdle is marketing of the produce.

Some of the bottlenecks in marketing are absence of testing or monitoring (certifying) agencies, lack of organized market, higher premium costs, low demand, and higher perishability of organic farm produce. The organic farming has in offing praiseworthy attributes to effectuate long term benefits to the lives on earth and their environment.

The potential for organic tea is increasing, which could be attributed to rising consumer purchasing power and concern for consuming healthy products. Certified organic teas tend to sell at higher prices than conventional teas.

Organic farming in tea has been observed to improve the faunal diversity as well. Mukhopadhyay *et al.* (2003) have found that variation in the arthropod faunal structure was numerically and taxonomically abundant in the organic tea as compared to the plot of conventional tea. Coleopterans were more diverse with largest number of families and recognizable taxonomic units (or morphospecies) in the organic tea plot. Faunal diversity study at ground level gave the clue that soil of the organic plantation was healthier than that of the conventional tea plot.

REFERENCES

Chaudhuri TC (1993) Recent developments in chemical residue aspects of tea. *Tea Science and Human Health, Proceedings of the International Symposium, Teatech*, Tea Research Association, Calcutta, 177-84.

Darmawijaya MI (1991) Organic farming of tropical tea. *Proceedings of the International Symposium on Tea Science*, August 26-29, 1991, Shizuoka, Japan, pp 712-716.

Dudeja V (1991) Growing tea and coffee- the organic way. The Planters' Chronicle 86(8): 347, 49, 51, 53.

Dudeja V (1997) Organic tea production - confusion confounded! *The Assam Review & Tea News* 86(5): 5-9.

Ganguly PK (2000) A few new concepts in organic tea cultivation. *Planters' Chronicle* 96(3): 119-123.

Hajra NG, Kabir SE (1993) Organic farming - an overview. *Indian Review of Life Sciences* 13: 121-131.

Hossain Md Zahid (2001) Farmer's view on soil organic matter depletion and its management in Bangladesh. *Nutrient Cycling in Agroecosystems* 61(1/2): 197-204.

Ipinmoroti RR, Adeoye GO, Iremiren GO (2007) Soil nutrient dynamics as influenced by organomineral fertilizers and tea seedling nutrient uptake in Nigeria. *International Journal of Tea Science* 6(3): 25-29.

Jayasinghe JMJK, Toyoda T (2004) Technical efficiency of organic tea small holding sector in Sri Lanka : a stochastic frontier analysis. *International Journal of Agricultural Resources, Governance and Ecology* 3(3/4): 252-265.

Lampkin NH, Padel S (eds.) (1994) The Economics of Organic Farming - An International Perspective, CAB International, U.K., p 468.

Lin ML, Shih JK, Tseng HK (1995) The organic farming of tea crop. *Taiwan Agriculture* 31(1): 123-128.

Mukhopadhyay A, Sherpa PW, Pradhan B (2003) Diversity of ground arthropod community at organic and chemically intensive tea plantation of Darjeeling terai. *Journal of Environmental Biology* 24(4): 471-476.

Nampoothiri KUK (2001) Organic farming - its relevance to plantation crops. *Journal of Plantation Crops* 29(1): 1-9.

Onduru DD, Diop JM, Werf E-van-der, Jager A-de (2002) Participatory on-farm comparative assessment of organic and conventional farmers' practices in Kenya. *Biological Agriculture and Horticulture* 19(4): 295-314.

Ramarethinam S (2004) Role of bio-pesticides in tea cultivation. *International Journal of Tea Science* 3(3/4): 111-115.

Sivapalan P (1990) Production of tea by organic farming systems. *Tea Bulletin*, 10(1): 3-14.

Sivapalan P (1993) Organic Farming of tea as an aid to maintaining environmental quality. *Tea Science and Human Health, Proceedings of the International Symposium, Teatech 1993*, Tea Research Association, Calcutta, 300-2.

Sud RK, Singh RD, Gulati Ashu, Singh Brajinder, Shekhawat RS, Prasad Ramdeen (2003) Organic tea farming in Himachal Pradesh. *International Conference on Impact of Science on Economy of Tea Industry*, Kolkata, November 20-22.

Verma DP (1995) Milestones in nutrient management of tea. *Planters' Chronicle* (August): 371-377.

Xu Yun Wen (2000) Prospects and key techniques for developing organic tea. *Journal of Tea* 26(1): 11-13, 18.

Zhao Qin, Tong QiQing, Tu YouYing, Luo YaoPing (2000) Effect of temperature and organic fertilizer on beta-glucosidase activity and contents of alcoholic aroma in fresh tea leaves. *Journal of Zhejiang Forestry College* 17(1): 28-31.

17

PLANT PHYSIOLOGICAL APPROACHES FOR ENHANCING PRODUCTIVITY IN TEA

S.K. Vats* and P.K. Nagar

CSIR-Institute of Himalayan Bioresource Technology,
Palampur (H.P.) 176 061, India. *E-mail:skvats@ihbt.res.in

INTRODUCTION

Tea or *Camellia sinensis* (L.) O.Kuntze, as a crop is grown in different parts of the world with diverse climatic conditions ranging from a Mediterranean-type climate to hot humid tropical and subtropical regions (Carr 1972). Out of about 82 species of *Camellia* (Banerjee 1992), the three major types known as *C. sinensis* (L.) O.Kuntze - the China type, *C. assamica* (Masters) Wight - the Assam type and *C. assamica* ssp. *lasiocalyx* (Planch ex Watt) Wight – the Cambod type. Tea is the oldest non-alcoholic beverage in the world (Mondal *et al.* 2004). The seed raised crop populations of tea are heterogeneous due to natural outcrossing. Hybrids like Assam, Cambod or China are most popular and related on the basis of their morphological proximity to the main taxa (Banerjee 1992). Selection and breeding programmes have contributed significantly towards developing improved varieties which are vegetatively propagated.

Despite its ability to grow under relatively diverse sets of environmental conditions, tea cultivation requires certain definite criteria for optimal productivity. Tea plant requires on average at least five sunshine hours per day and atmospheric humidity between 70 to 90% during the active stage of growth. Shoots tend to form dormant (*banjhi*) buds when the air is too dry. The harvesting

Ram 1990). The changing pattern in DMC in certain tea clones under agroclimatic conditions of Darjeeling was studied by Kumar *et al.* (1993). It was found that the higher amount is associated with dry period and lower values with relatively wet period. This may be attributed to the low water content of leaf and higher rate of transpiration during the dry period as compared to the high water content in leaf and low rate of transpiration during the wet period (Satyanarayana 1987).

Tea leaves attain self-sufficiency in photosynthetic capacity when they are fully expanded. Third leaf expanded from the bud is not fully dependent or independent, as it is in a transition phase (Marimuthu *et al.* 1994). Hence, plucking three leaves and a bud could improve the productivity, reaping the full benefit of sink induced photosynthesis. Three leaves and a bud definitely weigh heavier than the two leaves and a bud (Satyanarayan and Ilango 1993). This also leads to longer plucking interval without losing crop, which could save on labour inputs. However, the third leaf and the internodes are expected to dilute the quality through the additional fiber and loss of constituents (Owuor 1997).

Productivity in tea is influenced by time of plucking during the day. Sinks (growing shoots) are empty in the morning and metabolized fast for routine functions leaving less dry matter for accumulation. Plucking after 11 am gives higher percentage of recovery than morning plucking (Gogoi 1977), and thus adjusting the plucking times can influence the productivity. The recovery percentage is increased in afternoon plucked shoots than that of morning plucked ones in similar situations. The percentage of dry matter in plucked shoots in relation to the gross weight of the entire plant (total weight of the dry matter produced) is the harvest index of tea. A high harvest index does not necessary mean high yield but increasing the index helps in reducing the inputs without reducing the yield. In tea, the increment in photosynthesis, reduction in respiratory loss, mobilization of food materials to new shoots, and reduction in the formation of non-productive plant parts are important for increasing the harvest index.

WINTER DORMANCY

Tea cultivated near equator produces almost the same yield every month, but further from equator winter harvest gradually declines and at latitude beyond about 16° N or S, there is almost winter dormancy. Location wise, winter dormancy is almost negligible in south India and Sri Lanka, but tea plants enter into such a phase during

December to January in Assam, middle of November to end of March in Darjeeling and north-west of India. Earlier, Barua (1969) showed that tea bushes go dormant when the average day length remains approximately below about 11 hr for a period of about 6 weeks and this happens beyond about 16° N or S. In certain tea growing area at higher altitudes as in Georgia (42°N), Darjeeling, (27°N) and Iran (37°N), the minimum temperature in winter may drop below freezing point. Thus at high altitudes, short day length and low ambient temperature are encountered during the winter season when visible growth of the tea plant completely stops. But temperature prevailing in certain tea growing areas suggests that low temperature cannot be the sole reason. For example, mean range of maximum and minimum temperature in Tocklai during winter is 22-26°C and 10-16°C respectively where winter dormancy is experienced for 3-4 months, while at Kericho (Kenya), the temperature varies from 19-25°C and 8-9°C respectively and there is no dormancy. The cessation of growth with the onset of winter (dormancy) and its resumption in spring (bud break) are thought to result from an interaction between endogenous growth promoters and inhibitors (Wareing 1977). Quality of tea also varies with different time of harvest during the year. After winter dormancy, the first flush usually fetch higher price based the most desirable flavor profiles. Aroma component in leaves plucked during rainy season is considered less desirable and this tea is generally sold in bulk for use in blends.

DORMANCY AND PLANT GROWTH REGULATORS

Plant growth regulators (PGRs) have been used to boost the yield of horticultural and ornamental crops (Nickell 1992). Broadly stating, the response of tea plants to PGR is encouraging but are very often inconsistent due to fluctuating environmental conditions and highly heterogenetic nature of the plants (Ayyappan *et al.* 1987). It has been shown in tea that growth-inhibiting substances might determine the state of growth or dormancy of shoots (Barua 1969). Young leaves are considered to be site of cytokinin biosynthesis and nucleotides are known to be the primary products of cytokinin biosynthesis and their increased level at time of bud break may reflect resumption of cytokinin biosynthesis (Singh *et al.* 1992). The cytokinin levels, studied in tea shoots, were low during winter months and increased gradually prior to bud break (Nandi and Palni 1993). Abscisic acid (ABA) is also involved in the induction and maintenance of winter dormancy in tea. High levels of free ABA were detected in dormant shoots and the levels decreased with release from dormancy (Nagar 1996). It is suggested that some phenolic acids may participate in endogenous regulation of dormancy through their influence on hormonal metabolism and the role of high content of free phenolic acids in certain tissues during dormancy period is based on the inhibitory effect of some phenolics on cell division (Li *et al.* 1993).

Gibberellic acid (GA) activity studied in tea was reported to be low during dormancy but increased substantially at bud break (Nagar and Kumar 2000). Exogenous application of GA was quite effective in inducing early bud break (Nandi *et al.* 1995). However, response to GA has not been consistent from year to year on the same clone. Regarding field application, it appears that spraying GA under same situation may advance bud-break and increase early season crop at those latitudes where winter dormancy is experienced, provided the ambient temperature in late winter and early spring remains favorable and the tea bushes are not subjected to winter stress.

The role of polyamines that function as antisenescence agents (Galston and Kaur Sawhney1982) have also been studied in tea. Study on changes in endogenous polyamine levels in two different clones Tukdah-78 and UPASI-10, of tea shoots showed enhanced level of putrescine but low levels of spermidine and spermidine with imposition of dormancy, while high spermidine and spermine content in leaves were related with dormancy release (Kakkar and Nagar 1997). It appears that variations in endogenous levels of polyamine in tea during different growth stages control different developmental processes (Sood and Nagar 2004). Brassinosteroids (BRs), been reported to affect elongation, division and differentiation of plant cells in a broad range of test systems (Clouse 1997, Khripeach *et al.* 2000), have been reported from dormant shoots of tea plants (Gupta *et al.* 2004). At a very low concentration (5mM), BR are able to mobilize photosynthates out of the maintenance leaves at faster rates and thus contribute towards enhanced dry matter recovery and harvest index.

PLANT GROWTH REGULATORS FOR CROP IMPROVEMENT

Crop productivity, its clonal and seasonal responses are basically related to endogenous promoters and inhibitor levels. Imbalanced endogenous promote/inhibitor ratio may be moderated by the application of bio-stimulant, which in turn stimulate better growth, and metabolic activity of the plant. Despite its role, PGRs like auxins, gibberellins and cytokinins have found very limited use and that too in specific instances. The other PGRs of commercial interest are triacontanol (TRIA; a long chain fatty acid alcohol) and the products containing hydrolysed protein complexes. The response of tea plants to PGR is encouraging but very often inconsistent due to fluctuating environmental conditions and highly heterogenetic nature of the plants (Ayyappan *et al.* 1987). Earlier reports revealed that the response to applied biostimulants varies between jats/clones (Manivel *et al.* 1993). About 10-15% of the photosynthates are expected to go into the accumulation of dry matter (Manivel and Hussain 1982). Manivel *et al.* (1993) established that these regulators help particularly during drought season by enhancing the water use efficiency. It has been reported that bioregulators like brassinosteroids also elicit similar responses in tea at a very low concentration (Rajkumar *et al.* 1998). Besides mobilizing photosynthates these bioregulators enhances the dry matter recovery and harvest index. Due to favorable cell water content and turgidity, higher growth rate and leaf expansion is possible. Further; integrated use of growth promoters to augment mechanized harvesting in tea plantations is also a real possibility. On the contrary, certain growth retardants have also been used for young tea management in certain parts of India (http://www.upasitearesearch.org). Growth retardants are known to suppress apical dominance and induce the growth of latent buds without any negative role on the metabolic activities of the plant. It was found that growth retardants like paclobutrazol (500 ppm), ethaphon (1000 ppm) applied on the foliage, one month after planting was found effective in strengthening the root system as well promotion of lateral branching in young tea (Manivel *et al.* 1993).

Understanding the physiology of productivity in tea depends on the mode of dry matter production, which in turn is regulated by photosynthetic accumulation of stored carbon. The process of photosynthesis depends strongly on location specific environmental conditions, which may include drought or soil moisture status, temperature or light regimes, to invoke detrimental losses or limitation in terms of photorespiration or optimal harvesting of light energy. Studies have shown that optimal performance in tea do not depends on light, temperature and soil moisture conditions *per se* but the interaction of it. Given that maximum light-saturated photosynthetic rates of tea are below the average for most C_3 plants, the source capacity of tea is rather restricted. Tea is also sink-limited due to a continuous removal of young shoots. While environmental factors related to enhance productivity in tea appears to be highly location specific, other issues to be taken in account are the role of standard agronomic and cultural practices, conducive soil conditions, and pests and disease management, that can help minimize crop losses.

REFERENCES

Anandacoomaraswamy A, De Costa WAJM, Gawarammana MDW, Navaratna DMS (1996) Effect of soil moisture stress on physiological parameters of young tea. In: Proceedings of the 193rd Experiments and Extension Forum of the Tea Research Institute of Sri Lanka, Tea Research Board of Sri Lanka. pp. 17-35.

Anandacoomaraswamy A, De Costa WAJM, Tennakoon PLK, Van der Werf A (2002) The physiological basis of increased biomass partitioning to roots upon nitrogen deprivation in young clonal tea (*Camellia sinensis* (L.) O. Kuntz). *Plant Soil.* 238: 1-9.

Ayyappan P, Rajkumar R, Krishnamoorthy KK (1987) Effect of Miraculan on clonal and seedling teas. *Planters' Chronicle* 81: 225-227.

Banerjee B 1992. Botanical classification of tea. In: Willson KC, Clifford, MN (eds) *Tea: Cultivation to consumption.* Chapman and Hall, London, pp 25-51.

Barman TS, Baruah, U, Saikia JK (2008) Irradiance influences tea leaf (*Camellia sinensis* L.) photosynthesis and transpiration. *Photosynthetica* 46: 618-621.

Barman TS, Saikia JK (2005) Retention and allocation of ^{14}C assimilates by maintenance leaves and harvest index of tea (*Camellia sinensis* L.). *Photosynthetica* 43: 283-287.

Barua DN (1963) Use and abuse of clones. *Two and a Bud* 10: 3-6.

Barua DN (1969) Seasonal dormancy in tea (*Camellia sinensis* L.). *Nature* 224: 514.

Barua DN (1989) Production and partition of dry matter. In: Barua, DN (ed) *Science and Practice in Tea Culture* Tea Research Association, Kolkata, pp 369-395.

Baruah S, Hazarika M, Mahanta PK, Horita H, Murai T (1986) Effect of plucking intervals on the chemical constituents of CTC black teas. *Agric Biol Chem.* 50(4): 1039-1041.

Bonneure D, Willson KC (1992) Mineral nutrition and fertilizers. In: Wilson KC and Cliftord MN (eds) *Tea Cultivation to consumption.* Champman and Hall, London, pp 769.

Carr MKV (1972) The climatic requirements of tea: A review. *Exp Agric.* 8: 1-14.

Carr MKV (2000) Shoot growth plus plucking equals profit. TRIT Occasional Publication No. 1, Tea Research Institute of Tanzania, pp1-7.

Carr MKV, Stephens W (1992) Climate, weather and the yield of tea. In: Wilson KC and Cliftord MN (eds) *Tea Cultivation to consumption.* Champman and Hall, London, pp 87-135.

Clouse SD (1997) Molecular genetic analysis of brassinosteroid action. *Physiol Plantarum* 100: 702–709.

Costa MH, Yanagi SNM, Oliveira PJ, Ribeiro A, Rocha EJP (2007) Climate change in Amazonia caused by soybean cropland expansion, as compared to caused by pastureland expansion. *Geophys Res Lett.* 34: L07706, doi:10.1029/2007GL029271.

Davis JM, Loescher WH (1990) [^{14}C]-assimilate translocation in the light and dark in celery (*Apium graveolens*) leaves of different ages. *Physiol Plant.* 79: 656-662.

Galston AW, Kaur-Sawhney R (1982) Polyamines: are they a new class of plant growth regulators? In: Wareing PF (ed) *Plant growth regulators.* Academic Press, New York, pp 451-461.

Gamage AJ, Wijeratne MA, De Costa WAJM (2007) Role of shade trees in tea: A low-country perspective. In: *Proceedings of the 215th Meeting of the Experiments and Extension Forum.* Tea Research Institute of Sri Lanka, Talawakelle, Sri Lanka.

Gogoi BN (1977) A review of research on shade in tea. *Two and a Bud* 23: 67-73.

Gupta D, Bhardwaj R, Nagar PK, Kaur S (2004) Isolation and characterization of brassinosteroids from *Camellia sinensis* (L.) O. Kuntze. *Plant Growth Regul.* 43: 97-100.

Hansen P, Ryugo K (1979) Translocation and metabolism of carbohydrate fraction of ^{14}C-photosynthates in 'French' prune, *Prunus domestics* L. *J Am Soc Hortic Sci.* 104: 622–625.

Jain NK (1977) Technological gaps in tea research [India]. *Two and a Bud* 24: 1-5.

Joshi SC, Palni LMS (1998) Clonal variation in temperature response of photosynthesis in tea. *Plant Sci.* 137: 225–232.

Kakkar RK, Nagar PK (1997. Distribution and changes in endogenous polyamines during winter dormancy in tea *(Camellia sinensis* (L) O. Kuntze). *J Plant Physiol.* 151: 63-67.

Karunaratne PMAS, Mohotti AJ, Nissanka SP, Gunasinghe WADSK (2003) Effect of shade in minimizing photo-inhibition of photosynthesis of high grown tea (*Camellia sinensis* (L.) O. Kuntze). *Trop Agric Res.* 15: 133-143.

Khripach V, Zhabinskii V, deGroot A (2000) Twenty years of brassinosteroids, steroidal plant hormones warrant better crops for the XXI century. *Ann Bot.* 86: 441–447.

Kozlowski TT (1971) Growth and development of trees. vol. 1. New York and London: Academic Press, pp 443.

Fresh tea shoot is rich in polyphenols especially the flavanols (catechins) which may constitute up to 30% of the dry leaf weight. Other polyphenols include flavonols and their glycosides *viz.*, quercetin, myrcetin, and depsides such as chlorogenic acid coumarylq-uinic acid and, theogallin (3-galloylquinic acid, unique to tea). Caffeine is present at an average level of 3% along with very small amounts of the other methylxanthines, theobromine and theophylline. The amino acid theanine (5-N-ethylglutamine) is also unique to tea (Table 18.1).

Table 18.1. Chemical composition of fresh tea shoots

Compounds	% Dry weight	Contribution
Total Polyphenols	25 - 30	Astringency
Flavanols		
(-)Epigallocatechin gallate	8 – 12	
(-) Epicatechin gallate	3 – 6	
(-) Epigallo catechin	3 – 6	
(-) Epicatechin	1 – 3	
(+) Catechin	1 – 2	
(+) Gallocatechin	3 – 4	
Flavonols and flavonol glycosides	3 – 4	
Leuco anthocyanins	2 – 3	
Polyphenolic acids and depsides	3 – 4	
Caffeine	3 – 4	Briskness
Theobromine	0.2	
Theophylline	0.5	
Amino acids	4 – 5	Brothyness
Organic acids	0.5 – 0.6	
Monosaccharides	4 – 5	
Polysaccharides	14 - 22	
Cellulose and Hemicellulose	4 – 7	
Pectins	5 – 6	
Lignin	5 – 6	
Protein	14 – 17	

Tea leaf, apart from the normal cell enzymes, possesses a very active polyphenol oxidase which catalyzes oxidation of the catechins. This enzyme is responsible for the oxidation of catechins after the disruption of cellular structure of the shoot during black tea manufacture. The catechins are oxidized by the enzymatic oxidations to quinones which undergo condensation reactions to form compounds, including bisflavanols, theaflavins, epitheaflavic acids, and thearubigins. These compounds impart the characteristic taste and colour to black tea. Theaflavins and thearubigins complex with caffeine to give astringency and creaming property to black tea cup.

Tea is generally made from two main species of *Camellia viz., C. sinensis* (the China tea) and *C. assamica* (the Assam tea). These two differ in their composition of biochemicals. The broad leaf Assam tea has higher contents of flavan-3-ols and the flavonols present as in diglycosides, while the small succulent leaf China tea has lower flavan-3-ols content, and the flavonols are present as triglycosides and possesses higher content of volatile flavour components (VFC). Today due to the high yield and quality breeding strategies and clonal propagation, true China and Assam tea types are rare to find, instead hybrid varieties are found with characters that are a mixture of both the varieties. The one closer to *assamica* are named *assamica* hybrids and other *sinensis* hybrids.

POLYPHENOLS

Tea is a rich source of polyphenols which contribute to as high as 30% dry weight basis. There are six major catechins in tea. They are (+)-catechin, (+)-gallo-catechin, (-)epi-catechin, (-)epi-catechin gallate, (-)- epi-gallocatechin and (-)epi-gallocatechin-gallate. Catechins are reported to have the beneficial effects of tea.

The catechins in tea plant are synthesised by the shikimic acid pathway via chalchones, where phenyl alanine and tyrosine are its precursors. The catechins are first formed and are later epimerized and galloylated during storage in the epidermal tissue. The catechin biosynthetic pathway in tea has been worked out. Catechins are important quality attributes for green and black tea quality. The catechins are oxidized to theaflavins and thearubigins in black tea. Catechins show seasonal variations (Gulati and Ravindranath 1996). The increase in sythesis of EGCG and ECG with rainy season was observed by Bokuchava and Skobeleva (1980). In Australia grown fresh tea shoots, EGCG content was higher in warm harvesting season while higher levels of EGC were reported in cool months (Yao *et al.* 2005).

Estimation: A number of methods have been used for the determination of polyphenols. These include Folin-Denis reagent, vanillin reagent, Price-Butler (Ferricyanide and ferric chloride dip reagent), ceric sulfate and potassium permanganate and diazotized amines such as diazotized *p*-nitroaniline and sulfanilic acid methods. These reagents are nonspecific and can not distinguish flavanoid classes in general and best used for determination of total soluble polyphenols (Robards and Antolovich 1997). The flavanols and other polyphenols in tea were first separated by two-dimensional paper chromatography and located by ultraviolet light (Bhatia and Ullah 1968). Detection and identification of flavanoids was achieved with systems using fluorescence and UV illumination before or after ammonia treatment. Lately highly specific diazotized sulfanilamide reagent was synthesized for determination of tea catechins (Singh *et al.* 1999). The reagent is employed both as spray reagent for selective visualization of tea catechins on two-dimensional paper chromatograms and for their spectrophotometric quantification in the extracts of tea polyphenols. The formation of yellow colour between catechins and diazotized sulphanilamide is estimated. The yellow colour made the basis of spectrophotometric method for estmation of the total and individual catechins in different tea cultivars. Instrumental chromatographic methods such as gas liquid chromatography (GLC) and high pressure liquid chromatography (HPLC) have since been used for the separation and quantitative measurement of flavones. Due to their poor volatility, conversion to the trimethylselyl derivatives is prerequiste to analysis of the flavanols by GLC but separations can be made subsequently on a 5 feet x 4mm ID column packed with 3% OV-1 on diatomite. Today the method of choice for quantitative analysis of the flavanols and other non-volatile constituents of green leaf is C_{18} reversed phase HPLC, originally reported by Hoefler and Loggon (1976) who used isocratic elution with mobile phase consisting of acetic acid- methanol- dimethylformamide-water and UV detection at 254 nm. More recent work (Opie *et al*, unpublished data) has shown that gradient HPLC, using an acetonitrile-water-aceticacid mobile phase and 3 um particle C18 stationary phase Hypersil ODS, provided highly resolved flavanol peaks. In the analysis of catechins using HPLC, caffeine is simultaneously determined and minor catechins, like catechin gallate and gallo-catechin gallate were simultaneously determined along with major catechins like epigallocatechingallate (Goto *et al.* 1996) (Figure 18.1). In the HPLC study of catechins, many detection methods have been tried. Treutter (1989) used a post column reaction with 4-methylaminocinnamaldehyde and detection at an absorbance of 640 nm. Umegaki *et al.* (2001) detected catechins using an electrochemical detector with an applied potential of +550 mv. The sensitivity of this method was 1000 times superior to UV detection at

280 nm. Other studies mostly used a conventional UV detector of varying wavelengths i.e 275, 280, 231 nm. Bronner and Beecher (1998) mentioned the use of 210 nm to improve the signal-to-noise ratio,while Guo *et al.* (1994) used 231 nm to reduce the interference from other compounds.

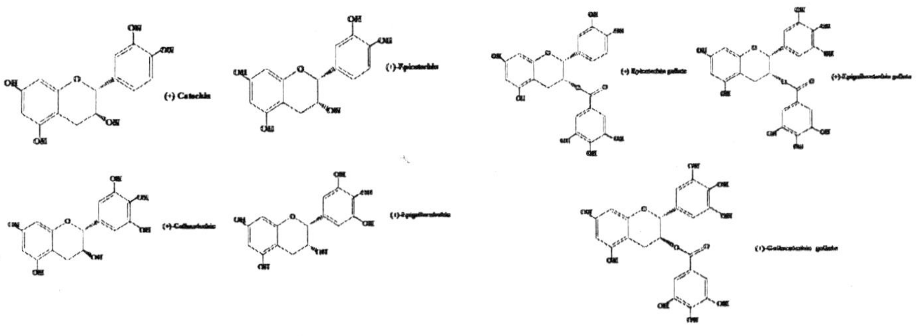

Fig. 18.1. Structures of catechins of tea

In most cases, separation was made using a C18 ncolumn, and in case of gradient separation, the concentration of acetonitrile or methanol was gradually increased. Wang *et al.* (2000) used isocratic elution system for the determinationb of catechins and caffeine using methanol/water/ortho-phosphoric acid but the run time was long (60 min) when compared to gradient elution. Sharma *et al.* (2005) used gradient elution system with mobile phase consisting of acetonitrile/ ortho phosphoric acid/water, and run time was reduced to 21 minutes with last peak eluting at 17 minutes. Dalluge *et al.* (1998) compared a variety of stationary phases which utilized ultrapure silica and found that maximized coverage of the silica support improved the separation.

L-THEANINE (N^5-ETHYL-GLUTAMINE)

In the late 1940s, a substance that was responsible for much of the green tea infusions infusion's flavour was isolated. It was later characterized as L-theanine and structure given by Sakato (1950). L-theanine (γ-glutamylmethylamine) is a unique and exclusive amino acid found in tea (Figure 18.2). It is also reported from the edible bay boletes mushroom *Xerocomus badius* (Casimir *et al.* 1960). It also exhibits several pharmacological effects i.e. antitumour and anti-hypertensive effects (Chu *et al.* 1998, Juneja *et al.* 1999, Kakuda *et al.* 2000, Egashira *et al.* 2004). Its ability to induce a state of relaxation is intriguing and suggests new possibilities for theanine as a functional food supplement in health foods and medicines (Chu *et al.* 1998, Kakuda, 2002).

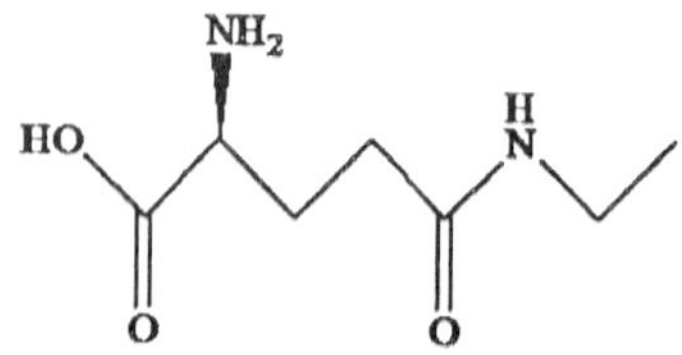

Fig. 18.2. Theanine (5-N-ethyl-gluatmine)

Theanine is one of the main amino acids in tea and usually constitutes about 2% of the dry weight of leaves and occupies around 50% of the total free amino acids in tea plant with little or no after taste. Theanine has a sweet taste and plays an important role in enhancing the quality of green tea. In general, higher grades of Japanese green tea contain high theanine content. Green tea of the highest quality, "Maccha" or "Gyokuro," contains on an average 1.5–2% of theanine per dry weight. Black and half-green teas (except for Formosa Oolong) also contain equal or higher theanine contents as green teas. The younger tea shoots (bud and first leaf) have a higher theanine content than mature leaves (second or third and fourth leaves) (Baptista *et al.* 2012).

The contents of theanine increased in tea roots with increasing levels of ammonium fertilizers (Selvendran and Selvendran 1972). High levels of theanine was observed in roots in the winter months when the bushes were dormant.

Studies on the factors affecting the theanine extraction rate including water volume, temperature and time of extraction showed that theanine slowly racemizes in aqueous solutions, undergoes hydrolysis, particularly at basic pH values. No correlation was found between the absolute concentration of theanine in tea and its enantiomeric composition (Ekborg-Ott *et al.* 1997). An inverse correlation was found between certain grades of tea (e.g., Pekoe, Flowery Orange Pekoe, etc.) and the percent of D-theanine present. Concentrations of theanine, and other amino acids in green tea infusion did not change even by heating at 120°C. Theanine content decreases with increase in degree of fermentation, and increase in drying temperature (Baptista *et al.* 2012).

In tea plants, L-theanine is bio-synthesized in the roots and stored in the leaves. Theanine is synthesized from glutamic acid and ethylamine by theanine synthetase [l-glutamate(Glu): ethylamine (EtNH2) ligase, EC 6.3.1.6] using ATP (Sasaoka *et al.* 1965) in the roots, and the synthesized theanine is transported to the leaves and accumulated there. Theanine synthetase (EC 6.3.1.6), is a special enzyme of tea plants (Feldheim *et al.* 1986). The overall amino acid content was found in order of roots > stems of new shoots > buds plus first leaves > second leaves > third leaves (Chakraborty *et al.* 1978). Roots contained ~2-9 times more amino acids than other organs.

The *N*-ethyl carbon of theanine was significantly incorporated into the phloroglucinol nucleus of catechins in tea shoots. This incorporation was controlled by light. Acetate-1-^{14}C was also incorporated into the phloroglucinol nucleus, but most of its activity was randomly distributed in other compounds and light had no effect. The physiological significance of theanine in catechin biosynthesis in tea plants has been shown by Kito *et al.* (1968).

CAFFEINE

Caffeine is an important component of both black and green tea. Caffeine contributes to the astringency along with the catechins and their derivatives in green tea and is responsible for briskness and creaming in the black tea infusion (Millin *et al.* 1969). Cream is a thick turbid layer that develops on the surface when a good cup is cooled. Cream is a complex of theaflavins, thearubigins and caffeine and is formed by weak hydrogen bonds. The complex formed modifies positively the taste characteristics of both caffeine and theaflavins (Millin *et al.* 1969, Sanderson 1972). The tender younger leaves of tea bushes contain higher caffeine contents than older leaves (Lin *et al.* 1996). Caffeine content decreases from about 4.7% in the bud to about 2.9% in the third leaf. Leaf stalks accumulate low levels of caffeine. Fine plucking standards ensure high caffeine content in brewed tea. There exists a positive and significant correlation established between caffeine levels in the green two and a bud shoot, and tasters' preferences of black teas. Seasonal variations have been observed in caffeine (Gulati and Ravindranath 1996). High relative humidity was the only climate factor significantly correlated with caffeine (Wang *et al.* 2011).

Extraction experiments of tea for caffeine extraction in the cup with loose tea and tea-bags at different temperatures showed that first order rate kinetics is followed. The activation energies were found to be 45.2 and 43.2 kJ moly1 for loose tea and tea-bag tea, respectively.

Studies conducted on variations in caffeine content during various stages of black tea manufacture revealed that the increase in caffeine content during the withering stage was time dependent, but independent of rate and degree of desiccation. The decrease in caffeine levels during fermentation was related to the time and temperature of fermentation. Minor amounts of caffeine were lost by sublimation during the drying process. Tea stored for 6 months contained more caffeine than the freshly manufactured product.

The ability to synthesize and accumulate caffeine in green leaf was clonal dependent (Owuor and Chavanji 1986, Robinson 1992). There is little seasonal variation in caffeine content in Kenya and central Africa (Owuor and Chavanji 1986), though high seasonal variations in caffeine content has been reported from India, Japan and China.

In tea, four-step caffeine biosynthetic pathway includes three methylation steps that utilise *S*-adenosyl-L-methionine (SAM) as the methyl donor. In the process SAM is converted to *S*-adenosyl-L-homocysteine (SAH) which in turn is hydrolysed to L-homocysteine and adenosine. Significant amounts of radioactivity from [methyl-^{14}C] methionine and [methyl-^{14}C] SAM were incorporated into theobromine and caffeine in young tea leaf segments, and very high SAH hydrolase activity was found in cell-free extracts from young tea leaves. Substantial amounts of radioactivity from [adenosyl-^{14}C]SAH were also recovered as theobromine and caffeine in tea leaf segments, indicating that adenosine derived from SAH is utilised for the synthesis of the purine ring of caffeine.

In addition, direct adenosine kinase-catalysed formation of AMP from adenosine may participate as an alternative minor route. The activity of two of the three *N*-methyltransferase activities involved in caffeine biosynthesis and part of the activities of SAH hydrolase, adenosine nucleosidase, adenine phosphoribosyltransferase and adenosine kinase were located in tea chloroplasts. In contrast, no detectable activity of SAM synthetase was associated with the purified chloroplast fraction.

In preparations from young leaves, N-3-methyltransferase was found associated with a purified chloroplast preparation obtained by using differential centrifugation and a discontinuous Percoll density gradient.

Ashihara *et al.* (1997) studied the purine alkaloid catabolism pathways in young, mature and aged leaves of tea by incubating leaf sections with ^{14}C-labelled theobromine, caffeine, theophylline and xanthine. They demonstrated that theobromine is the immediate precursor of caffeine, which accumulates in tea leaves because its conversion to theophylline is the rate limiting step in the purine alkaloid catabolism pathway.

The main fate of [U-^{14}C]-theophylline incubated with mature and aged leaves, and to a lesser extent young leaves, is conversion to 3-methylxanthine and onto xanthine which is degraded to $^{14}CO_2$ via the purine catabolism pathway. However, in young leaves, sizable amounts of [U-^{14}C]-theophylline were salvaged for the synthesis of caffeine via 3-methylxanthine to theobromine and caffeine. This was enhanced in the presence of 5 mM allopurinol which inhibits purine catabolism.

Caffeine synthase (**EC 2.1.1.160**) is an enzyme that catalyses the final two steps in the caffeine biosynthesis pathway. Caffeine synthase, comprises of 3-N-methyltransferase (3-NMT) and 1-N-methyl-transferase (1-NMT) activities, catalyses the final two steps of caffeine biosynthesis, when two methyl groups are added successively to 7-methylxanthine to produce theobromine and then caffeine (Suzuki, Takahashi 1975, Kato *et al.* 1999). Recently, the gene encoding caffeine synthase from young tea leaves of has been cloned (Kato *et al.* 2000).

Biosynthetic route of caffeine in tea is as described below:

SAM→SAH→adenosine→adenine→AMP→IMP→XMP→xanthosine→ 7-methylxanthosine→7-methylxanthine→theobromine→Caffeine

Decaffeination: Solvent extraction, column chromatography and hot water treatment have been used by researchers for selectively removing caffeine from different tea. Though organic solvents *viz.*, chloroform and methylene chloride remove caffeine from foods but their use has not been recommended due to the toxicity of these solvents. Hot water treatment has the disadvantage that it removes soluble solids and secondary metabolites along with the caffeine and is an expensive proposition. In addition, energy costs to make CO_2 to reach supercritical state are shown to be less than those associated with conventional solvent distillation. Efforts have been made to decaffeinate green tea and black tea using super critical CO_2 green tea (Hyong *et al.* 2007). As most of flavour volatile components were decreased during decaffeination process in green tea (Lee *et al.* 2007). The decaffeination process using SC-CO_2 reduced not only caffeine but also reduced some important volatile flavour components of tea (Lee *et al.* 2009).

In another study, attempt were made to produce a cultivar with low-caffeine content using RNAi technology. A partial fragment of caffeine synthase (CS) from the Kangra jat cultivar was used to design RNAi construct (pFGC1008-CS). Somatic embryos were transformed with this construct showed reduction in the levels of CS transcript expression as well as in caffeine content (Mohanpuria unpublished work).

LIPIDS

Lipid content increased with shoot maturity. It varied from 30 mg/g in bud to almost 72 mg/g in the fourth leaf. C6 aldehydes are obtained from C18 unsaturated fatty acids by lipid oxidase in leaf chloroplast.

The order of fatty acids (FA) content in the leaves and stalk of tea shoots is linolenic acid > linoleic acid > palmitic acid > stearic acid > oleic acid > palmitoleic acid, irrespective of the plucking standard or portion of the shoot. The stem has the least FA levels. Linolenic acid is the most dominant of the FA and increases with leaf-maturity and coarser plucking standard. Total FA, total unsaturated FA and linolenic and linoleic acids are higher in coarse pluck compared to standard pluck and increase with leaf maturity. That is why the quality deteriorates with higher amount of the Group I volatile flavour compounds arising from the coarse plucking standard.

Flavour of tea is dependent on the availability of precursors inherently present in tea shoot, stimulation of condition for their formation and retention of flavour component in the product. Flavour comprises of taste and aroma. Non-volatile components are generally responsible for the taste, while volatile components (VFCs) give aroma. Aroma of tea is a complex mixture of carbonyl, terpenoid and other volatile compounds (Table 18.2). Aroma is considered to have high impact on the flavour of food products, since the sense of smell (olfaction) is more developed than that of taste. The aroma characteristics of various teas i.e. green, black and oolong is influenced by various factors including the specific variety of tea plant (Cloughley *et al.* 1982), maturity of tea leaves, the growing environmental conditions as well as specific method of tea processing and manufacture like withering, rolling, fermentation and drying.

Table 18.2. Broad categorization of volatile compounds identified from tea aroma

		Compound	Number
I		Hydrocarbons	72
	1	Aliphatic	14
	2	Aromatic	25
	3	Terpenoid	33
II		Alcohols	89
	1	Aliphatic	51
	2	Aromatic	5
	3	Terpenoid	33
III		Aldehydes	68
	1	Aliphatic	45
	2	Aromatic	18
	3	Terpenoid	5
IV		Ketones	75
	1	Aliphatic	30
	2	Alicyclic	10
	3	Aromatic	16
	4	Terpenoid	3
	5	Ionone derivatives	16
V		Acids	69
	1	Aliphatic	63
	2	Aromatic	3
	3	Terpenoids	3
VI		Esters	82
	1	Aliphatic	52
	2	Alicyclic	3
	3	Aromatic	19
	4	Terpenoid	8
VII		Lactones	25
VIII		Phenolic compounds	22
IX		Oxygenated compounds	36
	1	Furanoid	17
	2	Aromatic	10
	3	Ionone related	5
	4	Others	4
X		Nitogenous compounds	86
	1	Pyrroles	12
	2	Pyridines	17
	3	Pyrazines	24

	4	Other nitrogenous compounds	33
XI		Sulfur compounds	14
	1	Total compounds	63

In tea, volatile components are present in very minute quantities *i.e.* 0.01% to 0.03% of the total dry weight. Morethan 600 volatile components have been identified so far (Yamanishi *et al.* 1973).

The composition, concentration and odour threshold of VFCs play an important role in the final flavour of tea and their valuation at the auction centers.

VFCs are classified into two groups based on either odour characteristic or the retention time of the aroma compounds during gas chromatographic analysis (Table 18.3).

Table 18.3 Classification of volatile flavour compounds in Group I and Group II

Compounds	Flavour
Group I	
1- Hexanol	Resinous
1-Hexenol	Fresh Green
2-Hexenol	Fresh Green
(*Z*)-3-Hexen-1-ol	Fresh Green
Heptanol	Fatty
(*Z*)-4- Heptenal	Biscuity
Octanol	Soapy
1-pentanol	Balsamic
Group II	
Benzyl alcohol	Sweet
Benzaldehyde	Almond
2-phenylethanol	Like honey
Phenylacetaldehyde	Hawthorne
Citronellol	Rosy
Geraniol	Rosy
Linalool	Lemon
Linalool oxides	
Nerol	Balsamic
α- terpineol	---
Nerolidol	---
Eugenol	---
Methyl salicylate	Peppermint
α-ionone	---
β-ionone	Like violet (flowery)
Methyl epijasmonate	flowery

Group I consists of mainly non-terpenoids, which imparts grassy aroma e.g. Hexenols, hexanals, hexenals, pentanol, 1-penten3-ol. Group II is mainly of terpenoids, which impart sweet flowery aroma to tea *viz.,* monoterpene alcohols like linalool and geraniol and also includes benzaldehyde, phenylacetaldehyde, methyl salicylate, benzyl alcohol and β-ionone. The presence of Group II compounds is highly desirable. The ratio of terpenoid (Group II) to non-terpenoid (Group I) the so called flavour index (FI) has been used to classify black teas for flavour quality. The tea quality increases with the increase in flavour index.

BIOGENESIS OF AROMA DURING MANUFACTURE

Black tea manufacture involves plucking, withering, rolling, fermentation and drying steps to give final product made tea. Various volatile compounds are liberated from their precursors after mechanical breakdown of leaf. Following types of mechanisms are involved in aroma formation:

Oxidative degradation

Carotenoids are converted into β-ionone derivatives by oxidative degradation during fermentation and drying process of tea manufacture. Carotenoid content reduces during the process (Hazarika and Mahanta 1984). They are present in minute quantity but contribute to the overall aroma of tea, as their threshold limit of perception is very low for e.g. methyl epijasmonate an important flavour compound that imparts flowery aroma, is produced by carotenoid degradation. (+) Z-Methyl epijasmonate is 400 times more flavoury than its isomer (-) Z-methyl jasmonate which is reported to be odourless (Sarkar *et al.* 1994). But methyl epijasmonate (cis) is quite unstable and is easily isomerised to methyl jasmonate (trans) by heat (Kobayashi *et al.* 1994).

Thermal generation of aroma compounds

Various aldehydes, furans, pyarzines and pyrroles are generated from sugars and amino acids during final step of tea manufacture i.e. drying. They five slightly roast aroma to the made tea. Short-chain volatile acids like propanoic, butanoic, hexanoic, heptanoic and various other acids that contribute to the tea aroma are thermal degradation products of fats or long chain fatty acids. A large number of lactones (olides) like butanolides contribute to the aroma of oolong

terpenoid compounds from leucine leads to the development of tea flavour, whereas intra-chloroplastidic fixation of CO_2, and the consequent derivation of terpenoids from acetate results in the production of teas without flavour. The leucine pathway for terpenoid biogenesis operates in conditions of climatic stress which limit CO_2 fixtation in the leaf, whereas the acetate pathway predominates in conditions of vigorous growth.

POLYPHENOL OXIDASE

Polyphenol oxidase (PPO) is an important copper-containing enzyme of tea which catalyses the conversion of catechins to theaflavins (Figure 18.3). This enzyme has been purified and characterised from tea leaves (*Camellia sinensis*). There are three isozymic forms of this enzyme (Haldar 1998). The active purified enzyme is 72 kD. The pH, temperature, and kinetic parameters were studied. The enzyme remains active in pH range is 4.03-7.00, optimum temperature 30°C. The highly purified enzyme was unable to oxidize monophenols, p-quinol but could oxidize catechol. Catechin was the best substrate with a K_m of 0.49 mM. The enzyme is inhibited by tropolone. The enzyme is localized in chloroplast and solubilized with Triton-X-100 (Haldar 1998). The ratio of PPO:catechins ratio has been found to play an important role in determining the quality potential of tea (Lopez *et al.* 2005). PPO activity is higher in tender leaves than in mature leaves (Takeo and Baker 1973, Thanaraj and Seshadri 1990). PPO activity reduces during withering and storage of the fresh leaves. Very hard wither or high reduction of moisture results in heavy loss of (50-55%) PPO activiy (Ullah and Roy 1982).

Variation in polyphenol oxidase (PPO) and phenylalanine ammonia lyase (PAL) activities with respect to season, different cultural and manufacturing processes and their effects on black tea quality have been studied. Variations exist in enzyme activities of different clones, as well as seasonal changes and shoot maturity. Field practices such as plucking standards and pruning have a great impact on enzyme activities. The enzyme activities positively correlated with tasters' scores. The extent of change in enzyme activities at different stages of manufacture differed widely. The loss of activity during withering could be restored by rehydration. Residual activity was observed in made tea. Supplementation of enzymes enhanced the black tea quality markedly in terms of cuppage and creaming properties.

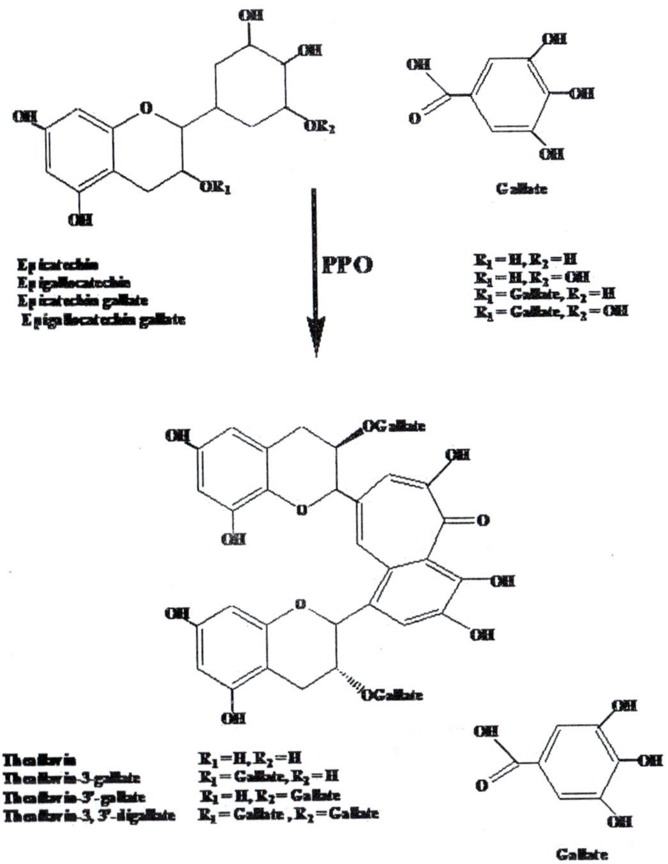

Fig. 18.3. Role of polyphenol oxidase enzyme

The relative amounts of chlorophyll, leucine, α-alanine and α-ketoisocaproic acid in tea leaves, and those of carotenoids, β-ionone, dihydroactinidiolide, theaspirone, β-amyrin, acetate, trans-2-hexenal and phenlacetaldehyde in processed black tea vary with climatic conditions, and indicate the mechanism of flavour development in tea. These variations are related to the degree of CO_2 fixation, and the extent of intra- as compared with extra-chloroplastidic biosynthetic reactions. Extra-chloroplastidic biogenesis of terpenoid compounds from leucine leads to the development of tea flavour, whereas intra chloroplastidic fixation of CO_2, and the consequent derivation of terpenoids from acetate results in the production of teas without flavour. The leucine pathway for terpenoid biogenesis operates in conditions of climatic stress which limit CO_2 fixtation in the leaf, whereas the acetate pathway predominates in conditions of vigorous growth.

Enzymes related to nitrogen assimilation: Studies of the activities of the enzymes nitrate reductase (NR), nitrite reductase (NIR), glutamate dehydrogenase (GDH), glutamine synthetase (GS) and glutamate synthase (GOGAT) in tea plants subjected to 100 to 300 kg N/ha showed NR and NIR activities negatively correlated with nitrogen application beyond 200 N kg/ha, while GDH activity was positively correlated up to 300 kg N/ha (TRA Annual Report). Potassium enhanced the nitrate reductase activity. Application of phosphate up to 50 kg P_2O_5 with 150–200 kg N/ha enhanced the activities of NR, NIR, GDH, GOGAT and acid phosphatase both in shoot and roots of tea plants. Higher doses of phosphate (>50 kg P_2O_5) reduced the activity of these enzymes. Application of higher doses of nitrogen (200-300 kg N/ha) showed accumulation of amide (mainly theanine) in feeder roots alongwith depletion of starch. When applied with adequate potassium, the accumulation of amide considerably reduced, nitrate reductase activity and starch reserve enhanced. However, continuous use of high rates of potash increased crude fibre. Application of NPK (330-60-290 kg/ha) adversely affected the synthesis of catechins in tea shoots thereby lowering the quality of black tea as theaflavins, thearubigins, brightness and total colour are all catechin dependent (Deb and Ullah 1968). However, caffeine the major fatty acid increased in the shoots resulting in deterioration in quality of black tea. (Fertilizer?).

REFERENCES

Alexopoulos N, Vlachopoulos C, Aznaouridis K, Baou K, Vasiliadou C, Pietri P, Xaplanteris P, Stefanadi E, Stefanadis C (2008) The acute effect of green tea consumption on endothelial function in healthy individuals. *Euro J Prev Cardio*. 15: 300-305.

Ament K, Kant MR, Sabelis MW, Haring MA Schuurink RC (2004) Jasmonic acid is a key regulator of spider mite-induced volatile terpenoid and methyl salicylate emission in tomato. *Plant Physiol*. 135: 2025–2037.

Anderson RA, Polansky MM (2002) Tea enhances insulin activity. *J Agric Food Chem*. 50: 7182-7186.

Batista J, Lima E, Paiva L, Andrade AL, Alves MG (2012) Comparison of Azorean tea theanine to teas from other origins by HPLC/DAD/FD. Effect of fermentation, drying temperature, drying time and shoot maturity. *Food Chem*. 32: 2181-2187.

Bhatia IS, Ullah MR (1968) Polyphenols of tea IV- Qualitative and quantitative study of the polyphenols of different organs and some cultivated varieties of tea plant. *J Sci Food Agric*. 19:535-542.

Bronner WE, Beecher GR (1998) Method for determining the content of catechins in tea infusions by high-performance liquid chromatography. *J Chromatogr A.* 805: 137-142.

Bokuchava MA, Skobeleva NI (1980) The biochemistry of tea manufacture. *CRC Crit Rev Food Sci Nutr.* 12: 303–370.

Casimir J, Jadot J, Renard M (1960) Separation and characterization of *N*-ethyl-γ-glutamine from *Xerocomus badius*. *Biochim Biophys Acta.* 39: 462-468.

Chakraborty S, Srivastava RAK, Choudhury MND (1978). Distribution of free amino acids and chlorogenic acids in different parts of "two leaves and a bud" and its relation to tea quality. *Two and a Bud* 25(1): 17-21.

Chu S, Herskowitz, I (1998) Gametogenesis in yeast is regulated by a transcriptional cascade dependent on Ndt80. *Mol Cell.* 1: 685–696.

Cloughley JB, Ellis RT, Pendlington F, Humphrey P (1982) Volatile constituents of some Central African black tea clones. *J Agric Food Chem.* 30: 842-845.

Dalluge JJ, Nelson BC, Thomas JB, Sander LC (1998) Selection of column and gradient elution system for the separation of catechins in green tea using high-performance liquid chromatography. *J Chromatogr A.* 793: 265-274.

Deb SB, Ullah MR (1968). The role of theaflavins (TF) and thearubigins (TR) in the evaluation of black tea. *Two and a Bud* 15: 101–102.

Dixon RA, Paiva NL (1995) Stress-induced phenylpropanoid metabolism. *Plant Cell* 7: 1085–1097.

Dooner HK, Robbins TP, Jorgensen RA (1991) Genetic and developmental control of anthocyanin biosynthesis. *Annu Rev Genet.* 25: 173–199.

Egashira N, Tanoue A, Higashihara F, Mishima K, Fukue Y, Takano Y, Tsujimoto G, Iwasaki K, Fujiwara M (2004) V1a receptor knockout mice exhibit impairment of spatial memory in an eight-arm radial maze. *Neurosci Lett.* 356: 195–198.

Ekborg-Ott KH, Taylor A, Armstrong D W (1997) Varietal differences in the total and enantiomeric composition of theanine in tea. *J Agric Food Chem.* 45: 353-363.

Feldheim W, Yongvanit P, Cummings PH (1986). Investigation of the presence and significance of theanine in the tea plant. *J Sci Food Agric.* 37(6): 527-534.

Fisher N, Nitz S, Drowert F (1987) Bound flavour compounds in plants. 2nd part: Free and bound flavour compounds in green and black teas. *Z. LebensmUntress Forsch.* 185: 195-201.

Goto T, Yoshida Y, Kiso M, Nagashima H (1996) Simultaneous analysis of individual catechins and caffeine in green tea. *J Chromatogr. A* 749: 295-299.

Gulati A, Ravindranath SD (1996). Seasonal variations in quality of Kangra tea (*Camellia sinensis)* O. Kuntze in Himachal Pradesh. *J Sci Food Agric*. 71: 231-236.

Gunata Z, Bitteur S, Brillouet JM, Bayonove C, Cordonnier R (1988) Sequential enzymic hydrolysis of potentially aromatic glycosides from grape. *Carbohydrate Res*. 184: 139-149.

Gunata Z, Blondeel C, Vallier MJ, Lepoutre JP, Sapis JC, Watanabe N (1998) An endoglycosidase from grape berry skin of cv. M. Alexandrie, hydrolyzing potentially aromatic dissacharide glycosides. *J Agric Food Chem*. 46:2748-2753.

Guo J, Han W, Wang M-H (2008) Ultraviolet and environmental stresses involved in the induction and regulation of anthocyanin biosynthesis: A review. *Afr J Biotechnol*. 7: 4966-4972.

Guo W, Hosoi R, Sakata K, Watanabe N, Yagi A, Ina K, Luo S (1994) (*S*)-linalyl, 2-phenylethyl and benzyl disaccharide glycosides isolated as aroma precursor from oolong tea leaves. *Biosci Biotech Biochem*. 58: 1532-1534.

Halder J, Tamuli P, Bhaduri AN (1998) Isolation and characterization of polyphenol oxidase from Indian tea leaf (*Camellia sinensis*). *J Nutr Biochem*. 9: 75-80.

Hatanaka A (1993) The biogeneration of green odour by green leaves. *Phytochem*. 34: 1201-1218.

Hazarika M, Chakravarty SK, Mahanta PK (1984) Studies on thearubigin pigments in black tea manufacturing systems. *J Sci Food Agric*. 35: 1208-1218.

Hoefler AC, Coggon P (1976) Reversed-phase highperformance liquid chromatography of tea constituents. *J Chr A*. 129: 460–463.

Hyong SP, Lee HJ, Shin MH, Lee KW, Lee H, Kim YS, Kim KO, Kim KH (2007) Effects of cosolvents on the decaffeination of green tea by supercritical carbon dioxide. *Food Chem*. 105: 1011–1017.

Juneja SC, Barr KJ, Enders GC, Kidder GM (1999) Defects in the germ line and gonads of mice lacking connexion. *Biol Reprod*. 60: 1263-1270.

Katiyar SK, Mukhtar H (1996) Tea in chemoprevention of cancer: Epidemiologic and experimental studies. *Int J Oncol*. 8: 221–238.

Kakuda R, Imai M, Yaoita Y, Machida K, Kikuchi M (2000) Secoirridoid glycosides from the flower buds of *Lonicera japonica*. *Phytochem*. 55: 879-881.

Kakuda T (2002) Neuroprotective effects of the green tea components theanine and catechins. *Biol Pharm Bull*. 25: 1513–1518.

Kato M, Mizuno K, Fujimura, T, Iwama M, Irie M, Crozier A, Ashihara H (1999) Purification and characterization of caffeine synthase from tea leaves. *Plant Physiol*. 120: 579–586.

Kato M, Mizuno K, Crozier A, Fujimura T, http://www.nature.com/nature/journal/v406/n6799/abs/406956a0.html - a2Ashihara H (2000) Plant biotechnology: Caffeine synthase gene from tea leaves. *Nature* 406, 956-957

Kito Y, Suzuki T, Azuma M, Sekoguti Y (1968) Absorption Spectrum of Rhodopsin denatured with Acid. *Nature* 218(5145): 955-957.

Kobayashi A, Kubota K, Joki Y, Wada E, Wakabayashi M (1994) (Z)-3-Hexenyl β-D-glucopyranoside in fresh tea leaves as a precursor of green odour. *Biosc Biotech Biochem.* 58: 592-593.

Koketsu M (1997) Antioxidant activity of tea polyphenols. In: Yamamoto T, Juneja LR, Chu DC, Kim M (ed) *Chemistry and Application of Green Tea*, CRC Press, London, U.K.: 37-50.

Lee SM, Lee HS, Kim KH, Kim KO (2009) Sensory characteristics and consumer acceptability of decaffeinated green teas. *J Food Sci.* 74 (3): S135-S141.

Lee S, Park MK, Kim KH, Kim YS (2007) Sensory and nutritive qualities of food effect of supercritical carbon dioxide decaffeination on volatile components of green teas. *J Food Sci.* 72 (7): S497-S502.

Lin YL, Juan IM, Chen YL, Liang YC, Lin JK (1996) Composition of polyphenols in fresh tea leaves and associations of their oxygen radical-absorbing capacity with antiproliferative actions in fibroblast cells. *J Agric Food Chem* 44: 1387–1394.

Lopez SJ, Thomas J, Pius PK, Raj Kumar R, Muraleedharan N (2005) A reliable technique to identify superior quality clones from tea germplasm. *Food Chem.* 91: 771–778.

Mahanta PK, Baruah S, Owuor PO, Murai T (1988). Flavour volatiles of Assam CTC black teas manufactured from different plucking standards and orthodox teas manufactured from different altitudes of Darjeeling. *J Sci Food Agric.* 45: 317-324.

Millin DJ, Swaine D, Disc PL (1969) Separation and classification of the brown pigments of aqueous infusion of black tea. *J Sci Food Agric.* 20: 296-302.

Moon JH, Watanabe N, Ijima Y, Yagi A, Ina K, Sakata K (1996) *Cis*- and *trans*-linalool 3,7-oxides and methyl salicylate glycosides and (*Z*)-3-hexenyl-β-D-glucopyranoside as aroma precursors from tea leaves for oolong tea. *Biosc Biotech Biochem.* 60: 1815-1819.

Moon JH, Watanabe N, Sakata K, Inagaki J, Yagi A, Ina K, Luo S (1994). Linalyl-ß-D-glucopyranoside and its 6'-*O*–malonate as aroma precursor from *Jasminum sambac. Phytochem.* 36: 1435-1437.

Moon JH, Watanabe N, Sakata K, Yagi A, Ina K, Luo S (1994). Trans-and cis-linalool 3,6 oxide 6-*O*- β-D-xylopyranosyl- β-D-glucopyranosides isolated as aroma precursor from leaves for oolong tea. *Biosc Biotech Biochem.* 58: 1742-1744.

Mukhtar H, Katiyar SK, Agarwal R (1994) Green tea and skin anticarcinogenic effects. *J Invest Dermatol*. 102: 3-7.

Nakachi K, Matsuyama S, Miyake S, Suganuma M, Imai K (2000) Preventive effects of drinking green tea on cancer and cardiovascular disease: Epidemiological evidence for multiple targeting prevention. *BioFactors* 13: 49–54.

Nishikitani M, Kubota K, Kobayashi A, Sugawara F (1996) Geranyl 6-*O*- β-L-arabinopyranosyl–β-D-glucopryranoside isolated as an aroma precursor from leaves of agreen tea cultivar. *Biosc Biotech Biochem*. 60: 929-931

Ogawa K, Ijima Y, Guo W, Watanabe N, Usui T, Dong S, Tong Q, Sakata K (1997) Purification of a ß-primverosidase concerned with alcoholic aroma formation in tea leaves (cv. Shuxian) to be processed to oolong tea. *J Agric Food Chem*. 45: 877-882.

Owour PO,Tushida T, Horita H, Murai T (1988). Effects of geographical area of production on the composition of the volatile flavour compounds in Kanyanclonal black teas. *Exptl Agric*. 24: 227-235.

Owour PO, Chavanji AB (1986) Caffeine content of clonal tea, seasonal variations and effect of plucking standards under Kenyan conditions. *Food Chem*. 20: 225–233.

Rietveld A, Wiseman S (2003) Antioxidant effects of tea: Evidence from human clinical trials. *J Nutr.* 133: 3285S-3292S.

Robards Kevin, Antolovich Michael (1997) Analytical chemistry of fruit bioflavonoids- A review. *Analyst* 122: 11R–34R.

Robards K, Li X, Antolovich M, Boyd S (1997). Characterization of citrus by chromatographic analysis of flavonoids. *J Sci Food Agric*. 75: 87-101.

Robinson JM, Owuor PO (1992) Tea aroma. In: Willson KC, Clifford MN (ed) *Tea Cutivation to Consumption*. Chapman & Hall, London. pp 603-626.

Sakanaka S, Kim M, Taniguchi M, Yamamoto T (1989) Antibacterial substances in Japanese green tea extract against Streptococcus mutans, a carcinogenic bacterium. *Agric Biol Chem*. 53: 2307-2311.

Sakata K, Guo WF, Moon JH (1995). Molecular basis of alcoholic aroma formation in Oolong tea. In: Proceedings of 95th International Tea Quality- Human Health Symposium in Shanghai. Hangzhou Tea Society of China. pp 175-187

Sakato Y (1950) The chemical constituents of tea: III. A new amide, theanine. *Nippon Nogei Kagaku Kaishi* 23: 262-264.

Salt SD,Twzen S, Kuc J (1986) Effects of β-ionone and abscicic acid on growth and resistance to blue mold. *Physiol Mol Plant Pathol*. 28: 287-297.

Sarkar A, Layania SC, Pandey DN, Pant MC (1994) Changes in blood lipid profile after administration of Tulsi (*Ocimum sanctum*) leaves in the normal albino rabbits. *Indian J Physiol Pharmacol*. 38 (4): 311-312.

Sanderson, GW (1972) Recent Advances in Phytochemistry. Runeckles, VC and Tso, TC (eds). Academic Press, New York.5: 247-316.

Sasaoka K, Kito M, Onishi Y (1965) Some properties of the theanine synthesizing enzyme in tea seedlings. *Agricultural and Biological Chemistry* 29: 984-988.

Selvendran RR, Selvendran S (1972) Chemical changes in young tea plant (*Camellia sinensis* L.) tissues following application of fertilizer nitrogen. *Ann Bot.* 37: 453-461.

Seskar, M, Shulaev V, Raskin I (1998) Endogenous methyl salicylate in pathogen inoculated tobacco plants. *Plant Physiol.* 116: 387-392.

Sharma V, Gulati A, Ravindranath SD, Kumar V (2005) A simple and convenient method for analysis of tea biochemicals by reverse phase HPLC. *J Food Comp Anal.* 18: 583-594.

Singh HP, Singh C, Ravindranath SD (1999) Analysis of tea shoot catechins: Spectrophotometric quantitation and selective visualization on two dimensional paper chromatograms using diazotized sulphanilamide. *J Agric Food Chem.* 48: 1041-1045.

Suzuki T, Takahashi E (1975) Biosynthesis of caffeineby tea-leaf extracts. Enzymic formation of theobromine from 7-methylxanthine and caffeine from theobromine. *Biochem J.* 14: 87–96.

Takeo T, Baker JE (1973). Changes in multiple forms of polyphenol oxidase during maturation of tea leaves. *Phytochem.* 12: 21–24.

Takeo T (1981) Production of linalool and geraniol by hydrolytic breakdown of bound forms in distrupted tea shoots. *Phytochem.* 20: 2145-2147.

Takeo T (1982). *Nippon Nogeikagaku Kaishi* (in Japanese). 56: 455-457.

Takeo T (1984) Production of linalool and geraniol by hydrolytic break down of bound forms in distrupted tea shoots. *Phytochem.* 20: 2145-2147.

Takeo T, Mahanta PK (1983). Comparison of black tea aroma of orthodox and CTC manufacture and black tea made from different varieties. *J Sci Food Agric.* 34: 307-310.

Thanaraj SNS, Seshadri R (1990) Influence of polyphenol oxidase activity and polyphenol content of tea shoot on quality of black tea. *J Sci Food Agric* 51: 57–69.

Treutter D (1989) Chemical reaction detection of catechins and proanthocyanidins with 4-dimethylaminocinnamaldehyde. *J Chromatogr A.* 467: 185-193.

Ullah MR, Roy PC (1982) Effect of withering on the polyphenol oxidase level in the tea leaf. *J Sci Food Agric* 33: 492–495.

Umegaki K, Sugisawa A, Yamada K, Higuchi M (2001). Analytical method of measuring tea catechins in human plasma by solid phase extraction and HPLC with electrochemical detection. *J Nutr Sci Vitaminol.* 47: 402-408.

Wang HF, Helliwel K, You X (2000) Isocratic elution system for the determination of catechins, caffeine and gallic acid in green tea using HPLC. *Food Chem.* 68: 115-121.

Wang LY, Wei K, Jiang YW, Cheng H, Zhou J, He W, Zhang CC (2011) Seasonal climate effects on flavanols and purine alkaloids of tea (*Camellia sinensis* L.) *Eur Food Res Technol.* 233: 1049–1055.

Yamanishi T, Shimojo S, Ukita M, Kawashima K, Nakatani Y (1973) Aroma of roasted green tea (Hoji-cha). *Agric Biol Chem.* 37: 2147-2153.

Yano M, Okada K, Kubota K, Kobayashi A (1990) Studies on the precursors of monoterpene alcohols in tea leaves. *Agric Biol Chem.* 54: 1023-1028.

Yao L, Caffin N, D'Arcy B, Jiang Y, Shi J, Singanusong R, Liu X, Datta N, Kakuda Y, Xu Y (2005) Seasonal variations of phenolic compounds in Australia-grown tea (*Camellia sinensis*). *J Agric Food Chem.* 53: 6477–6483.

19

TEA MANUFACTURE

Ashu Gulati

CSIR-Institute of Himalayan Bioresource Technology
Palampur (HP)- 176 061, India. E-mail: ashugulati@ihbt.res.in

Tea is one of the most widely consumed drinks in the world. In the Far East (particularly in China and Japan), tea is consumed mainly as a hot infusion of either "unfermented" tea (green tea) or partially fermented tea (oolong tea) whereas in most other countries the beverage is prepared from predominantly fermented (black) tea. The fermentation is a misleading term as there are no microbes involved in tea manufacture. Now the term fermentation has been replaced by oxidation but in tea terminology, fermentation is still the jargon in use. Hence, in this chapter term fermentation will be used. The major classification in the three types of manufactured teas is the degree of oxidation of catechins. Of the approximately 4.0 million metric tons of dried tea manufactured, only 25% is green tea and less than 3% is oolong tea. Chemical composition of green tea is very similar to that of the fresh leaf except for a few enzymatically catalyzed changes which occur extremely rapidly following plucking. New volatile substances are produced during the drying stage. Oolong tea is intermediate in composition between green and black teas.

Quality of made tea is determined by the chemical composition of the shoot, the environmental conditions in which it grows and process conditions. The different methods of manufacture of tea account for the marked differences in the chemical compositions of green, oolong and black teas.

GREEN TEA

Green tea is unfermented or unoxidised tea and manufactured in such a manner that the oxidation of green leaf polyphenols is prevented. If the shoots are left as such then the polyphenols are acted upon by the enzyme polyphenol oxidase and are oxidized to quinines. To prevent this oxidation of polyphenols, it is essential to 'fix' or inactivate the enzyme polyphenol oxidase. The major steps therefore in green tea manufacture are:

Inactivation of enzyme polyphenol oxidase (PPO), cooling and rolling, partial drying, second rolling, drying, and sieving and polishing.

The most important step in green tea manufacture is inactivation of the enzyme PPO. Tea shoots immediately after plucking are either steamed or roasted for a short while so as to inactivate the enzyme polyphenol oxidase. The Japanese steam the shoots to fix the enzyme while the Chinese parch or roast tea shoots for green tea manufacture.

In the Japanese way of manufacture, tea shoots immediately after plucking are steamed in a steamer for 3-5 minutes and left on the floor for extra moisture to dry. This also cools and stabilizes the shoot components and makes it suitable for rolling. The shoots are then rolled for 10 minutes at ambient temperature and half-dried at 70-85°C so as to reduce the leaf-moisture to about 50-55%. The shoots are again rolled under pressure to give proper shape and dried in hot pans at 90°C to decrease the moisture level to 2-3%.

Popular green teas of Japan are Sen-cha, Ban-cha, Gyukuro-cha. The high quality Sen-cha is produced from the first crop of spring. This crop has high amino acid content specially L-theanine and low contents of flavanols. There exists a positive correlation between taste of Sen-cha and amino acid content and negative correlation with flavanol content. The Japanese use extremely high content of nitrogen fertilizers for this purpose.

In the Chinese way of manufacture, the shoots are roasted in big pans. Roasting is accomplished at high temperatures as high as 230°C, for 8-10 minutesand moisture level is reduced to around 60%. The shoots are then cooled as is done for Japanese tea. The cooled shoots are rolled at ambient temperature for 10 minutes and pan fired at 150°C for 10-25 minutes and the moisture level of the shoots is brought down to about 20%. The shoots are rolled again under pressure and finally dried in pans at 80°C, for 40-60 minutes to bring the moisture content at 3-5%.

The popular Chinese green teas are Kamaira-cha, Chun Mee and Sow Mee.

The teas produced by the above procedures are sifted and refined to remove stalk and fibre. In Japan, teas are polished to give proper shape and shine. Teas are sorted and graded before they are sold in the market.

Recently, inactivation of polyphenol oxidase by microwave heating, and oven heating were shown to give enhanced quality green tea and better shelf-life. Microwave inactivated and dried teas showed highest levels of total phenols and catechins, their infusion was bright in colour and sweet in taste with subtle pleasant odour (Gulati *et al.* 2003). The microwave inactivated teas had enhanced preserved quality of green tea in terms of vitamin C and chlorophylls (Huang *et al.* 2007). Tea made using microwave had lower levels of pesticide residues (Sood *et al.* 2003).

The quality attributes of green tea are polyphenols, amino acids, caffeine and volatile flavour components. The astringency of green tea brew has been ascribed to polyphenols in tea, mainly the flavanols. The brothy taste of the brew has been correlated to amino acids, especially to L-theanine. Caffeine contributes to the astringency, through complexation with flavanols. Best quality green tea (Sen-cha) is made from chinary tea shoots with high amino acid content and low flavanol content. The content of ascorbic acid is taken as an index of freshness for green tea. Ascorbic acid content is higher in young shoots than in mature shoots, and more in unshaded than in shaded tea plants. Quality of green tea deteriorates with fall in ascorbic acid content. Recently, two new methylgallates *viz*., (-)-epigallocatechin-3-(3-*O*-methylgallate) and (-)-epigallo-catechin-3-(4-*O*-methyl gallate) have been found in Chinese green tea

Systematic studies on changes in PPO activity due to fixation time, polyphenols, amino acids, and soluble sugars due to fixation temperature, and polyphenols, amino acid, chlorophyll contents, soluble sugars and total sugars during green tea manufacture have been conducted. Variations in chemical components of roasted and steam roasted Korean green teas have also been reported in respect of crude lipid, total nitrogen and water soluble protein. Chemical and organoleptic tests revealed that total nitrogen gave the best positive correlation between the quality of green tea and the chemical constituents.

Green tea producing countries prefer the tea varieties with low catechin content and high amino acid contents (6-8 %) on dry weight basis. The ceremony teas of Japan are produced from fresh leaves of the first crop plucked from shaded tea plants with rice straw. The levels of theanine and caffeine in the first crop are high due to shade. The contents of caffeine and theanine decrease with progress of season. The flavanol contents are low throughout the season.

The fresh and greenish aroma in green tea is due to (Z)-3-hexen-1-ol, and its esters. Most important roast aroma compounds 2,5-dimethyl pyrazine and 1-ethylpyrrol-2-aldehyde are present in Chinese green tea. 2,3-diethyl-5-methyl pyrazine and 2 ethyl-3,5-methyl pyrazine were the major aroma components in Chinese Kaimari-cha. The superior roast aroma of best quality teas are obtained by using leaves with high amino acid contents.

Seasonal variations in green tea aroma and effect of shade by rice straw have also been studied by Japanese workers. Dimethylsulphide formed by degradation of S-methyl methionine is one of the characteristic flavour compound in these teas. Lesser proportions of geraniol and linalool are present in total aroma in teas grown in shade with rice straw compared to teas grown in unshaded areas. Sen-cha manufactured during spring season is characterized by the presence of 4-mercapto 4-methyl pentanone as the characteristic aroma component along with α-damascenone. The content decreases with progress in season. Seasonal clonal variations in green tea aroma are influenced by liberation of monoterpene alcohols, especially linalool and geraniol, liberated from glycosides in tea shoots. Terpene index, based on geraniol and linalool contents, was constituted for flavoury teas. Recently, primeverosides of linalool, linalool oxides, geraniol and methyl salicylate have been found in fresh tea shoots and the enzyme primeverosidase, responsible for flavour generation characterized

The contributors to the differences in the odors of different grades of green tea were selected, and the relations to the odors of their infusions showed D-nerolidol, 6-methyl-alpha-ionone, methyl jasmonate, coumaran, indole, and coumarin were possible contributors to a typical green tea odor. 3-Hexenoic acid and methyl jasmonate along with aliphatic alcohols and aldehydes provided fresh and brisk odors. 3-Hexenoic acid, 3,5,5-trimethyl-2(5H)-furanone, and 5,6,7,7a-tetrahydro-4,4,7a-trimethyl-2(4H)-benzofuanone gave deep and mild odors. Linalool, 2,6-dimethyl-1,3,7-octatrien-6-ol, benzeneacetaldehyde, and 3-hexenyl hexanoate afforded floral and fruity odors. 1-Ethyl-1H-pyrrol-2-carboxyaldehyde, 3-ethyl-4-methyl-1H-pyrrole-2,5-dione, 3-ethyl-3-methyl-2,5-pyrrolidinedione, coumaran and coumarin provided burnt and sweet odors. 2-Methylbenzonitrile, 2,5-diethylphenol, and 2-

ethyl-4,5-dimethylphenol afforded woody and green odors. D-Limonene, acetic acid, linalool, 2,6-dimethyl-1,3,7-octatrien-6-ol, phenyl-acetaldehyde, and 3-hexenyl hexanoate gave harsh odors, while (E)-2-hexenal and (E,Z)-2,4-heptadienal gave green odors. 3-Hexenic acid and D-nerolidol provided a deep odor.

Delayed light emission of dependent on the moisture content was used as a measure of drying tea. The intensity decreased inversely with temperature from 12.5 to 45.4°C and became weak at temperatures greater than 40°C. Delayed light emission characteristics of crude and fresh tea were different.

For the purpose of the objective quality evaluation of Japanese green tea (Sen-cha), chemical sensors have been introduced. An enzyme electrode, which consists of an oxygen electrode and a biocatalytic membrane containing α-amino acid oxidase, is used to measure amino-acid concentrations in tea infusions. The values determined by this method are highly correlated to the values determined by the conventional colorimetric method. Since high-grade teas show high total amino-acid values, this enzyme electrode can be used for the practical quality evaluation of green tea. The aroma of green tea delicately reflects quality deterioration. An odorant sensor is used for a simple analysis of the aroma of tea leaves. Teas with bad aroma, judged from sensory tests, show higher odor values than teas of normal aroma with this sensor.

OOLONG TEA

Oolong tea is a semi-fermented type of manufacture or partially oxidized manufacture. It is an intermediate between the green and black tea. The plucked tea shoots are withered under the sun for 6-8 hours followed by indoor withering for 8-10 h, with soft rolling by hand once in about 30-60 min. The shoots are then pan fired at 230-250°C for 10-15 minutes to inactivate the enzyme polyphenol oxidase and terminate the oxidation reaction of catechins. The degree of oxidation is approximately 15-18%. The shoots are rolled and partial dried to moisture level of 40%. The shoots are cooled and rolled again. This is followed by pan drying to bring the moisture level of 3-5%. As the enzyme polyphenol oxidase is not fixed at the initial stages, some amount of polyphenols are oxidized to give theaflavins. Therefore, the quality attributes for oolong tea are theaflavins, catechins, caffeine and volatile flavour components.

The effect of tea cultivars, processing technology and seasons on the quality of Oolong tea from China was investigated. The free and bound alcoholic aromas in fresh tea leaves varied in different seasons and processing technologies. The free aroma content was highest in spring, followed by summer and autumn; while that of bound aroma was highest in summer, followed by autumn and summer. During the manufacturing process, the free and bound aromas showed different trends of dynamic change and the leaf polyphenol content decreased gradually. Free amino acid content increased continuously at the initial withering stage and decreased in the drying stage, while the opposite was observed for the soluble sugar content. Results of the organoleptic evaluation indicated that oolong tea made from tea cv. Zhuye, harvested in late spring or early summer and with heavy shaking showed distinctive tea quality, and typical tea aroma and taste.

A direct qualitative and quantitative determination of the glycosides of tea aroma compounds at the four stages of the oolong tea manufacturing process (plucking, solar withering, indoor withering, and oolong tea product) was carried out by a capillary gas chromatographic-mass spectrometric analysis after trifluoroacetyl derivatization of the glycosidic fractions. Sixteen glucosides and primeverosides were identified and quantified in cv. Chin-shin-oolong and cv. Chinhsuan-oolong. A comparison of the glycosides in dried fresh leaves between the two cultivars showed significant differences. During the manufacturing process, the amounts of most of these glycosides increased from the solar-withering stage, reaching the highest level at the final stage of oolong tea production. It was noted that no glycoside decreased in its content during the manufacturing process, this being quite different from the manufacture of black tea. In addition, the contents of these alcoholic aroma compounds in the free aroma concentrate from each cultivar remained almost unchanged or slightly decreased, and they constituted only about 12 and 17% in amount of the whole oolong tea aroma compounds. However, jasmine lactone and indole were markedly higher in the final oolong tea products.

The other semi-fermented tea is Pouching tea. The processing is similar to the oolong tea but the difference is in the fermentation time. The degree of fermentation is low in pouching tea. The fermentation duration shows in the flavour profile of the two semi-fermented teas. Oolong tea has higher levels of geraniol, benzyl alcohol, 2-phenylethanol and B-ionone while pouching has higher levels of jasmine lactone, nerolidol and methyl jasmonate.

The formation of cream in Paochung tea was investigated at various extraction temperatures, extraction times, pH values and leaf/water ratios. The primary components of Paochung tea cream were catechins (30%), caffeine (20%) and protein (16%). (-)-Epigallocatechin gallate and (-)-epicatechin gallate were the major catechins precipitated during creaming, constituting 19% and 5% of the tea cream respectively. The amount of tea cream produced and its composition were influenced by extraction temperature and pH. The tea leaf/water ratio determined the amount of tea cream formed but did not affect the cream composition. Catechins were considered to be the key component in tea cream. They interacted with caffeine and protein to induce tea cream formation.

BLACK TEA

Black tea production is characterized by oxidation reactions so that most of the catechins are oxidized. Quality of tea is determined by the presence or absence of chemical compounds which impart colour, briskness, brightness, strength and flavour in the black tea infusion. Majority of the chemicals imparting quality of black tea are produced during processing of the tea leaves. Biogenesis of such precursors is influenced either by the genetic and environmental factors which cannot be controlled or by the cultural practices as well as conditions of processing, which can be carefully manipulated. Though biochemical changes start immediately after plucking the shoots, the precise changes required for quality start from withering onwards.

The major steps in the black tea manufacture are (i) withering, (ii) rolling, (iii) fermentation, (iv) drying and (v) sorting.

Withering

Withering a very important step in black tea manufacture where the leaf is forced to loose turgidity, becomes flaccid and fit for rolling, the next step of the manufacture.

In the true sense withering of tea shoot starts immediately after plucking. It is the first stage of tea manufacture and quite labour intensive time consuming step in black tea manufacture. The harvested tea flush is brought to the factory from the field and is evenly spread on the withering troughs which are normally 4 to 5 feet wide and 50 to 70 feet long. During the process hot and cold air is blown through the leaves to remove the moisture of fresh leaf from 75-80 per cent down to 50-55 per cent. This is achieved in 16 to 18 hours.

The prime object of withering is to make the turgid leaf flaccid and prepare the leaf for next stage and facilitate rolling.

The decrease in water content of the shoots increases the cell membrane permeability. There is also increase in the level of cell solutes mainly the catechins. Besides the physical wither, by virtue of separation from bush, many biochemicals including enzymes undergo metabolic readjustment. The major changes that take place during withering have been described below:

Breakdown of complex molecules into simpler molecules *viz.,* due to prolonged storage under mild desiccating the membrane lipids get partially affected to generate fatty acids, complex proteins into simpler proteins and peptides, complex sugars into simple sugars and breakdown of chlorophylls into pheophytin and pheophorbides. There is also an increase in caffeine content, inorganic phosphate contents. The development of flavour volatiles from glycosidic precursors also starts during withering. Organic acids also increase which decrease the pH of the leaf to make it more acidic. This decrease in pH tends to favour the formation of theaflavins from catechins by polyphenol oxidase enzyme. These changes are due to weakened organelle barriers that allow for nearly free exchange of cytoplasmic components.

The factors affect withering are standard of plucking, type of leaf, drying capacity of air and period of wither

Black tea quality, as measured by chemical quality parameters and sensory evaluations, declines with coarse plucking standards. The extent of the decline varies with clone. Studies conducted in Kenya show that Clone 6/8, with high polyphenol content in the green leaf had a higher decline than clone S15/10 with low polyphenol content results suggest a possible variation of recommended plucking standard for different clones.

A good planting material with good plucking standards would make good tea while good planting material with poor plucking standards would mar the quality of tea. With increase in mechanization of farm operations, one has to be careful of plucking standards. The coarse leaf is rich in fibre and low in polyphenols compared to fine pluck, and therefore would take more time to wither. It then becomes imperative to sort the leaves before putting in withering troughs so as to have as far as possible uniform wither. The thermo-sensitive nature of many biochemical reactions imposes an upper limit to which air used could be heated. Uniformity of shoot size, plucking standard, shoot succulency, amount of moisture in the shoots and presence of surface moisture due to precipitation determine the total amount of moisture present in the shoot. Relative humidity of the air used for withering would determine the duration for which the shoots need to be withered. Normally, shoots have 75% moisture and require 16-20 hours of withering for orthodox and 14-16 hours for CTC (crush-tear-curl) tea.

The succulent bud and first leaf would wither at a faster rate while the coarse plucked shoot would take a much longer time to wither. This makes some portions over-withered and some under-withered.

Studies conducted for CTC tea showed that very hard wither affected polyphenol oxidase activity and the formation of theaflavins, but the formation of thearubigins continued, producing teas with body and without adequate briskness and brightness. Unwithered CTC teas produced higher theaflavins and lower thearubigins resulting in bright, brisk and thin liquor. Natural withering produces better quality teas than leaf withered at 37.8°C in a chamber for 13 hours. During withering polyphenol oxidase activity was depressed. The theaflavins, thearubigins and brightness in tea were observed higher in 18 hours wither than those of teas of less wither.

Leaf maceration/ rolling

Major chemical changes mainly oxidation reactions take place in this step to give the tea pigments theaflavins and thearubigins. Apart from this flavor compounds are also formed by glycolysis, lipoxygenases and oxidation. The importance of this step is cellular structure broken to provide proper milieu for the enzyme mainly Polyphenol oxidase and substrate catechins to react. There are two kinds of black teas the orthodox and the CTC (crush-tear-curl). The two types of black tea differ in the way the shoots are rolled after withering and will be deált later in the chapter under rolling.

Plant cells have different organelles to carry out different metabolic processes and have special compartments to store secondary metabolites. Polyphenols are situated in the vacuoles while the enzyme polyphenol oxidase that work on them are either located in the cytoplasm or bound to the membrane. It is therefore necessary that the enzyme and substrate are brought together to start the chemical reactions and produce the quality determining chemicals. Evenness of wither is crucial for generation of right type of chemicals.

gallocatechin (trihydroxy) dimerise to produce theaflavins, which are orange-red substances that contribute significantly to astringency, briskness, brightness and colour of tea beverage. Theaflavins comprise 0.3 to 2.0% of the dry weight of black tea. Theaflavins is a group of comprising a number of fractions namely theaflavin (formed from (-)-epicatechin and (-)-epigallocatechin), theaflavin monogallate (formed from either (-)-epicatechin and (-)-epigallocatechin gallate or from (-)-epicatechin gallate and (-)-epigallocatechin) and theaflavin digallate (from (-)-epicatechin gallate and (-)-epigallocatechin gallate), epitheaflavic acid comes from (-)-epicatechin and (+)-gallic acid. Isotheaflavin comes from reaction of (-)-epicatechin and (+)-gallo catechin. As the gallation increases the astringency also increases and the proportions of theaflavin fractions present in black tea depend upon the method and conditions of manufacture.

Both theaflavins and sensory evaluations vary in a quadratic manner with fermentation duration in the process of black tea manufacture. The rate of formation of total (Flavognost) theaflavins varies with clones, and within a clone the individual theaflavins form at different rates with resultant changes in the theaflavin ratios as fermentation progresses. The astringency of black tea also varies with composition and total amounts of the theaflavins, as individual theaflavins have different astringencies. The measurement of total theaflavins may therefore not correctly predict total astringency and hence value of tea. When macerated tea leaf is fermented at lower pH, the resultant black tea contains increased levels of theaflavin, an important quality marker in black tea. In vitro oxidation experiments using polyphenol oxidase (PPO) from fresh tea leaves, horseradish peroxidase (POD), and tea catechins, precursors for theaflavins, were carried out. In vitro oxidation experiments using crude tea PPO resulted in higher content of theaflavins at pH 4.5 in comparison with pH 5.5, the normal pH of the macerated tea leaf. When purified PPO was used in the in vitro system, surprisingly a reversal of this trend was observed, with more theaflavins being formed at the higher pH. A combination of pure tea PPO and POD led to an observation similar to that with the crude enzyme preparation, suggesting a possible role for POD in the formation or degradation of theaflavins. POD was observed to oxidize theaflavins in the presence of H_2O2 leading to the formation of thearubigins, another black tea pigment. This paper demonstrates that tea PPO, while oxidizing catechins, generates H_2O_2. The amount of H_2O_2 produced is greater at pH 5.5, the optimum pH for PPO activity, than at pH 4.5. Hence, an observed increase of theaflavins in black teas fermented at pH 4.5 appears to be due to lower turnover of formed theaflavins into thearubigins.

Drying

The chemical reactions are stopped to increase the shelf-life of the product. It is a very important step in tea manufacture. During drying the moisture of the fermented leaves which is about 60-75% is brought down to 2-3% by exposing the fermented leaf to a stream of hot air. During this process some of the flavour compounds like pyrazines and ionones are developed due to heat as the leaves loose moisture which give characteristic aroma to the teas. The heating also inactivates the enzymes and ends other processes which are taking place during fermentation thereby increasing the storage and preservation quality of the manufactured tea.

Sorting

Sorting is a mechanical process and is done to remove fibre and grade the teas according to the size.

Chemical composition of black tea

Table 19.1 provides list of chemicals present in black tea. Thearubigins (TR) are complete condensation products of oxidised catechins with theaflavins (TF). Together with high polymerised substances (HPS), TR contributes to the colour, mouth feel and body of the tea liquor.

Caffeine is relatively a stable molecule and is a direct stimulant of the central nervous system. Along with TF it imparts briskness to the tea liquor. High levels of caffeine indicate a good leaf standard. Caffeine decreases with the maturity of the crop shoots.

In addition to the biochemical constituents that gives the tea its unique taste, black tea is also well known for its characteristic aroma. Volatile flavour constituents *viz*., terpenoid glycosides and amino acids degrade to produce linalool, geraniol, phenyl acetaldehyde, phenyl ethanol and methyl salicylate which contribute to desirable aroma of tea.

During black tea manufacture, lipids are degraded to yield volatile constituents. *Cis*-3-hexenal is present in appreciable amounts in the various parts of fresh shoots and decreased in the second leaves during manufacture. There was a simultaneous increase in *trans*-2-hexenal. Linalool and methyl salicylate also increased appreciably during rolling and fermentation. Most of the volatiles were lost during the firing process. The C_6-aldehydes present in the headspace of withered shoots increased significantly following mechanical damage. The major fatty acids of the lipids in the various parts of the shoots were linolenic, linoleic, palmitic, oleic and stearic acids. The ratio of linoleic to linolenic acid in the stems was much higher than that of the leaves or buds and this was reflected in its higher potential for formation of hexanal. During withering and rolling of the second leaves, the unsaturated fatty acids showed substantial losses compared with the saturated acids. It is suggested that the enzymic breakdown of membrane lipids initiate the formation of volatile carbonyl compounds which are partly responsible for the flavour of black tea.

Table 19.1 Biochemical compounds responsible for colour of black tea

Compounds	Colour
Theaflavins	Yellowish brown
Thearubigins	Reddish brown
Flavonol glycosides	Light yellow
Pheophorbide	Brownish
Pheophytin	Blackish
Carotene	Yellow

Biochemical compounds responsible for taste

Compounds	Taste
Polyphenols	Astringent
Amino acids	Brothy
Caffeine	Bitter
Theaflavins	Astringent
Thearubigins	Ashy and slight astringent

Principal components of black tea beverage

Components	Concentration (g/100g)
Catechins	3.00
Theaflavins	3.00
Thearubigins	12.00
Flavanols	6.00
Phenolic acids and Depsides	10.00
Amino acids	13.00
Methylxanthines	8.00
Carbohydrates	10.00
Protein	0.80
Mineral matter	8.00
Volatiles	0.05

CTC teas were made from shoots plucked at 5,7 9 and 11 day interval. Crude fibre and theaflavins and thearubigins increased while total soluble solids and organoleptic evaluation decreased with increase in length of plucking interval. The total volatile flavour components increased from 5 to 7 day and decreased thereafter. The quality of green tea and yield of summer crop improved by 75 % with shading for 2-3 weeks.

Quality of Uva black tea improved with tat-withered leaf than with trough wither leaf and morning plucked leaf rather than the afternoon pluck.

Deep pruning and root pruning improved the quality in Japan. Similar results have been obtained in Kangra region in north-west India and in Assam. Years after prune had inverse effect on the quality. Quality decreased with increase in time from pruning.

By contrast, thearubigins are found to be a highly heterogeneous mixture of pigments, generally characterized by a very broad hump when analyzed by reverse-phase HPLC (Bailey *et al.* 1992). At present there is little firm knowledge on their composition and structure. Theaflavins can be synthesized from selected pairs of catechins *via* chemical oxidation (Takino *et al.* 1964).

However, the yield of the expected theaflavins is low, and a complex mixture of products is often produced. Bailey *et al.* (1993, 1994) studied oxidation of individual catechins, using potassium hexacyanoferrate (III), which has been widely employed to effect oxidations normally catalysed by the enzymes, peroxidase, laccase and polyphenoloxidase (Geissman and Crout 1969). They demonstrated that, although in the cases of catechin, epicatechin and epigallocatechin-3-O-gallate the oxidation products were also a complex mixture, epicatechin-3-O-gallate (ECG) gave one major product (Bailey *et al.* 1993).

Natural withering of plucked tea shoots for 16 to 18 hours was considered adequate for achieving both physical and chemical wither, though natural wither for 12 hours qualified the requisite chemical wither for CTC teas. Natural wither beyond 20 hours appeared detrimental to tea quality. In artificial withering 16 to 18 hours period was found adequate for attaining necessary chemical wither equivalent to that in natural withering as indicated by breakdown of protein and synthesis of caffeine in withered shoot. During withering polyphenol oxidase activity was depressed. Very hard wither affected PPO activity and the formation of theaflavins, but the formation of thearubigins continued, producing teas with body and without adequate briskness and brightness. Unwithered CTC teas produced higher theaflavins and lower thearubigins resulting in bright, brisk and thin liquor. Natural withering produce better quality teas than leaf withered at 37.8°C in a chamber for 13 hours. The theaflavins, thearubigins and brightness in tea were observed higher in 18 hours wither than those of teas of less wither.

In another *in vitro* model fermentation system containing purified catechins and partially purified polyphenol oxidase (EC 1.14.81.1) from green tea shoots, has been used to determine the efrect of catechin mixtures of different concentration and proportions on the formation of theaflavin and thearubigin. Increases in total catechin concentration, 25% above that typical in green tea shoots of Malawi-grown bushes, inhibited polyphenol oxidase activity and, consequently, depressed theaflavin levels. Individual or combined concentrations of epicatechin gallate and epigallocatechin gallate in excess of 110 mM were shown to be responsible for enzyme inhibition, whereas epicatechin and epigallocatechin had no effect. Fermentation of a catechin mixture, containing the four major catechins, epicatechin, epigallocatechin, epigallocatechin gallate and epicatechin gallate, at equal individual concentrations (55 mM), produced, after 30 minutes, total theaflavin levels 68% higher and thearubigin levels only 25% higher than those from a standard catechin mixture fermented under similar conditions. Continued fermentation of this mixture produced no further theaflavin, but the thearubigin fraction increased significantly, due to subsequent oxidation of the excess of simple catechins. A new catechin mixture was, therefore, calculated to give a similar level of theaflavin to that of the previous mixture without leaving an excess of unoxidized simple catechins. The catechin proportions and concentrations of the latter mixture agree well with those of the green shoots of quality Kenyan teas or similar quality Malawi teas grown during the dry cold season. The results indicate that a high ratio of simple to gallocatechins will facilitate a high theaflavin-thearubigin ratio in black tea.

Main polyphenol oxidation products in black tea infusion are thearubigins and theaflavins, which amount to 30-60% of infusion solids and predominant chemical components of tea cream, which is a coloured precipitate in suspension, formed as a strong black tea cools down. Theaflavins content is positively correlated with tea quality or market tea price. The creaming ability of black tea is also an indicator of black tea quality. However, creaming down is also a major problem in production of cold water-soluble iced tea because it makes the iced tea become a turbid liquor, which is not favoured by consumers. Theaflavins and thearubigins have acidic properties. Acidification of tea infusion by sulfuric acid results in partial precipitation of the coloured oxidation products and is therefore a means for encouraging a considerable degree of separation of the oxidation products from unchanged flavanols in black tea and for obtaining a cold water-soluble instant tea for iced tea production. Tea contains about 300 - 450 g/kg of extractable solids, but the yield of instant tea is only 200 g/kg tea in production scale. H^+ level affects the equilibrium concentration of theaflavins and more theaflavins are liberated when tea is extracted at low pH. As pH of an infusion is raised, the infusion becomes dark and loses its creaming ability. Thus, behaviour of tea extracted at extreme pH levels might be of great interest in both controlling the creaming down of cold water-soluble instant tea and increasing yield of normal drinking instant tea.

The cream of black tea contains a relatively higher amount of thearubigins TR1, and a greater ethyl acetate soluble part than the decreamed portion, which in turn is rich in TR3 and theaflavins. This shows that TR1 has a strong tendency to precipitate, and its presence contribute significantly to the dull colour of the brew. Further, the briskness is reported to be related to the condensation index and is lowered with an increasing amount of TR1. The large amount of TR3 is expected to be precipitated only when small levels of TR1 is present, thereby increasing briskness. It is noted that the diminished cream in black tea manufactured from fresh leaves by the CTC process can be attributed to its low TR1 content. This leads light liquor characteristics to a tea brew, as observed in orthodox tea.

Volatile components in unfermented green tea, semi-fermented Oolong tea and fully fermented black tea were analyzed by gas chromatography (GC) and gas chromatography-mass spectrometry (GC-MS). For differentiating, three tea categories based on their volatile components, unsupervised and supervised pattern recognition techniques were applied to the resulting GC data. Three distinct clusters each corresponding to green tea, Oolong tea and black tea were observed in the dendrogram and the principal component (PC) score plot. However, a subcluster of Oolong tea was observed in the vicinity of black tea cluster in both the dendrogram and the PC plot. The first and second PC corresponded to the fermentation products and aroma components originally contained in tea leaves, respectively. Both the partial least squares (PLS) analysis and linear discriminant analysis correctly differentiated tea samples into the three categories. (E)-2-Hexenal, the major fermentation product from unsaturated fatty acids, was the most efficient for the discrimination. Although three teas are produced from the same plant species, pattern recognition clarified the existence of the apparent quality difference among their volatile component profiles.

Foliar application of zinc sulphate and manganese sulphate (1-2%) enhanced the nitrogen uptake, protein synthesis and nitrate reductase activity in tea shoot and no adverse effect on catechin concentration and thus on theaflavins and thearubigins content of black teas.

Theaflavins are catechin dimers that contribute to the yellow-orange color and astringency of black tea liquors. Theaflavins are formed during the fermentation stage of black tea manufacture through the oxidative condensation of a catechin quinone (dihydroxy B-ring) e.g. epicatechin (EC) and a gallocatechin quinone (trihydroxy B-ring) e.g. epigallocatechin (EGC). However, the theaflavins are not stable and can be further oxidized to thearubigins, which are a heterogeneous group of catechin polymers, of unknown structure, which can also be formed directly from catechins. It has been reported that EC is required as an intermediate for the enzymic oxidation of theaflavin (Bajaj *et al.* 1987, Opie *et al.* 1993).

Goodsall *et al.* (1996) reported the factors influencing theaflavin formation, using pure tea components, in a model fermentation system. In these experiments, theaflavin was formed very readily from its precursors (EGC and EC), and continued to accumulate until EGC was exhausted, whereupon theaflavin oxidation commenced. All oxidized EC was quantitatively accounted for in theaflavin formation and there was no theaflavin oxidation whilst EGC was present. Hence, theaflavin accumulation was not a balance of simultaneous synthesis and further oxidation, rather theaflavin was formed whilst EGC was present and then degraded after EGC had been consumed.

Experiments investigating the factors that control theaflavin turnover, and the mechanism whereby EGC prevents theaflavin oxidation by CSIR-IHBT confirmed that theaflavin was not a direct substrate for tea PPO, but readily oxidized in the presence of EC. Variations in the initial EC concentration demonstrated the absence of stoichiometric relationship between TF and EC oxidation. This indicated that EC was not combining directly with TF to give catechin trimers, rather EC quinone was acting as an intermediate in enzymic TF oxidation. All simple catechin quinones (C, EC, ECG) could enable enzymic TF oxidation, whilst gallocatechins (EGC, EGCG, GC) could not. This was probably due to the relative redox potentials of the different species. The results confirmed that TF was not a substrate for tea PPO, but that catechin quinones could act as intermediates for enzymatic theaflavin oxidation. These results explain the profile of theaflavin formation seen in the EGC + EC + tea PPO incubation (Goodsall *et al.* 1996). After 30 minutes oxidation, EGC and theaflavin were present at approximately equal concentration, thus EC quinone was equally likely to interact with either, yet no theaflavin oxidation was seen. It was suggested that EC quinone may oxidize theaflavin to a reactive intermediate, but that this intermediate was, in turn, able to oxidize EGC and thus regenerate theaflavin. In this way the rate of EGC oxidation was increased and apparent EC oxidation reduced.

Variation of chemical wither time by up to 18 hours in the manufacture of black tea, from shoots of seedling tea, caused minimal changes in plain tea quality parameters. Thus, in a two-stage wither, plain black teas can be withered for between 0 and 18 hours without loss or improvement in quality. Both the Group I VFC and Group II VFC decreased with chemical wither time, but the decrease in Group I VFC was much higher, resulting in improvement of flavour index with chemical wither time. The changes in the VFC necessarily brought significant change in the tasters' evaluation which seemed to follow the plain tea quality parameters.

LEAF PRECONDITIONING MACHINE FOR WITHERING

Concept

Recently CSIR-IHBT discovered that it was possible to accelerate both chemical and physical wither of green leaf by creating control stress to trigger chemical wither. On the basis of this concept a machine was designed and develop in association with M/s. Mesco Equipment (P) Ltd (Figure 19.1).

It has been claimed that the machine has changed the accepted theory that 12 to 16 hours are essential for chemical wither in order to maximize the quality property of black tea, and it is possible to cut the trough area requirement by 27 to 40% by pre-treatment of the green leaf.

Fig. 19.1. Leaf pre-conditioning machine for withering

Scientific basis

(a) Controlled physical stress to trigger chemical wither.

(b) Cooling at ambient temperature by cold air circulation.

(c) Physical as well as chemical wither is achieved in $4^{1}/_{2}$ - $5^{1}/_{2}$ hours for CTC and $5^{1}/_{2}$ - 6 hours for orthodox.

(d) The process works in all weather conditions.

(e) Extra surface water on leaf will have to be removed before setting the process. However, if the process start with wet surface leaf it may become necessary to prolong application of cold air.

(f) The quality of tea through this process is at par or superior to normal process machine.

System Management

Successful wither with this processing machine will very much depend on the system management. Therefore, while going for this machine one has to change the manufacturing process programme for conversion of wither leaf to black tea.

There are five important reasons for preferring this machine over the existing system:

(a) Saving of energy by substantial reduction of number of trough fan used.

(b) No reddening of bruised leaf during wither and thereby enhancement of quality.

(c) Saving of fuel as hot air is not required.

(d) Manufacture during cooler period which always gives better control and quality.

(e) Labour saving.

Specification of the machine

(a) Compact with space requirement of 3.22 m x 1.58 m.

(b) Flexibility of stroke and pressure in relation to horizontal speed of movement.

(c) Throughput 2500 – 4000 kg green leaf per hour.

(d) One machine can cope with green leaf of a 10 lakh kg factory by adjustment of manufacturing time.

(e) Easily adaptable to a continuos process from withering to drying.

Studies carried out on this machine at UPASI (Ramamurthy *et al.* 2009) showed reduction of 2-5 hours in the time of physical wither depending upon the relative humidity and trough load. The laboratory analysis for the quality parameters indicated that the requisite quality was achieved in almost half of the normal withering time. There was further reduction in the withering time when the controlled stress on the leaves was enhanced by increasing the number of pressings. In addition, theaflavin content increased from 3 to 23% over the untreated control, depending on the grade of tea. UPASI recommended that the present output of the machine must be increased from 800 kg to 10000 kg/hour if this machine has to be use on a commerial scale. Volatile flavour compounds of orthodox teas were also analyzed and found a remarkable improvement in the Group II flavour compounds of the machine treated tea. An overall improvement of 43% in the flavour index was recorded.

OTHER DEVELOPMENTS IN TEA TECHNOLOGY

Recently, efforts are being made to automate the green as well as black tea manufacture especially the withering step since withering is a very time consuming process in black tea manufacture. This includes the development of sensors, electronic nose and electronic vision systems, fuzzy logic to detect proper withering and a preconditioning machine that reduces the time of withering.

Green tea manufacture involves inactivation of leaf indigenous enzymes immediately after plucking so as to prevent the oxidation of polyphenols and other biochemicals by these enzymes. Using fuzzy logic, the manufacture of high grade Sen-cha has been attempted. Since the leaf moisture and air temperature are critical for green tea manufacture, the sensors are based on the temperature and moisture content of the air inlet and leaves. Green tea manufacture involves different steps from leaf inactivation to drying with intermitant rolling and cooling steps. Six different sensors on different machines were used for controlling the different steps of green tea manufacture.

In tradional factories, the termination of withering is judgeded by experience and judgement of the tea maker. Replacing human experience by a machine is a very tough task. Withering involves both physical as well as chemical wither. The parameters controlling withers are mainly dependent on the flush i.e. type of leaf, leaf size, age, standard of plucking, thickness of spread in the withering trough, drying capacity of air, condition of leaf, period of wither, condition of withering room, temperature of leaf, texture etc. The decision to terminate the withering process has directly bearing on the colour and quality attributes of the final product as well as the overall energy consumption of the process. Therefore, the different automation systems target to monitor the relative humidity and temperature of the inlet air, monitor the leaf moisture and weight of the leaf to be withered, switching the fan on/off, forward or reverse along with hot air or cold air are essential components for ensuring the even wither. Most manufacturers of automated systems recommend making a chart for correlating the weight loss in shoots to wither percentage and moisture content so as to determine the wither percentage at any point of time (Yoshitimi *et al.* 2000).

Apart from withering automation of other steps of black tea manufacture *viz.,* rolling, fermentation and drying have also been automated using fuzzy logic. Recently, chemical sensors using metal oxide sensors have been developed for optimizing fermentation step (Bhattacharyya *et al.* 2007). These are based on the formation of linalool, geraniol as volatiles during fermentation and takes into account the concept of second peak in black tea manufacture.

REFERENCES

Bailey RG, Nursten HE, McDowell I (1992) Isolation and analysis of a polymeric thearubigin fraction from tea. *J Sci Food Agric.* 59(3): 365-375.

Bailey RG, Nursten HE, McDowell I (1993) The chemical oxidation of catechins and other phenolics - A study of the formation of black tea pigments. *J Sci Food Agric.* 63(4): 455-464.

Bailey RG, Nurnsten HE, McDowell I (1994) A comparison of the hplc, mass spectra, and acid degradation of theafulvins from black tea and proanthocyanidin polymers from wine and cider. *J Sci Food Agric.* 64(2): 231-238.

Bajaj KL, Anan T, Tsushida T, Ikegaya K (1987). Effects of (-)epicatechin on oxidation of theaflavins by polyphenol oxidase from tea leaves. *Agric Biol Chem.* 51: 1767–1772.

Bhattacharyya N, Seth S, Tudu B, Tamuly P, Jana A, Ghosh D, Bandyopadhyay R, Bhuyan M (2007). Monitoring of black tea fermentation process using electronic nose. *J Food Eng.* 80: 1146-1156.

Geissman TA and Crout DHG (1969) Organic chemistry of secondary plant metabolism. Freeman, Cooper and Co., San Francisco, p 592.

Goodsall CW, Parry AD, Safford R (1996) Polyphenols Communications 1996. XVIIIth International Conference on Polyphenols, Bordeaux, France, July 15-18, pp 287-288

Gulati A, Rawat R, Singh B, Ravindranath SD (2003). Use of microwave for quality green tea manufacture. *J Agric Food Chem.* 51: 4764-4768.

Huang Y, Sheng J, Yang F, Hu QH (2007). Effect of enzyme inactivation by microwave and oven heating on preservation quality of green tea. *J Food Eng.* 78: 687-692.

Opie SC, Clifford MN, Roberson A (1993). The role of (-)epicatechin and polyphenol oxidase in the coupled oxidative breakdown of theaflavins. *J Sci Food Agric.* 63: 435-438.

Ramamoorthy G, Senthilkumar RS, Muraleedharan N (2009) A technical report on the evaluation of pre-withering machine. *Planters' Chronicle* 105: 6-14.

Sood C, Jaggi S, Kumar V, Ravindranath SD, Shanker A (2004). How manufacturing processes affect the level of pesticide residues in tea. *J Sci Food Agric.* 84: 2123-2127.

Takino Y, Imagawa H, Harikawa H, Tanaka A (1964). Studies on the mechanism of the oxidation of the catechins. Part 111, Formation of a reddish-orange pigment, its spectral relationship to some benzotropolone derivatives. *Agric Biol Chem.* 28:64-71.

Yoshitomi H, Nakano F, Tanaka S, Sumikawa O (2000). Automatic control of tea manufacturing Process (Part I), *J Soc Agr Machin.* 62(4): 127-136.

20

DIVERSIFIED TEA PRODUCTS

H.P. Singh, Brajinder Singh and Mahesh Gupta*

CSIR-Institute of Himalayan Bioresource Technology,
Palampur-176 061, H.P., India.*E-mail: mgupta@ihbt.res.in

Tea is consumed worldwide, although in greatly different amounts, it is generally accepted that, next to water, tea is the most consumed beverage in the world, with per capita consumption of ≈120 mL/day (Katiyar and Mukhtar 1996). The total amount of tea produced and consumed in the world, 78% is black, 20% is green, and <2% is oolong tea.The natural conditions affect the quality of green leaf during harvesting. In general every tea processing unit is facing problem of disposing of its 30-40% sub-standard produce. This creates the need for diversified use of such teas. These teas contain all the essential characteristics of quality tea but in lesser amounts. Such teas can be very easily converted into quality diversified products having all the nutraceutical values.

Both black as well as green tea can be converted into value added products which have ready world market. China has come up in a big way in such products. These products are already liked in one form or other in western and American countries.

There is an explosion of consumer interest in the health-promoting and disease-preventive effects of foods. These foods have been given several names, such as "functional foods, "nutraceutical," "healthy foods," "healthful foods," and "wellness foods." The Institute of Medicine's Food and Nutrition Board defined functional foods as any food or food ingredient that may provide a health benefit beyond the traditional nutrients it contains. Another working definition is that functional foods are foods or food components that provide a health benefit beyond basic nutrition. A shorter definition is a food that has medicinal effects. Health benefits of foods are not a new discovery -Hippocrates recommended 2,500 years ago, "Let food be thy medicine and medicine be thy food."

TEA AS MAJOR NUTRACEUTICALS BASE

Tea as a major base and well defined store-house of nutraceutical chemicals can play a lead role in defining shape of nutraceuticals industry. Research on the effects of tea on human health has been fuelled by the growing need to provide naturally healthy diets that include plant-derived polyphenols. In line with this, there is need to elucidate how known functional components in foods could expand the role of diet in disease prevention and treatment (Mandel *et al.* 2006).

Although the overall value of the market continues to decline, the popularity of specialty, herbal and fruit teas sectors have done much to curtail a dramatic slump, experiencing a value rise of 50% and 30% respectively between 2002 and 2004. These sectors have been fuelled not only by consumers' tendency to trade up to premium specialty teas, but also by their increasingly healthy lifestyles and concerns over long-term health, which can be enhanced by tea drinking, according to manufacturer claims. There is already growing evidence that tea polyphenols reduce the risk of heart diseases and cancer in humans (Vanessa and Williamson, 2004). In some studies, tea has been associated with antiallergic action (Yamamoto *et al.* 2004) and antimicrobial properties (Paola *et al.* 2005). Further studies have demonstrated that the co-administration of drugs with catechins (EC and EGCG) inhibits glucoronidation and sulfonation of orally administered drugs thereby increasing the bioavailability of such drugs (Hang *et al.* 2003). Moreover some epidemiological studies have associated consumption of tea with a lower risk of several types of cancer including those of the stomach, oral cavity, oesophagus and lungs (Cabrera *et al.* 2003). Therefore, tea appears to be an effective chemopreventive agent for toxic chemicals and carcinogens.

Bottled green tea is by far the most popular soft drink in Japan. We consumed 1.5 billion gallons last year, compared with 700 million gallons of the next most popular drink, carbonated soft drinks. Beverage companies are capitalizing on this popularity to create value added products. Adding functionality is one way to do it. The most successful product launch in functional green tea was Kao's Healthy Green Tea. It sold $100 million a year in 2002.

While Ready-to-Drink (RTD) tea beverage consumption is increasing, the typical RTD consumer demands differently like - a younger male who prefers sweeter, flavored drinks - is rarely the same consumer who prefers high quality, loose leaf or bagged tea. Increasing consumption of higher quality flavored and unflavored teas are a developing trend much like the specialty coffee market of 10 years ago.

High doses of ECG and other catechins are particularly effective in preventing inflammation and skin cancer, especially if used as a topical lotion. Researchers discovered that topical ECG reduced the release of inflammatory prostaglandins, which play a role in generating free radicals and promoting tumor growth (Pashka *et al.* 1998).

Green tea weight loss supplements contain both caffeine and the chemical epigallocatechin gallate (EGCG). When these two substances in green tea react to each other, metabolism rates increase along with the body's 24-hour energy expenditure. EGCG of green tea also triggers the release of hormone noradrenaline, an appetite-suppressant.

Major tea growing countries like Japan, China and India have already branched out in a big way in this direction e.g. Indian tea major, Tetley Tea has branched out into the functional tea category in the UK with the launch of a new range of tea and herb blends which promise various physical and mental benefits. Japanese companies are selling Functional Green Teas with Green Tea as a base added or admixed with neutraceuticals important herbs or their extracts *viz*., Green tea; more than antioxidants, Green tea with Glucosamine, and Green tea with French Pine Bark Extract.

Tokuho (FOSHU) approved Green Tea for Diabetics. Green tea with soy isoflavones, caffeine and citrus are already in the Japanese market. Collagen, seramide, CoQ10, omega 3 might be the next line of green tea fortification.

CSIR-IHBT ROLE IN DEVELOPING TEA NUTRACEUTICALS

Under the programme on product diversification from tea, following technologies are developed by CSIR-IHBT for commercia-lization:

1. Ready-to-drink (RTD) Tea Concentrates (both black and green tea) and formulations to RTD's
2. Tea wines and tea based Himalayan berries herbal health wines.

3. Tea polyphenols free from caffeine, amino acids and soluble sugars.
4. High grade bioprocessed theaflavins.

Theanine

The Chinese and Japanese consume green tea while the rest of the world prefers the black tea. Tea is rich in polyphenols and contains theanine, an amino acid very specific to tea. There is a spurt in research activity to find foods that have nutraceutical values or functional foods. Theanine (5-N-ethyl glutamine) an amino acid has been reported to have the following activities:

- Promotes relaxation without causing drowsiness
- Reduces stress and anxiety
- Improves learning ability
- Supports the immune system
- Lowers cholesterol

Theanine is synthesized in tea roots and translocated to different parts of the plants, mainly the growing shoots. Efforts at CSIR-IHBT are to develop eco-friendly green technology for extraction of theanine from tea and developing systems for producing theanine *in vitro*. Work has been initiated on biotransformation of theanine employing the following techniques:

1. Developing hairy root cultures and comparison of theanine content *in vitro* and *ex vitro* induced tea roots, callus and cell suspension cultures
2. Microbial biotransformation employing suitable microbes and glutamine/glutamic acid as substrates
3. Enzymatic biotransformation employing the enzyme theanine synthase and glutamine/glutamic acid as substrates

Driving Factors Increased Tea Consumption

- Scientific validation of healthy attributes being covered in major U.S. media on a consistent basis
- An endless diversity of flavors, scents, and accessories to enhance the consumer's tea experience
- Increased education by trade associations, tea houses, and cafes regarding proper tea preparation and serving is resulting in better quality tea offerings

- Broader worldwide tea supply keeps cost per serving generally lower than coffee with its fluctuating prices.

FUTURE TRENDS

The list of ingredients making the rounds in the natural products industry is quite extensive. And customers on all levels are demanding more accountability relative to safety, quality and efficacy of new formulations. This leads to a more complicated process for formulators of products containing numerous ingredients from the vast realm of compounds beyond the boundaries of vitamins and minerals. Common health benefits of various fruits, analysis of the market for functional specialty beverages and highlights market sizes and trends, key players and significant product activity. Functional and specialty drinks, energy drinks and other functional drinks (including health and wellness drinks, nutraceutical drinks, enriched drinks, near-water drinks).

There is also a trend towards more exotic, fruity flavours, in addition to caffeine-free hot beverages which is boosting the popularity of fruity herbal infusions. In fact, it can be argued that the tea market is becoming increasingly diversified and segmented, as manufacturers attempt to re-vitalize the market. To sum up with the emerging trend is in:

1. A proliferation of flavored and unflavored Green teas and their use in new consumer products
2. Specialty ethnic teas.
3. Nutrients and herbal concentrates being added to tea to fortify human health.

REFERENCES

Cabrera C, Gimenez R, Lopez CM (2003) Determination of tea components with antioxidant activity. *J. Agric Food Chem.* 51: 4427-4435.

Dulloo AG, Duret C, Rohrer D, Girardier L, Mensi N, Fathi M, Chantre P, Vandermander J (1999) Efficacy of a green tea extract rich in catechin polyphenols and caffeine in increasing 24-h energy expenditure and fat oxidation in humans. *Amr J Cli Nutr.* 70 (December): 1040-1045.

Functional and Specialty Beverages: Market, Regulations, Processing, Formulation and Health Benefits. WORLDNUTRA Symposium June 23-24, 2006. Orange County Center, Orlando, Florida, USA.

Hang L, Meng X, Chuvan I, Sang S, Patten C, Sheng S, Hung J, Winnik B, Yang S (2003) Glucoronides of Tea Catechins: Enzymology of Biosynthesis and Biological Activities. *Drug Metab Dispos*. 31: 452- 461.

Katiyar SK, Mukhtar H. (1996) Tea in chemoprevention of cancer: epidemiologic and experimental studies. *Intr J Oncl*. 8: 221–38.

Landau JM, Yang CS (1997) The effect of tea on health. *Chemistry and Industry*. (November 17) 22: 904-906.

Leung L K, Su Y, Chen R, Zhang Z, Huang Y, Yu Chen Z (2001) Theaflavins in black tea and catechins in green tea are equally effective antioxidants. *J Nut.* 131: 2248-2251.

Mandel S, Packer L, Moussa YBH, Weireb O (2006) Proceedings from the 3rd International Conference on Mechanism and Action of Nutraceuticals. *J Nutr Biochem*. 16: 513-520.

Miyagawa C, Wu C, Kennedy DO, Nakatani T, Ohtani K, Sakanaka S, Kim M, Matsui-Yuasa I (1997) Protective effect of green tea extract and tea polyphenos against the cytotoxicity of 1,4-naphthoquinone in isolated rat hepatocytes. *Biosci Biotech Biochem*. 61(11): 1901-1905.

National Convention, South India Teas. 2006. Tea for health In: Sarronwala BK (ed) Proceedings, "Tea for Health. Tea - A Broad Spectrum Curative". Sarronwala BK, March 10-12, Coimbatore, India.

National Workshop IV: Value Addition to Foods-Tea 2005. In Proceedings, Organised by Association of Food Scientists & Technologists (India), November 5, Kolkata, India.

Pashka AG, Butler R, Young CY (1998) Induction of apoptosis in prostate cancer cell lines by the green tea component ECG. *Cancer Lett*. 130: 1-7

Paola RD, Mazzon E, Muia C, Genovese T, Menegazzi M, Zaffini R, Suzuki ., Cuzzocrea S (2005) Green Tea Polyphenols Attenuates Lung Injury in Carrageean-Induced Pleurisy Injury in Mice. *Respir. Res*. 6: 1465-9921.

Vanessa C, Williamson G (2004) A Review of the Health Effects of Green Catechins *in vivo* Animal Models. *J Nutr*. 134: 3431- 3440.

Wiseman SA, Balentine DA, Frei B (1997) Antioxidants in tea. *Crit Rev Food Sci*. 37 (8): 705-718.

Yamamoto MM, Inagaki N, Kitaura J, Chikumoto T, Kawahara H, Kawakami Y, Kawakami T, Nagai H (2004) O-Mmethylated Catechins from Tea Leaves Inhibit Multiple Protein Kinases in Mast Cellscells. *J Immu*. 172: 4486-4492.

Yamamoto T, Juneja LR, Chu DC, Kim M (1997) *Chemistry and applications of green tea*. Boca Raton CRC Press, pp.160.

21

SOME RECENT INNOVATIONS IN TEA BIOTECHNOLOGY

Anil Sood, Madhu Sharma, Amita Bhattacharya Sanjay Kumar, R.K. Sharma, P.K. Nagar, Om Prakash and P.S. Ahuja*

CSIR-Institute of Himalayan Bioresource Technology
Palampur (H.P.)- 176 061, India. *E-mail: asood@ihbt.res.in

INTRODUCTION

Improvement of tea through conventional breeding methods is a difficult and time consuming task, as in other highly heterozygous, long-lived and cross-pollinated crops. Thus, biotechnological improvement programs including production of transgenics, somaclonal variants, somatic hybrids, artificial seeds, cryopreservation of elite clones and embryo rescue etc. offer alternative routes to crop improvement. While molecular fingerprinting helps in evaluation of existing diversity, rationalization of gene banks, identification of parental groups and authenticating important germplasm. Cloning of useful genes adds to the crop improvement programs. A large number of reports are available on tea biotechnology, yet the significant strides that have been made by the CSIR-IHBT deserve special mention and are highlighted.

The genus *Camellia* comprises of a large number of species both cultivated and non-cultivated (ornamental) (Jain 1999). The cultivated taxa have evolved out of a large-scale gene introgression among the various species of *Camellia* because of its out breeding nature. A large number of the cultivation at the national level are derived from the selections of existing open pollinated populations and, hence highly heterozygous. Thus, this has affected the identity of original progenitors in tea and makes it difficult to identify discrete taxonomic groups (Wickramaratne 1981). Further, the selection of high productive elite clones in wide range of tea plantations has narrowed down the genetic variability and resulted in loss of valuable genetic resources. As a result, the expansion of germplasm repositories, conservation and maintenance of germplasm with distinct geographical identities becomes necessary. In this regard, application of molecular marker based approaches necessarily provide a valuable tool for evaluating genetic diversity and distribution of tea genetic resources in wide range of genetic pools in tea. Especially because most of the vegetative characteristics are influenced by the environmental factors and thus, show a continuous variation and high degree of plasticity and may not reflect true genetic diversity. Considering, various properties, traditional breeding coupled with molecular markers can be alternative approaches for evaluating genetic diversity and genetic improvement of economically important crops, such as tea.

Of the various types of genetic markers that have been used in the past for evaluating tea genetic diversity, such as morphological (Banerjee 1992), metal content (Caceres *et al.* 2001, Pineiro *et al.* 2003) and distribution of minerals and flavonoids in tea (Ferrara *et al.* 2001, Borse *et al.* 2002), the utilization of molecular markers has been found to be promising tool for genetic characterization of tea.

Molecular markers have revolutionized the entire scenario of biological sciences by detecting polymorphism at DNA level especially because of the simultaneous occurrence of a trait within the same population of two or more discontinuous variants or genotypes (Joshi *et al.* 1999). DNA based molecular markers have also acted as a versatile tool for assessing genetic variation in plant breeding programs (Paterson *et al.* 1991), phylogenetic evaluation (Wang *et al.* 1998, Blair *et al.* 1999), gene localization (Wang *et al.* 1998, Jong *et al.* 1999), tagging and marker aided selection of agronomically important genes, linkage map construction (Yamagishi *et al.* 2002) and also for molecular diagnosis. Though, tremendous progress has been made in genomic technologies in crop plants, availability of molecular tools, such as molecular marker resoureces has been limited in tea. The first such markers were reported by Matsumoto *et al.* (1994) who used phenyl-ammonia lyase (PAL) cDNA probes to differentiate various tea cultivars and found considerable PAL variations by RFLP among different clones of Assam hybrids. On this basis, the Japanese green tea clones could be grouped into five discrete classes. Wachira *et al.* (1995a) detected genetic diversity through RAPD markers for the first time in tea among the 38 accessions, representing the three varietal types and reported an average of 62% variability. Wachira *et al.* (1997) could also detect genetic diversity among 28 cultivars of *C. sinensis* and its wild relatives by using RAPD and organelle-specific sequence tagged sites (STS) primers. Their findings suggested that organelle genome were remarkably conserved with polymorphism detected in only 1 of 4 non-coding regions in the chloroplast and mitochondrial genomes. Mondal (1999, 2002) carried out similar RAPD, ISSR studies on 27 tea cultivars of wide geographical origin including two ornamental types and recorded high level of intra-population variability. Paul *et al.* (1997) used AFLP markers to detect diversity and genetic differentiation between Indian and Kenyan tea cultivars and recorded high level of genetic variability (79%) within population and low level (21%) between populations. Lai *et al.* (2001) studied the genetic relationships in 37 cultivated tea clones and native wild tea in Taiwan using RAPD and ISSR markers and recognized the three major groups i.e., cultivars of China tea and cultivars developed in Taiwan from hybridization and selection; Assam tea and native Taiwan wild tea. Balasaravanan *et al.* (2003) used AFLP for the 49 south Indian tea cultivars wherein all the cultivars were distinguished into Chinary, Assam and Cambod, however, limited intermediates were also reported. Among the tested germplasm China types were found to be the most diverse (37.76%) due to the popularity of certain cultivars among commercial breeders. Recently, AFLP based analysis of Indian tea germplasm at the major gene banks revealed high level of heterozygosity in tea cultivated at National level (Sharma *et al.* 2010, Raina *et al.* 2012).

Recently, ribosomal RNA gene(s) was also used to study the organization and their utilization for better understanding of the phylogenetic relationships between accessions for their use as molecular markers (Singh *et al.* 2001).

Tea glutamine synthetase : gene isolation, characterization and exploration

Glutamine synthetase is a very important enzyme of ammonium assimilation or GS/GOGAT pathway in plants. Its detailed biochemical and molecular characterization has been conducted in several plants. However, CSIR-IHBT cloned a cDNA encoding glutamine synthetase (*CsGS*) from *Camellia sinensis*. The recombinant protein purified from *Escherichia coli* using Ni–NTA affinity chromatography was confirmed by blotting with anti-His antibodies. Catalytic parameters of the protein were determined using glutamate and ATP as substrates. The observed Km was 9 mM and Vmax was 93 U/mg protein with glutamate as substrate, while with ATP Km and Vmax values were 6 mM and 70 U/mg protein, respectively. Purified enzyme showed pH optima at 8. Cations were found to show enhancing effect on the activity of GS enzyme and Mg^{2+} ion exhibited maximum enhancing effect among the various ions used in this study. This enzyme activity increased by 25% in presence of DTT and decreased by 18% when incubated with PMSF. Transcript analysis in tea bud (youngest leaf) showed that *CsGS* gene expression is stimulated in response to abscisic acid (ABA), salicylic acid (SA), and hydrogen peroxide (H_2O_2), while gibberellic acid (GA_3) has no influence on its expression (Rana *et al.* 2008a). Further the expression and activity of *CsGS* was increased by ammonium and nitrate and also by cadmium and salt stress but decreased by copper, aluminium, cold and heat stress (Rana *et al.* 2008b).

Structural analysis of *CsGS* has been conducted employing computational techniques. This was conducted to compare its structural aspects with other known structures of GS. The disordered residues and their distribution in *CsGS* are in close comparison to earlier reported GS. The 3-D structure of *CsGS* also showed high degree of similarity with the only known crystal structure of GS from *Zea mays*. The Km values observed with recombinant *CsGS* for all the three substrates are higher compared to rice, Arabidopsis, maize and human. This suggests lower affinity of *CsGS* for substrates. Further, kinetic mechanism of *CsGS* catalysis was investigated using initial velocity analysis and product inhibition studies. Initial velocity data eliminate the possibility of pingpong mechanism and favor the random mechanism of catalysis. Through product inhibition studies, ADP was found to be a competitive inhibitor with respect to ATP and noncompetitive inhibitor versus both glutamate and ammonium. Taken together, these observations are consistent with a random catalysis mechanism for the *CsGS* where the binding order of certain substrates is kinetically preferred by the enzyme (Yadav 2009). The regulation of a *CsGS* during developmental stages and light/dark conditions on the utilization of nitrate and ammonia was also conducted. The *CsGS* expression levels decreased during dormancy compared to non-dormancy phase of growth. Different leaf positions present different ages of the leaf and *CsGS* expression level was highest in apical bud (youngest leaf) and lower in mature fourth leaf,

Thus, attempts were made to understand the physiological basis of WD. Tea bushes experience WD when the winter day is shorter than a critical length of about 11 hour 15 minutes and minimum temperature falls below 13°C for at least six weeks. The duration of the dormant period increases with the number of short days (Laycock 1969). Barua and Barua (1969) found the hastening of bud break and drastic reduction of flowering when tea bushes are exposed to 13 hours photoperiod during the winter season at Tocklai (27°N). Barua (1969) concluded that winter dormancy in tea is due to short day length (or long night length). Short days alter the balance of the endogenous plant growth regulators (PGRs) in favor of dormancy and long days in the favor of growth.

Higher levels of free ABA were detected in dormant shoots and the levels decreased with release from dormancy suggesting that ABA is involved in the induction and maintenance of WD in tea (Nagar 1996). With increase in free ABA, the bound ABA levels increased in shoots prior to dormancy release suggesting of metabolic inter-conversions of ABA during these periods. The levels of acidic and residual (bound) phenols varied with the state of dormancy and these phenols could possibly play some role in dormancy. Free gibberellin like activity was extremely low at the initiation of WD and remained so during the dormancy period. Conjugated gibberellin like activity (ficin hydrolysably and β–glucosidase hydrolysable compounds) remained high during the dormancy period. The decrease in both ficin and β–glucosidase hydrolysable gibberellin activity with a concomitant increase in free gibberellin activity during the later stages of dormancy was observed by Nagar and Kumar (2000) thereby, suggesting metabolic inter-conversions. Nandi and Palani (1992) reported higher cytokinin glucoside levels during WD with increase in their free levels during bud-break.

Various PGRs, namely, NAA, ABA, GA_3 and BAP were studied in relation to WD. It was reported that The BAP was not particularly effective in inducing bud-break, it stimulated shoot growth during spring and in the active growth period (Nandi *et al.* 1995). ABA and NAA inhibited shoot growth, whereas, GA_3 induced bud-break earlier by about 2 weeks. GA_3 also had a beneficial effect by significantly increasing amino acids, protein and caffeine content.

High polyamines (especially putrescine and low spermine and spermidine) were reported to be associated with onset of WD in tea (Kakkar and Nagar 1997). Conversely, high spermidine and spermine were related with dormancy release. An inverse relationship between free and conjugated polyamines was observed only for putrescine. In general genotypic differences for polyamine levels were observed in the shoots of high yielding clones.

Decline in temperature during periods of WD is one of the most obvious environmental factors experienced by tea plants. It is known that exposure of plants to abiotic factors including low temperature causes oxidative stress (Bowler *et al.* 1992) during which an increased production of reactive oxygen species (ROS) is expected. Low temperatures coupled to light lead to the increased the levels of reactive oxygen species (ROS) in plants mainly because of chilling-induced photoinhibition (Wise 1995), where the molecular oxygen acts as an electron acceptor in place of $NADP^+$. ROS reacts with lipids, proteins and DNA, and cause immense cellular damage. Therefore, plants contain an array of enzymatic and non-enzymatic mechanisms to protect themselves from increased accumulation of ROS (Inzé and Van Montagu 1995). Hypothesis by Allen (1995) and transgenic studies by McKersie *et al.* (1999) and Samis *et al.* (2002) showed that over expression of ROS scavenging enzymes had marked effect on yield enhancement. This suggested that the management of oxidative stress was key to improve plant performance under stress conditions.

Oxidative stress mediated by low temperatures has been implicated in cellular and metabolic damage, and thus affecting growth and development of plants (Fryer *et al.* 1998). Therefore, it was pertinent to ask if tea also experienced cellular damage in response to low temperature and that if there was a correlation between the degree of damage and the tea clones exhibiting variable periods of WD. What happens to the accumulation of ROS in response to low temperature? How various enzymes involved in ROS scavenging respond to low temperature? Studies of Vyas (2003), Vyas and Kumar (2005) and Vyas *et al.* (2007) on several tea clones with varied dormancy periods exhibited significant reduction in the rate of bud growth during winter months with concomitant decrease in net photosynthesis rates (P_N) and *Fv*/*Fm* ratios, in general; though clonal variation for these parameters was evident. Another related parameter, superoxide anion increased in all the clones under study, though the clonal variation did exist. The anti-oxidative enzymes superoxide dismutase (SOD), ascorbate peroxidase and glutathione reductase also showed enhancement during the onset of dormancy and during peak dormancy period as compared to the period of active growth and the period of dormancy release. Among the three enzymes under study, only SOD showed qualitative differences during different months. Data showed that the clone which exhibited maximum enhancement in the activity of anti-oxidative enzymes accumulated the least superoxide anions. A notable point was that the clone that accumulated the least superoxide anions exhibited least dormancy and *vice-versa*. Data thus suggested that modulation of oxidative stress should be central to manage WD in tea. Further detailed studies lead to the proposal that Mn-SOD was closely linked to WD in tea (Vyas and Kumar 2005).

Concomitantly, efforts were also made to understand the global changes in gene expression during WD. For this purpose, expression messages were displayed on a poly-acrylamide gel (Figure 21.2) and differentially expressed sequences were cloned and analysed using differential display of mRNA (Kumar *et al.* 2002).

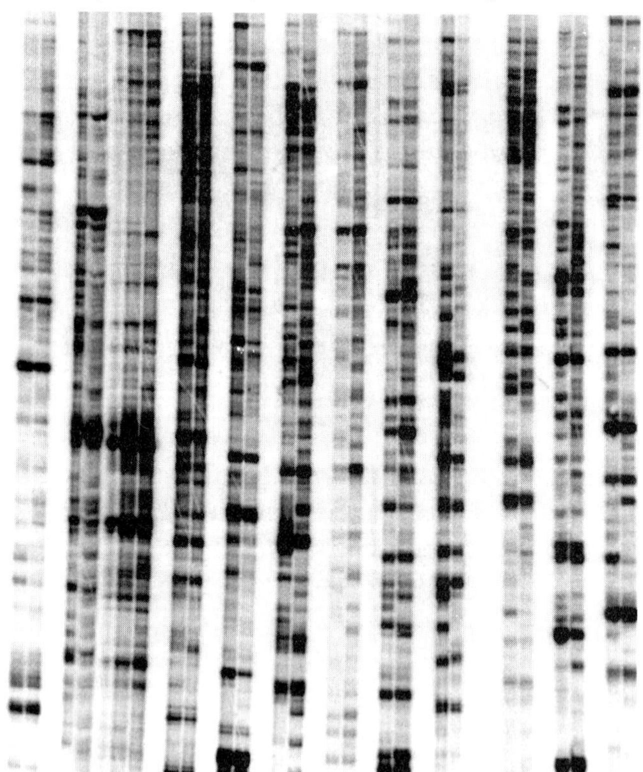

Fig. 21.2. A typical DD gel exhibiting expression messages in dormant and non-dormant buds

DD identified QM like protein gene (*CsQM*) associated with WD in tea (Singh *et al.* 2009a). *CsQM* encodes for ribosomal protein L10, which is implicated in tumor suppression, transcription factor regulation, and ribosome stability in yeast and mammals. During active-growth period of tea, higher expression was observed in apical buds that decreased gradually with increasing age of the leaf. During dormancy season, the expression of *CsQM* gene was severely down-regulated in all the leaves studied. *CsQM* transcript was found to be down regulated in response to drought stress and abscisic acid treatment but up-regulated by gibberellic acid treatment. A positive association of *CsQM* transcript abundance with active cellular growth suggested its role in plant growth and development.

Yet another gene *Histone H3* (*CsH3*) was cloned from tea (Singh *et al.* 2009b). Histone proteins are integral part of chromatin and their expression is typically linked to DNA replication in the S phase of cell cycle. Histone H3 is one of the four histones, along with H2A, H2B and H4, which forms the eukaryotic nucleosome octomer core. During active-growth period of tea, higher expression was observed in apical buds that decreased gradually with increasing age of the leaf. During dormancy season, the expression of *CsH3* was severely down-regulated in all the leaves studied. *CsH3* was found to be down regulated in response to drought stress and ABA treatment and was up-regulated by GA3 treatment. A positive association of *CsH3* abundance with active cellular growth suggested its role in plant growth and development, in general.

In addition to DD, a comprehensive approach of transcriptome analysis was adopted to study WD in tea. Genes associated with WD were studied by analyzing two subtractive cDNA libraries, forward library (FL) and reverse library (RL) with a total of 1102 sequences (Accession number: FF682697 to FF682833, GH454303 to GH454326 and FE942774 to FE943102). Out of 602 sequences obtained in the FL (genes up-regulated during WD), 468 (77.74%) sequences were clustered into 27 contigs and 134 (22.26%) as singletons, generating 161 unigenes. From RL [(genes up-regulated during the period of active growth (PAG)], 232 (46.4%) sequences were clustered into 61 contigs and 263 (53.6%) into singletons, yielding a total of 329 unigenes. Removing the redundancy, a total of 960 clones were analysed by macro-array and later reverse transcriptase (RT)-PCR was used to confirm the expression patterns of putative differentially expressed ESTs during PAG, WD and dormancy release (DR) in TAB and the mature leaf (ML) in two subsequent years. The results indicated, up-regulation of stress responsive genes, and down-regulation of gene involved in cell cycle and DNA processing during WD. The induction of ESTs for *MADS-box*, *MYB*, *nam*, and *eIF5A* suggested change of transcription and synthesis of specific set of

Molecular basis of catechin biosynthesis in tea

The characteristic astringency and bitterness to black tea is imparted by catechins which constitute 25- 30% of the dry mass of young tea leaves (Singh *et al.* 1999). Catechins are synthesized through the flavonoid (FL) pathway (Figure 21.3) wherein 4-coumaroyl CoA, the substrate for the first committed step of the FL pathway, is supplied by the phenylpropanoid (PP) pathway (Rani *et al.* 2009).

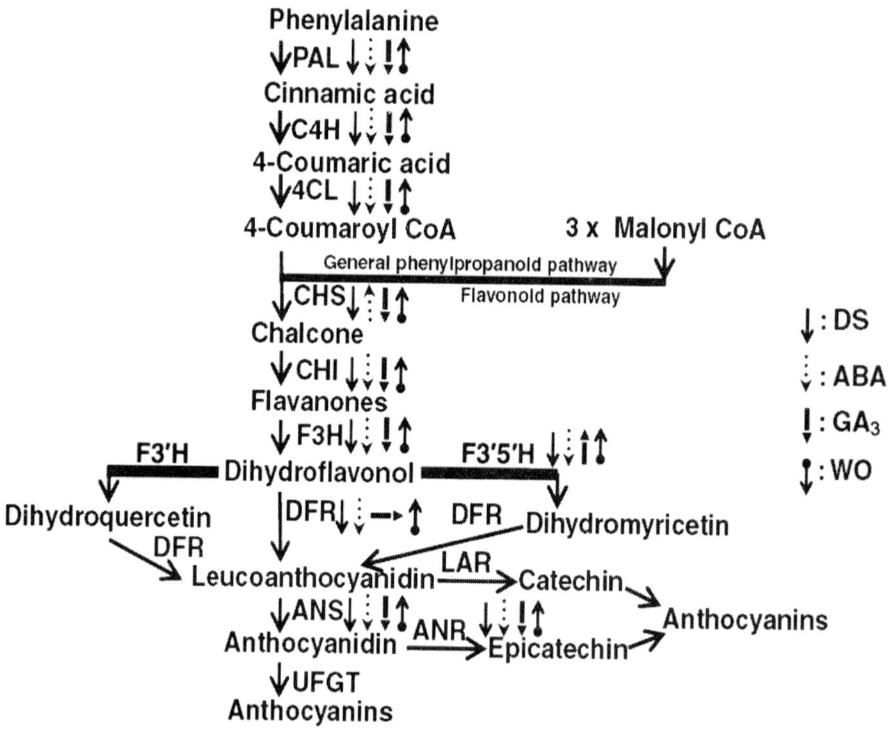

Fig. 21.3. General phenylpropanoid and flavonoid pathway involved in the biosynthesis of catechins and other flavonoids (adapted from Rani *et al.*, 2012). Arrows on right side of the gene indicate the change in expression in response to (DS) Drought, ABA, GA3 and WO (Wounding). Arrows pointing upward, downward and horizontal indicate increase, decrease and no change in expression, respectively. ANR, anthocyanidin reductase; ANS, anthocy-anin synthase; 4CL, 4-coumarate :CoA ligase; C4H, cinnamate 4-hydroxylase; CHI, chalcone isomerase; CHS, chalcone synthase; DFR, dihydroflavonol reductase; F3H, flavanone 3ß-hydroxylase; F3′H, flavonoid 3′-hydroxylase; F3′5′H, flavonoid 3′5′-hydroxylase; LAR, leuacoanthocyanidin reductase; PAL, phenylalanine ammonia-lyase; UFGT, UDP-glucose flavonoid 3-O-glucosyl transferase.

Apart from the role on PP and FL pathways in catechins biosynthesis, these pathways function in providing both structural and signaling molecules. Catechins are synthesized through the joint activity of phenylpropanoid (PP) and flavonoids (FL) biosynthetic pathways (Figure 21.3). PP pathway includes three enzymes, phenylalanine ammonia-lyase (PAL), cinnamate 4-hydroxylase (C4H) and 4-coumaroyl-CoA ligase (4CL). The substrate for PAL catalyzed reaction phenylalanine is derived from the shikimate pathway that begins with the condensation of phosphoenol pyruvate and erythrose 4-phosphate. PAL catalyzes the synthesis of cinnamic acid from phenylalanine and provides a link between primary metabolism and the PP pathway. Cinnamic acid is converted into 4-coumaric acid via aromatic hydroxylation, which is further transformed to 4-coumaroyl CoA by the action of 4CL.

FL pathway (Figure 21.3) begins with the sequential condensations of one molecule of *p*-coumaroyl CoA with three molecules of malonyl CoA to form a C15 chalcone intermediate naringenin chalcone (4,2,4'6'-tetrahydroxychalcone). The reaction is catalyzed by chalcone synthase (CHS), the first enzyme of FL pathway. Chalcone isomerase (CHI) provides the first flavonoid naringenin by catalyzing the stereospecific cyclization of naringenin chalcone to naringenin. Hydroxylation of naringenin at 3' position is catalyzed by flavanone 3-hydroxylase (F3'H) to yield dihydroflavanols.

Dihydroflavonols serve as biosynthetic intermediates in the formation of flavonols, catechins, proanthocyanidins and anthocyanidins. Dihydroflavonols are reduced by dihydroflavonol 4-reductase (DFR) catalyzed reaction at position 4 to yield leucoant-hocyanidins (flavan 2,3-*trans*-3,4-*cis*-diols). Leucoanthocyanidins either serve as immediate precursors for the synthesis of 2, 3-*trans*-(+)-catechins in an enzyme catalyzed reaction called leucoanthocyanidin reductase (LAR) or converted into anthocyanidin by anthocyanidin synthase (ANS). Anthocyanidins are reduced to 2,3-*cis* (-)-epicatechin in the presence of anthocyanidin reductase (ANR) or polymerize to give anthocyanins (Harborne 1967, Harborne 1988).

The majority of the enzymes of catechin biosynthesis as described above are the members of three classes of enzymes found in all organisms: the oxoglutarate-dependent dioxygenases, F3H and ANS; NADPH-dependent reductases, DFR and ANR; and cytochrome P450 hydroxylases, F3'H and F3'5'H (Dixon and Steel 1999).

ption factor(s) regulating the pathway that will open path for metabolic engineering of catechin. Also, the availability of genes and a knowledge on their regulatory role can be utilized in pathway transplantation in heterologous system.

Tea micropropagation

Tea is traditionally propagated through cuttings but a large number of existing plantations are likely to be of seed origin. Seed grown plants show a high degree of variability, therefore, the alternative choice is propagation through vegetative means so as to generate uniform stands of selected elites of superior quality and higher yields. However, the growth of the vegetatively propagated cuttings is slow, season-dependent and the rooting ability varies amongst clones. Micropropagation helps in circumventing these problems. Moreover, this technique can be commercially effective and utilized for introducing improved clones in a relatively shorter time.

Successful micropropagation of tea has been reported by different workers (Chen and Liao 1983, Phukan and Mitra 1984, 1990, Kato 1982, 1986, 1989, 1990, 1992, Arulpragasam and Latiff 1986, Nakamura 1987a,b, 1988, 1989, 1990, Das and Barman 1988, Sarathchandra *et al.* 1988, Banerjee and Agarwal 1990, Jain *et al.* 1991, Agarwal *et al.* 1992, Tahardi and Shu 1992, Vieitez *et al.* 1992, Kuranuki and Shibata 1993, Sood *et al.* 1993, Bag *et al.* 1997, Mondal *et al.* 1998, Xin *et al.* 2008, Ghanati *et al.* 2009, Esan *et al.* 2009). Various explants have been used for *in vitro* regeneration of different tea cultivars such as zygotic embryos, immature and mature cotyledons (Iddagoda *et al.* 1988, Jha and Sen 1992) epidermal layers of stem segments, stem segments without epidermal layer and intact stem segments (Kato 1985), flower stalks and leaf petioles (Sarwar 1985), shoot tips and nodal segments of either juvenile or adult plants (Vieitez *et al.* 1992, Guo *et al.* 2008), leaf explants (Sandal *et al.* 2005, Seran *et al.* 2007) for the induction of adventitious buds. All these reports deal with micropropagation using agar gelled medium. The only report on shoot proliferation using liquid medium was by (Sandal *et al.* 2001). This protocol has a good potential for commercial application as use of liquid medium with extended subculture periods is cost and labour effective in terms of agar and volume of medium.

The genetic fidelity of micropropagated plants was also assessed using RAPD and microsatellite markers (Borchetia *et al.* 2011)

Rooting, hardening and field transfer of tissue culture raised plants

Effective rooting and hardening of tissue culture raised plants have eluded several workers and it is in this context that the work of CSIR-IHBT gains prominence. Rooting has been reported to be a major limitation in the micropropagation of tea wherein reduced salts of MS, light and a low pH (4.5-4.6), IBA treatment (100 mg/l) followed by hormone free medium favored rooting (cf. Mondal *et al.* 2004). Many papers have appeared on micropropagation of *Camellia* spp. but complete and systematic information on root emergence and hardening are lacking (Wu *et al.* 1981, Arulpragasm and Latiff 1986, 1988, Carlisi and Tores 1986, Samartin *et al.* 1986, Beretta *et al.* 1987, Das and Barman 1988, Zhuang *et al.* 1988, Kato 1989, Vieitez *et al.* 1989 a, b, San-Jose and Vieitez 1990). There are reports on root induction from micropropagated tea shoots on auxin containing media under *in vitro* conditions (Banerjee and Agarwal 1990, Jain *et al.* 1991, Nakamura 1991, Jha and Sen 1992, Jain and Newton 1993). Das *et al.* (1990) and Jain *et al.* (1993) reported rooting of IBA treated microshoots prior to direct planting in the soil mix. *Ex vitro* rooting was considered to be better than *in vitro* rooting (Jain *et al.* 1993).

Transfer of *in vitro* plantlets to the greenhouse is the most critical phase of micropropagation. Different methods have been employed for the transfer of tissue culture raised plantlets to green house/field conditions (Tian Ling 1982, Kato 1985, Banerjee and Agarwal 1990, Agarwal *et al.* 1992, Jain *et al.* 1993). However, relevant detailed information on the conditions of hardening was not mentioned in any of the reports. In this context, IHBT played a pivotal role in developing a functional protocol for hardening of tea microshoots (Sharma *et al.* 1999). The time of harvesting (spring and early summer) of microshoots, shoot size, soil pH (5.4), plant growth regulator treatment (IBA; 500 mg/l, 30 min), CO_2 enrichment (9.09/10 x 10^{-5} mol to 10.22/10 x 10^{-5} mol to 80/13 x 10^{-7} mol) and light (15 μmol $m^{-2}5^{1}$) conditions were optimized in specially designed hardening chambers (Figure 21.4).

Significant impact on the percent success for hardening i.e., up to 71.60% root induction, 73% survival and successful field transfer were achieved by the group (Table 21.3).

With a view to save the period of raising plants under nursery and to ensure availability of early planting material of tea, the effect of various auxins in combination with phenolics were examined for rooting of four and eight node cuttings (Prakash *et al.* 2002). A combination of NAA and phenolics was superior to a combination of IBA and phenolics in both types, since 100% rooting was evident in the former. The rooting of cuttings took only ten months for transfer to field as compared to single node cuttings which takes 15-16 months.

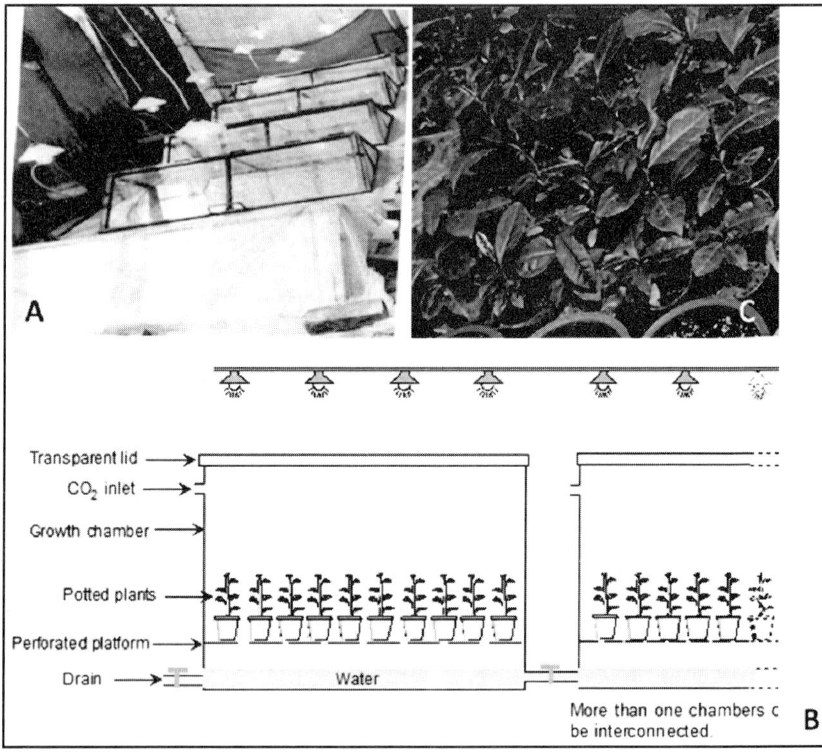

Fig. 21.4. Hardening chambers (A & B) and hardened tea plants (C)

Micro-grafting has been successfully reported in a range of horticulture plants, as a means to obtain clones free from viruses and virus-like diseases and also to detect graft incompatibilities at an early stage. Some important successes are related to fruit trees like citrus (Oyiaima 1992, Starrantino 1992, Obeidy and Smith 1991, Jonard *et al.* 1987), kiwifruit (Ke *et al,* 1993) stone fruit (Cupidi 1992), apple (Richardson *et al.* 1996) and grapes (Martino 1992). Similar experiments have also been reported on forest species like *Larix deciduous* (Ewald and Kretzschmar 1996) and *Picea* spp. (Ponsonby and Mantell 1993). In most of these reports, *in-vitro* raised scions have directly been transplanted on *in-vitro* raised stocks under sterile conditions. CSIR-IHBT has the distinction to have reported the first successful grafting of *in vitro* grown tea micro-shoots for increased vigour and simultaneous hardening for field transfer in *C. sinensis*

(Prakash *et al.* 1999). Tea microshoots excised from well established multiple shoot cultures grown *in vitro* and 8-week old, three to five leaved seedlings from the Kangra Chinary stock (Banuri 96) and UPASI-9 (from south India) were selected as scions and root stocks, respectively, for grafting (Figure 21.3).

Table 21.3. Comparative growth performance of vegetative (single node cutting) and micropropagated tea plants in the fields

Type of cuttings survival (%)	Season of root	Days to root induction	Time of growth establishment & sustainable Management of Nursery	Hardening period in mild shade		Time taken for establishment (from rooting to field)	Growth under field conditions after one year establishment			Average survival (%) under field conditions
				Gradually in Open tunnel Conditions	Natural conditions		Average Height (cm)	Average no. of leaves	Average stem thickness at Collar point (mm)	
Single node Cuttings Av. L. 2.5-3.0 cm Below soil 2.0-2.5 cm Above soil 0.5 cm Dia. 4 mm	Autumn (10.10.95 to 20.12.95)	70-85 days (3 months)	210 days (7 months)	90 days (3 months)	120 days (4 months)	510 days (17 months)	32.50 ± 0.41	8.60 ± 0.16	12.50 ± 0.71	91.2%
Microshoots Av. L. 3.0-4.0 cm ('C' category of shoots) Below soil 0.5-1.0 cm Above soil 0.5-3.0 cm Dia. 1.5 mm	Spring (13.3.95 to 14.5.95)	60 days (2 months)	210 days (7 months)	90 days (3 months)	150 days (5 months)	510 days (17 months)	20.00 ± 0.21	16.00 ± 0.23	4.5 ± 0.16	73.3%

Soil pH = 5.4 Transplanting time=14.2.97; 'c' category of microshoots= 3.0 – 4.0 cm long with short internodes (0.5-1.0cm) and broad leaves (1.0 cm). Number of single node cutting and microshoots=25.

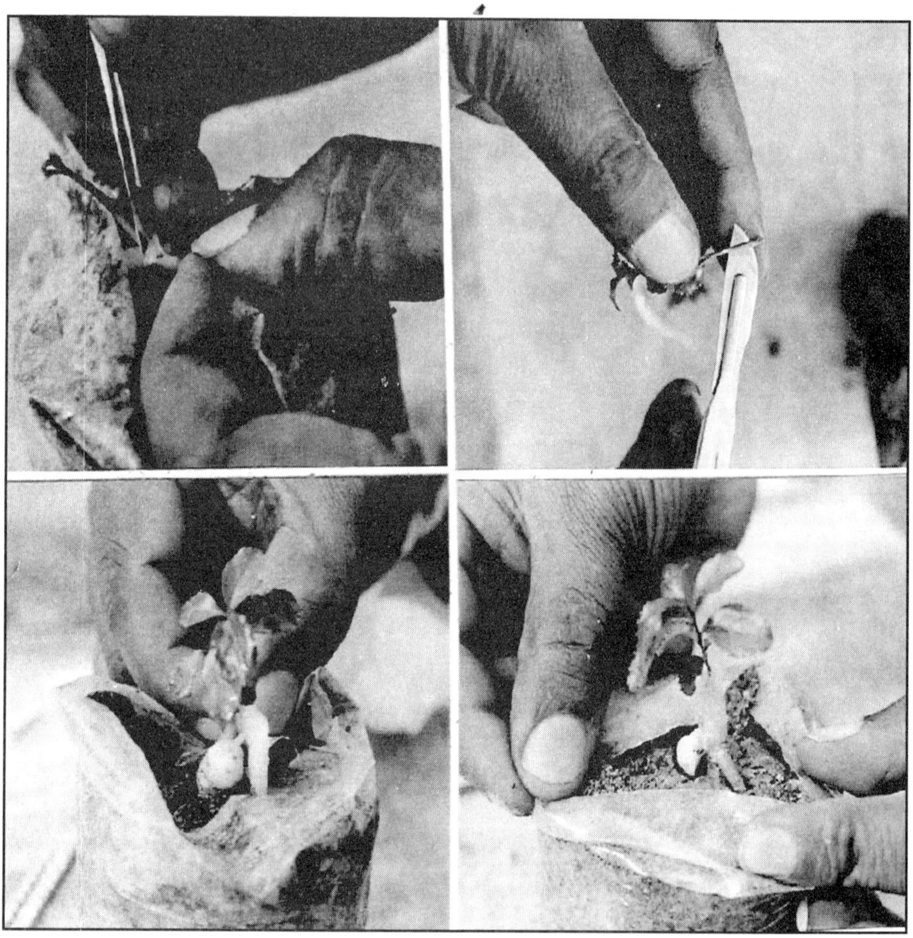

Fig. 21.5. Grafting of tea micro-shoots on superior root stock

In addition, 4 and 12 months old seedlings of Banuri-96 were also used as root stocks. Cut ends of root stocks and scions were pretreated with varying concentrations of BAP and NAA for 10 minutes. A treatment of BAP (5 mg/l) and NAA (5 mg/l) solution to both scion and stocks renewed foliar development at a relatively early stage (40-60 days). The grafted plants were kept in CO_2 enriched air and at high light intensity in specially designed hardening chambers. No significant difference was observed between the auto-graft (scion and root stock of Banuri clone) and the hetero-graft (scion of the Banuri clone and root stock of UPASI-9). Of the three types (in terms of age) of seedling-raised- rootstocks employed, grafts on young tea (4-month old) performed the best (88.33%) (Table 21.4). Grafts made in early summer established relatively faster with a higher rate of success (Table 21.4).

Table 21.4. Effect of age of the root stock on the graft union with microshoots of tea

Age of root stock grown from seed	Number of grafts	Number of successful	Percentage establishment grafts
2 Months			
Banuri-96	500	355	71.00
UPASI-9	250	150	60.00
4 Months			
Banuri-96	180	159	88.33
UPASI-9	200	160	80.00
12 Months			
Banuri-96	55	26	47.00
UPASI-9	120	87	72.50

χ^2 Cal: for Banuri-96 stocks= 41.4627; χ^2 tab, 2 *df* at 1% = 9.21;
χ^2 Cal: for UPASI-9 = 21.6082; χ^2 tab, 2 *df* at 1% = 9.21

Table 21.5. Seasonal effect on the survival of grafted microshoots of tea (scion: local China hybrid; root stock: UPASI-9)

Season	Number of grafts	Established grafts made	Percentage establishment
Winter (1st week of Jan-1st week of Feb)	200	149	74.50
Spring (2nd week Feb-1st week of April)	200	184	92.00
Early summer (April-1st week of May)	200	167	83.50
Peak summer	200	15	7.50

χ^2 Cal:389.46; χ^2 tab, 3 df at 1% = 11.34

The percentage survival of plants transferred to the field was 88.33%. Moreover, success could be achieved within 10 months as compared to the usual 20-24 months required for field transfer of cuttings.

In a totally different approach, the native isolates of *Pseudomonas fluorescens*, *Azospirillum brasilense*, and *Trichoderma harzianum* were used for rooting and acclimatization of *in vitro* grown shoots and plantlets (Thomas *et al.* 2010). They found that while *in vitro* bacterization of *P. fluorescens* failed to establish, both *T. harzianum* and *A. brasilense* retarded shoot growth. Nutrient uptake and consequent plantlet survival was promoted when rooted plantlets were acclimatized in soil amended with bioinoculants. Similarly, a consortia of arbuscular mycorrhizal fungi (AMF) from natural and cultivated tea rhizospheres) were used to improve the growth and quality of plants obtained from seedlings and cuttings in acid soils under net-house (Singh *et al.* 2010).

Somatic embryogenesis

Plant regeneration via somatic embryogenesis is an alternative pathway that offers a great potential for use in mass propagation besides direct or indirect shoot multiplication (Garvelli and Senaratna 1994). Somatic embryogenesis has served as an efficient and essential tool in the propagation and improvement programs of tea. The first report on somatic embryogenesis was by Wu *et al.* (1981) followed by a number of other workers (Nakamura 1985, Sarathchandra *et al.* 1988, Kato 1996, Abraham and Rahman, 1986, Paratasilpin 1990, Bano *et al.* 1991, Rajkumar and Ayappan 1992, Jha *et al.* 1992, Sood *et al.* 1993, Wachira and Ogado 1995, Ponsamuel *et al.* 1996, Bag *et al.* 1997, Akula and Dodd 1998, Akula *et al.* 2000, Balasubramanian *et al.* 2000, Haridas *et al.* 2000, Mondal *et al.* 2000, Aoshima 2005, Ozudogru *et al.* 2006). Arulpragasam *et al.* (1988) defined the medium supplements for callus initiation, embryoid development, maintenance and plant regeneration from cotyledons of *C. sinensis*. Abraham and Rahman (1986) obtained somatic embryogenesis on immature embryos and cotyledons of six cultivars of tea on a hormone free medium. In all these reports, emphasis was laid on manipulating nutrient composition, growth regulators, physical conditions during incubation including stress treatments for the induction of somatic embryogenesis.

In our laboratory, Sood *et al.* (1993) achieved somatic embryogenesis on half strength MS medium with sucrose (2%), BAP (2.0 mg/l) and IAA (0.2 mg/l) which germinated when 0.2 mg/l GA_3 was also present in the medium. However, none of these reports dealt with the induction of synchronized secondary somatic embryos in the form of number and developmental stages (Vieitez 1995). Therefore, a systematic assessment was made by Mondal *et al.* (2001a) on the type and concentration of sulphate and nitrate salts in the basal medium to elicit an optimum and synchronous secondary embryogenic response (Figure 21.6).

Fig. 21.6. Secondary somatic embryos (A) Globular, (B) Heart shaped, (C) Germinated somatic embryo, and (D) Somatic embryo-derived plants

They reported that nitrate salts of potassium (9.39 mmol) and ammonium (10.31 mmol) with only 1.5 mmol potassium sulphate produced the maximum number of synchronous secondary embryos (i.e. 20-25 secondary embryos per primary embryo in 91.6% responsive explants). Of the different factorial combinations of L-glutamine, BAP and IBA, maximum number of synchronous secondary embryos developed on a medium supplemented with 8.88 mmol BAP, 0.98 mmol IBA and 10 mmol L-glutamine. Germination of the embryos (52%) was achieved only when sulphate salts of potassium were removed from the medium. While stimulatory effects of GA_3 for complete plantlet production and that of BAP for somatic embryogenesis was observed by Abraham and Rahman (1986 a,b), Raj Kumar and Ayyappan (1992) reported somatic embryogenesis and embryo germination from mature

Surface – disinfection of mature sinker seeds

Germination on 1/2 MS + 30 g/l sucrose + 0.7 agar
2 weeks

Induction of somatic embryos from cotyledon slices on EIM
6-8 weeks

Multiplication of primary embryos on EMM
6 weeks

Infection of globular embryos with *Agrobacterium* suspension
(O.D. 600 = ~0.610-9 cell/ml)
20 min
Co-cultivation (EMM, pH = 5.6) in darkness
5 days
Washing 2 times with sterile water and 2 times with liquid EMM
containing 400 mg/l sporidex and and blot drying

Transfer of embryos to EMM + sporidex (250 mg/l) + carbenicillin (250 mg/l)

2 weeks
Selection on above medium + kanamycin (50 mg/l)
8-12 weeks
Transfer to MM + kanamycin (75 mg/l)
4 weeks
Transfer to germination medium + kanamycin (75 mg/l)
8 weeks
Multiplication of transgenic shoots on 1 mg/l TDZ + 2 mg/l NAA
8 weeks
Micrografting on seedling rootstocks and growth in green house

Fig. 21.8C. Protocol for *Agrobacterium* mediated genetic transformation of tea

Although Mondal *et al.* (2001c) was successful in developing transgenic plants from somatic embryos of the cultivar, Kangra jat, most of the commercially important cultivars of tea remain outside the host range of *Agrobacterium*. Particularly, the leaves of several commercially important tea cultivars are recalcitrant to *Agrobacterium* mediated transformations. This was either due to the variations in surface characteristics and/or biochemical constituents of the tea leaves of different cultivars (Kumar *et al.* 2005) or the presence of high contents of bactericidal polyphenols (Biao *et al.* 1998, Sandal *et al.* 2009, 2011, MeiQin *et al.* 2008). However, this was overcome by the use of autoclaved L-glutamine by Sandal *et al.* (2009, 2011) or by a combination of filter sterilized L-glutamine and glutamic acid.

Genetic transformation using *Agrobacterium rhizogenes* was attempted by several workers. Zehra *et al.* (1996) used a wild strain of *A. rhizogenes* and confirmed the stable integration of transgene in root callus by mannopine assay. Later Konwar *et al.* (1998) used the wild strain of *A. rhizogenes* for root induction in 4-6 months old *in vitro* tea shoots. In 2009, John *et al.* reported that the *A. rhizogenes* strain (MTCC 532) was suitable for hairy-root cultures from leaf explants and accumulation of phenolic compounds. They used the rol C gene primers for PCR confirmation and HPLC analysis for detection of higher levels of catechins in the hairy roots.

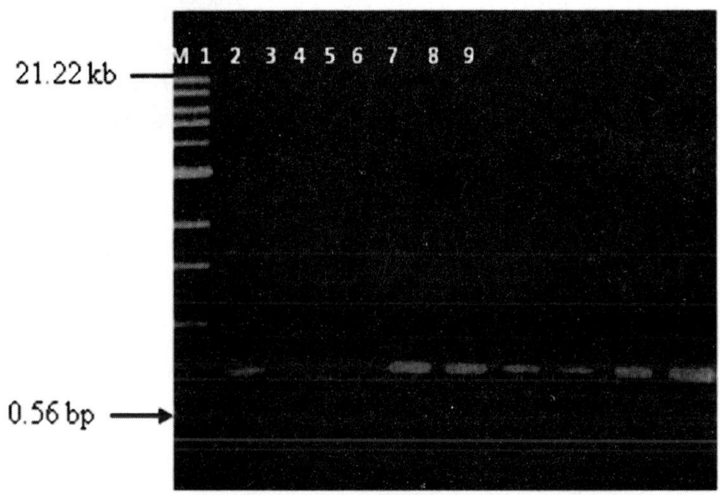

Fig. 21.8D. PCR amplification product of gus gene in putative transformants where Lane M is 1 kb ladder, Lanes 2 & 3 are DNA from untransformed tea plants; Lanes 4-9 are DNA from transformed tea plants obtained from *Agrobacterium* mediated transformation of somatic embryos

Biolistic mediated genetic transformation was also attempted by Saini *et al.* (2012) and Sandal *et al.* (2009). The factors crucial for successful transformation were osmotic pretreatment of leaf explants with 0.25 M sorbitol or hormone free, distances (gap, microcarrier flight and target), burst pressure and DNA concentration. About 40% of the GUS positive transformants (Figure 21.8 D) that survived the lethal dose of kanamycin showed PCR amplification for both *npt*II and *gus* primers (Figure 21.8 D).

Southern hybridization further confirmed the stable transformation. However, these plants were incapable of transgene expression probably due to inappropriate localization of transgene(s) in the genome. Delivery of transgenes into an appropriate locus in the plant genome is governed by the parameters that affect the process of gene delivery (Romano *et al.* 2003). Moreover, intact and single copy integration as compared to multiple copies of rearranged transgene inserts is considered to be crucial for stable expression (Thomas *et al.* 2001, Gelvin 2003).

In view of these, Saini *et al.* (2012) made further modifications in the combinations of target distances and burst pressures. These modified parameters were used to bombard somatic embryos with *osmotin* (source: Prof. KC Bansal, IARI, New Delhi) and *chitinase* (chi11) (source: Prof. Muthukrishnan, Kansas State University, Kansas) genes for biotic and abiotic stress tolerance. Although a total of 99 lines were produced in the various parameter combinations, only 33% were found to have intact *osmotin* inserts in Southern hybridization, and only 15% of these yielded stable expression in Northern hybridization (Figure 21.9 A & B).

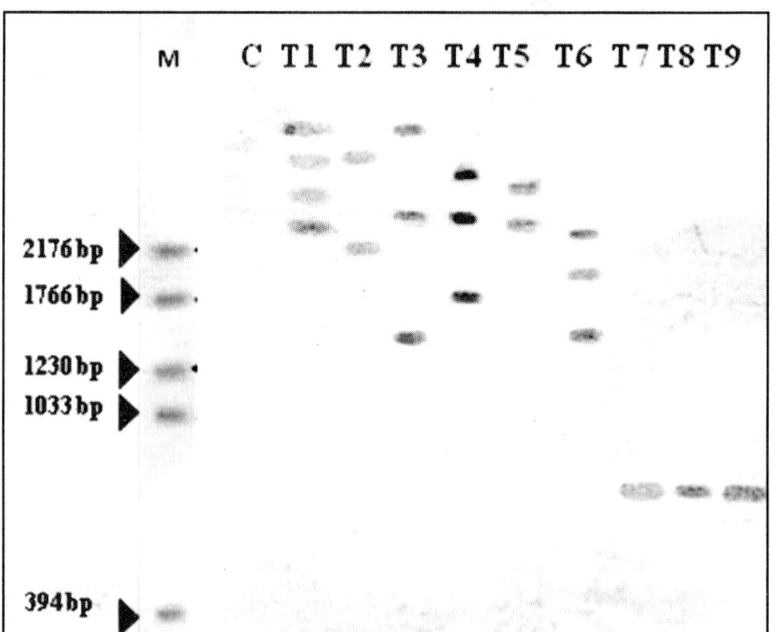

Fig. 21.9A. Southern blot analysis showing *osmotin* gene integration patterns in PCR positive (T1-T9) lines. DIG-labeled, PCR amplified product of osmotin was used as probe. M: molecular marker, T1-T9: transgenic lines, C: Control

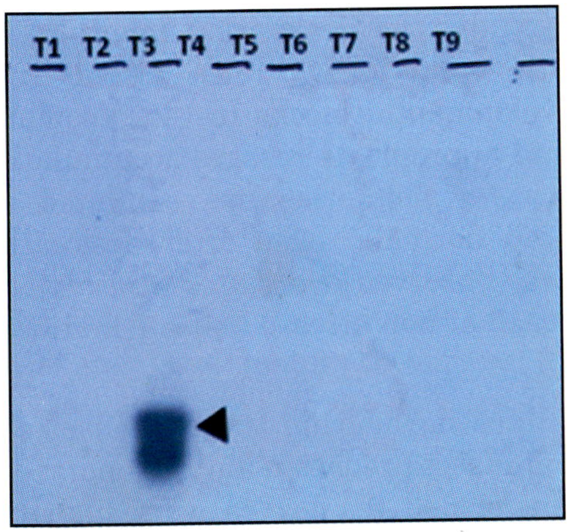

Fig. 21.9B. Northern blot of Southern positive transgenics (T1-T9), showing strong hybridization signal in T2 when α-P^{32} labeled PCR amplified osmotin fragment was used as probe

Interestingly, all these lines were produced by using a combination of a target distance of 9 cm and burst pressure of 1100 psi. These lines were healthier and showed higher multiplication rates. In contrast, the other lines were poor performers with staggered multiplication rates. When these lines were assessed for stress tolerance, increased osmotic adjustment with the ability to tolerate osmotic stresses imposed by polyethylene glycol was observed (Figure 21.10).

Fig. 21.10. Response of control and transgenic shoots to (A) osmotic stress imposed by 20% PEG and (B) their recovery in water

Although healthy vegetative growth of the plants growing under contained polyhouse conditions was maintained, their reproductive ability was found to be highly suppressed.

Arulpragasam PV, Latiff R, Seneviratne P (1988) Studies on the tissue culture on tea (*Camellia sinensis* (L.) O. Kuntze). 3. Regeneration of plants from cotyledon callus. *Sri Lanka J Tea Sci.* 57(1): 20-23.

Arulpragasum PV, Latiff R (1986) Studies on the tissue culture of tea (*Camellia sinensis* (L.) O. Kuntze). 1. Development of a culture method for the multiplication of shoots. *Sri Lanka J Tea Sci.* 5(1): 44-47.

Bag N, Palni LMS, Nandi SK (1997) Mass propagation of tea using tissue culture methods. *Physiol Mol Biol Plants* 3: 99-103.

Balasaravanan T, Pius PK, Kumar RR, Muraleedharan N, Shasany AK (2003) Genetic diversity among south Indian tea germplasm (*Camellia sinensis*, *C. assamica* and *C. assamica* spp. *lasiocalyx*) using AFLP markers. *Plant Sci.* 165: 365-372.

Balasubramanian S, Marimuthu S, Rajkumar R, Haridas V (2000) Somatic embryogenesis and multiple shoot induction in *Camellia sinensis* (L.) O. Kuntze. *J Plant Crops* 28: 44-49.

Ballester A, Janeiro LV, Vieitez AM (1997) Cold storage of shoot cultures and alginate encapsulation of shoot tips of *Camellia japonica* L. and *Camellia reticulata* Lindl. *Sci Hortic.* 71: 67-78.

Banerjee B (1992) Botanical classification of tea. In: Wilson K C and Clifford M N (ed) Tea cultivation to consumption Chapman and Hall, London, pp 25-51.

Banerjee M, Agarwal B (1990). *In vitro* rooting of tea *(Camellia sinensis.* (L). O. Kuntze). *Ind J Exp Bot.* 28: 936-939.

Bano Z, Rajaratnam S, Mohanty BD (1991) Somatic embryogenesis in cotyledon culture of tea (*Thea sinensis* L.) *J Hortic Sci.* 66: 465-470.

Barua DN (1963) Classification of the tea plants. *Two and a Bud* 10: 3-11.

Barua DN (1969) Seasonal dormancy in tea (*Camellia sinensis* L.). *Nature* 224: 514.

Barua DN (1989) Science and practice in tea culture. Tea Research Association, Calcutta. p 509.

Barua DN, Barua SC (1969) Seasonal dormancy in tea. *Two and a Bud* 15: 41-45.

Beratta D, Vanoli M, Eccher T (1987) The influence of glucose, vitamins and IBA on rooting of *Camellia* shoots *in vitro*. Abstr. Symposium on Vegetative Propagation of Woody Species, Pisa, Italy, p 105.

Bhattacharya A, Saini U, Ahuja PS (2006) Transgenic Tea. *Intl J Tea Sci.* (Special Issue) 5(1-2): 39-52.

Bhattacharya A, Saini U, Sharma P, Nagar PK, Ahuja PS (2006) Osmotin-regulated reserve accumulation and germination in genetically transformed tea somatic embryos: A step towards regulation of stress tolerance and seed recalcitrance. *Seed Sci Res.* 16: 203-211.

Biao Xi, Toru K, Jian Xu, Yongyan B (1998) Effect of polyphenol compounds in tea transformations. Abstr. no. 314. In: American Society of Plant Physiologists, Plant Biology.

Blair MW, Panaud O, McCouch SR (1999) Inter-simple sequence repeats (ISSR) amplification for analysis of microsatellite motif frequency and fingerprinting in rice (*Oryza sativa* L.). *Theor Appl Genet.* 98: 780-792.

Borchetia S, Das SC, Handique PJ, Das S (2011) High multiplication frequency and genetic stability for commercialization of the three varieties of micropropagated tea plants (*Camellia* spp.) *Sci Hortic.* 120(4): 544-550 DOI: 10.1016/j.

Borse BB, Rao LJM, Nagalakshmi S, Krishnamurthy N (2002) Fingerprint of black teas from India: identification of the regio-specific characteristics. *Food Chem.* 79: 419-424.

Bowler C, Montagu MV, Inze D (1992) Superoxide dismutase and stress tolerance, Annu. Rev. *Plant Physiol Plant Mol Biol.* 43: 83–116.

Braun AC, Mandle RJ (1948) Studies on the inactivation of the tumor inducing principle in crower gall. *Growth* 12: 255-269.

Caceres PLF, Martin MJ, Pabios F, Gonzalez AG (2001) Differentiation of tea (*Camellia sinensis*) varieties and their geographical origin according to their metal content. *J Agric Food Chem.* 49: 4775-4779.

Carlisi JC, Torres KC (1986) *In vitro* shoot proliferation of *Camellia* 'Purple Dawn. *Hort Sci.* 21: 314.

Chen Z, Liao H (1983) A success in bringing out tea plants from the anthers. *China Tea* 5: 6-7.

Cupidi A (1992) *In vitro* micrografting of stone fruits. *Petria* 2 (Suppl.): 7-15.

Das SC, Barman TS (1988) Current State and future potential of tissue culture in tea. Proceedings of 30th Tocklai Conference. TRA, Jorhat, pp 90-94.

Das SC, Barman TS, Singh R (1990) Plant regeneration and establishment in the nursery. *The Assam Review and Tea News* 79: 24-27.

Dhiman B (2003) Development of species specific PCR primers for some common plants adulterants of tea. Ph.D thesis submitted to Guru Nanak Dev University, Amritsar, India.

Dixon RA, Steel CL (1999) Flavonoids and isoflavonoids a gold mine for metabolic engineering. *Trends Plant Sci.* 4: 394-400.

Dudeja V (1992). *Tea Today* 13: 1-5.

Esan EB, Muyiwa AA, Lawal JO (2009) Performance of tea (*Camellia sinensis* L.) on culture media modified with locally sourced substitutes. *J Animal Plant Sci.* 4(1): 298-303.

Ewald D, Kretzschmar U (1996) The influence of micro-grafting *in vitro* on tissue culture behaviour and vegetative propagation of old European larch trees. *Plant Cell Tissue Org Cult.* 44: 249-252.

Ferrara L, Montesano D, Senatore A (2001) the distribution of minerals and flavonoids in the tea plant (*Camellia sinensis*). *IL FARMACO* 56: 397-401.

Fryer MJ, Andrews JR, Oxborough K, Blowers DA, Baker NR (1998) Relationship between CO_2 assimilation photosynthetic electron transport and active O_2 metabolism in leaves of maize in the field during periods of low temperature. *Plant Physiol.* 116: 571–580.

Gelvin SB (2003) *Agrobacterium*-mediated plant transformation, the biology behind the gene-jockeying tool. *Microbiol. Mol. Biol. Rev.* 67: 16-37.

Ghanati F, Ishka MR (2009) Investigation of the interaction between abscisic acid (ABA) and excess benzyladenine (BA) on the formation of shoot in tissue culture of tea (*Camellia sinensis* L.). *Intl J Plant Product.* 3(4): 7-14.

Grandmaison J, Ibrahim RK (1996) Evidence for nuclear protein binding of flavonol sulfate esters in Flaveria chloraefolia *Journal Plant Physiol.* 147: 653-660.

Guo YQ, Xiong LZ, LiuXin L, TengHui L, Jing C (2008) *In vitro* preservation of germplasm of oolong tea trees (*Camellia sinensis*) in Fujian: I. Establishment of aseptic lines from adult stem nodes. *J Fujian Agric Forestry* 37(6): 587-591.

Harborne JB (1967) Comparative biochemistry of the flavonoids. Academic Press, New York.

Harborne JB (1988) The flavonoids Advances in research since 1986. Chapman and Hall, New York.

Haridas V, Balasaravanan T, Rajkumar R, Marimuthu S (2000) Factor influencing somatic embryogenesis in *Camellia sinensis* (L.) O. Kuntze. In: Muraleedharan N, Rajkumar R (eds) *Recent Advances in Plantation Crops Research.* Allied Publishers Ltd, India, pp 31-35.

Hausman JF, Neys O, Kevers C, Gaspar T (1994) Effect of *in vitro* storage at 4°C on survival and proliferation of poplar shoots. *Plant Cell Tissue Org Cult.* 38: 65-67.

Holford P, Newbury HJ (1992) the effect of antibiotics and their break-down products on the *in vitro* growth of *Antirrhium majus*. *Plant Cell Rep.* 11: 93-96.

Hrazdina G, Jensen RA (1992) Spatial organization of enzymes in plant metabolic pathways Annu. Rev. Plant Physiol. *Plant Mol Biol.* 43: 241-267.

Hutzler P, Rischbach, R, Heller W, Jungblut TP, Reuber S, Schmitz R, Veit M, Weissenbck G, Schmitzler JP (1998) Tissue localization of phenolic compounds in plants by confocal laser scanning microscopy. *J Exp Bot.* 49: 953-965.

Iddagoda N, Kataeva NN, Butenko RG (1988) *In vitro* clonal micro-propagation of tea (*Camellia sinensis* L.) 1. Defining the optimum condition for culturing by means of a mathematical design techniques. *Ind J Plant Physiol.* 31(1):1-10.

Inzé D, Van Montagu M (1995) Oxidative stress in plants. *Curr Opin Biotechnol.* 6: 153–158.

Jain NK (1999) "Global advances in Tea Science" 1st Edn, Published by Aravali Books International, W-30 OKHLA Industries (p) Ltd. Area phase II, New Delhi, pp 333-358.

Jain SM, Das SC, Barman TS (1991) Induction of roots from regenerated shoots of tea (*Camellia sinensis* L.). *Acta Hort.* 289: 339-340.

Jain SM, Das SC, Barman TS (1993) Enhancement of root induction from *in vitro* regenerated shoots of tea (*Camellia sinensis* L.). *Proc Ind Natl Sci Acad.* B59 (6): 623-628.

Jain SM, Newton RJ (1993) Prospects of biotechnology for the improvement. *Proc Ind Nat Sci Acad.* B-56: 441-448.

Jha T, Sen SK (1992) Micropropagation of elite Darjeeling tea clone. *Plant Cell Rep.* 11: 101-104.

Jha TB, Jha S, Sen SK (1992) Somatic embryogenesis from immature cotyledons of an elite Darjeeling tea clone. *Plant Sci.* 84: 209-213.

Jian Z, Hao C, Yuan WL (2008) Factors affecting the germination rate and the form in the culture of tea (*Camellia sinensis*) young embryos. *Southwest China J Agric Sci.* 21(2): 440-443

John KMM, Joshi SD, Mandal AKA, Kumar SR, Kumar RR (2009) *Agrobacterium rhizogenes* mediated hairy root production in tea leaves (*Camellia sinensis* (L.) O. Kuntze) Ind. *J Biotechnol.* 8(4): 430-434.

Jonard R, Soedharma I, Villemur P (1987) Analysis of the influence of various factors on the improvement of successful micrografts in citrus. *C R Acad Sci.* 305: 45-49.

Jong JHD, Fransz P, Zabel P (1999) High resolution FISH in plants-techniques and applications. *Trends in Pln Sci.* 4 (7): 258-263.

Joshi PS, Ranjekar PK, Gupta VS (1999) Molecular markers in plant genome analysis. *Curr Sci.* 77(2): 230-241.

Kakkar RK, Nagar PK (1997) Distribution and changes in endogenous polymines during winter dormancy in tea (*Camellia sinensis* (L.) O.Kuntze). *J Plant Physiol.* 151(1): 63-68.

Kato M (1982) Result of organ culture on *Camellia japonica* and *Camellia sinensis*. *Jap J. Breed.* 32(2): 267-277.

Kato M (1985) Regeneration of plantlets from tea stem callus. *Jap J Breed.* 35: 317-322.

Kato M (1986) Micropropagation through cotyledon culture of *Camellia japonica* L. and *C. sinensis* L. *Jap. J. Breed.* 36: 31-38.

Rani A, Singh K, Ahuja PS, Kumar S (2012) Molecular regulation of catechin biosynthesis in tea (*Camellia sinensis* (L.) O. Kuntze). *Gene* 495: 205-210.

Rani A, Singh K, Sood P, Kumar S, Ahuja PS (2009) p-Coumarate: CoA ligase as a key gene in the yield of catechins in tea [*Camellia sinensis* (L.) O. Kuntze]. *Funct Integr Genomic* 9:271–275. DOI: 10.1007/ s10142-008-0098-3.

Richardson FVM, Saoir SMA, Harvey BMR (1996) A study of the graft union in *in vitro* micrografted apple. *Plant Growth Regul.* 20: 17-23.

Romano A, Raemakers K, Bernardi J, Visser R, Mooibroek H (2003) Transgene organization in potato after particle bombardment mediated (co) transformation using plasmids and gene cassettes. *Transgenic Res.* 12: 461-473.

Saini, U., Kaur, D., Bhattacharya, A., Kumar, S., Singh, R.D. and Ahuja, P.S. (2012) Optimising parameters for biolistic gun-mediated genetic transformation of tea [*Camellia sinensis* (L.) O. Kuntze]. *J. Hort. Sci. Biotechnol.* (In Press).

Samartin A, Vieitez AM, Vieitez E (1986) Rooting of tissue cultured camellias. *J. Hort. Sci.* 61: 113-120.

Samis K, Bowley S, McKersie BD (2002) Pyramiding Mn-superoxide dismutase transgenes to improve persistence and biomass production in alfalfa. *J Exp Bot.* 53: 1343–1350.

Sandal I, Bhattacharya A, Ahuja PS (2001) An efficient liquid culture system for tea shoot proliferation. *Plant Cell Tissue Org. Cult.* 65: 75-80.

Sandal I, Bhattacharya A, Ahuja PS (2009) Transgenic tea through biolistic using leaf explants. Patent no. 7129394, dated 31/10/06, USA.

Sandal I, Bhattacharya A, Saini U, Kaur D, Sharma S, Gulati A, Kumar JKK, Kumar N, Dayma J, Das P, Singh B, Ahuja PS (2011) Chemical modification of L-glutamine to alpha-amino glutarimide on auto-claving facilitates *Agrobacterium* infection of host and non-host plants: A new use of a known compound. *BMC J. Chemical Biol.* 11:1 doi:10.1186/1472-6769-11-1

Sandal I, Kumar A, Bhattacharya A, Sharma M, Shanker A, Ahuja PS (2005) Gradual depletion of 2,4-D in the culture medium for indirect shoot regeneration from leaf explants of *Camellia sinensis* (L.) O. Kuntze. *Plant Growth Regul.* 47(2-3): 121-127.

San-Jose MC, Vieitez AM (1990) *In vitro* regeneration of *Camellia reticulata* cultivar 'Capitain Rawes' from adult material. *Sci Hortic.* 43: 155-162.

Sarathchandra TM, Upali PD, Wijeweardena RGA (1988) Studies on the tissue culture of tea (*Camellia sinensis* (L.) O. Kuntze) 4. Somatic embryogenesis in stem and leaf callus cultures. *Sri Lanka J Tea Sci.* 52 (2): 50-54.

Sarma KS, Evans NE, Selby C (1995) Effect of carbenicillin and cefota-xime on somatic embryogenesis of Sitka Spruce (*Picea sitchensis* (Bong. Carr.). *J Exp Bot.* 46: 1779-1781.

Sarwar M (1985) Callus formation from explanted organs of tea (*Camellia sinensis* L.). *J Tea Sci.* 54 (1): 18-22.

Seran TH, Hirimburegama K, Gunasekare MTK (2007) Production of embryogenic callus from leaf explants of *Camellia sinensis* (L.). *J. Natn. Sci. Foundation Sri Lanka* 35(3): 191-196.

Sharma M, Sood A, Nagar PK, Prakash O, Ahuja PS (1999) Direct rooting and hardening of tea microshoots in the field. *Plant Cell Tissue Org. Cult.* 58: 111-118.

Sharma P, Kumar S (2005) Differential display mediated identification of three drought-responsive expressed sequence tags in tea (*Camellia sinensis* (L.) O. Kuntze). *J Bioscience.* 30: 231-235.

Sharma P, Pandey S, Bhattacharya A, Nagar PK, Ahuja PS (2004) ABA associated biochemical changes during somatic embryo development in *Camellia sinensis* L. (O.) Kuntze. *J Plant Physiol.* 161(11): 1269-1276.

Sharma RK, Negi MS, Sharma S, Bhardwaj P, Kumar R, Bhattachrya E, Tripathi SB, Vijayan D, Baruah AR, Das SC, Bera B, Rajkumar R, Thomas J, Sud RK, Muraleedharan N, Hazarika M, Lakshmikuma-ran M, Raina SN, Ajuja PS (2010) AFLP-Based Genetic Diversity Assessment of Commercially Important Tea Germplasm in India. *Biochem Genet.* 48:549-564.

Singh D, Singh M (2001) Organization of 5S ribosomal RNA genes in tea (*Camellia sinensis*) *Genome* 44: 143-146.

Singh HP, Ravindranath SD, Singh C (1999) Analysis of tea shoot catechins Spectrophotometric quantitation and selective visualization on two dimensional paper chromatograms using diazotized sulphanilamide. *J Agri Food Chem.* 47: 1041-1045.

Singh K, Kumar S, Ahuja PS (2009b). Differential expression of Histone H3 gene in tea (*Camellia sinensis* (L.) O. Kuntze) suggests its role in growing tissue. *Mol Biol Rep.* 36: 537-542.

Singh K, Kumar S, Rani A, Gulati A, Ahuja PS (2009) Phenylalanine ammonia-lyase (PAL) and cinnamate 4-hydroxylase (C4H) and catechins (flavan 3-ols) accumulation in tea. *Funct Integr Genomic.* 9: 125–134. DOI 10.1007/s10142-008-0092-9.

Singh K, Kumar S, Yadav SK, Ahuja PS (2009) Characterization of Dihydroflavonol 4-reductase (DFR) cDNA in tea (*Camellia sinensis* (L.) O. Kuntze). *Plant Biotechnol Rep.* 3: 95-101.

Singh K, Paul A, Kumar S, Ahuja PS (2009a) Cloning and differential expression of human Wilm's tumor suppressor homologue, QM-family protein (L10) isolated from tea [*Camellia sinensis* (L.) O. Kuntze]. *Mol Biol Rep.* 36: 921-927.

Singh K, Rani A, Kumar S, Sood P, Mahajan M, Yadav SK, Singh B, Ahuja PS (2008) An early gene of flavonoid pathway, flavanone 3-hydroxylase, exhibits a positive relationship with catechins content in tea (*Camellia sinensis* (L.) O. Kuntze). *Tree Physiol.* 28: 1349–1356.

Singh K, Rani A, Paul A, Dutt S, Joshi R, Gulati A, Ahuja PS, Kumar S (2009) Differential display mediated cloning of anthocyanidin reductase gene from tea (*Camellia sinensis*) and its relationship with the concentration of epicatechins. *Tree Physiol.* DOI: 10.1093/ treephys/tpp022.

Singh S, Pandey A, Kumar B, Palni LMS (2010) Enhancement in growth and quality parameters of tea (*Camellia sinensis* (L.) O. Kuntze) through inoculation with arbuscular mycorrhizal fungi in an acid soil. *Biology and fertility of soils* 46(5): 427-433; Doi: 10.1007/s00374-010-0448.

Smith EF, Townsend CO (1907) A plant tumour of bacterial origin. *Science* 25: 671-673

Sood A, Palni LMS, Sharma M, Rao DV, Chand G, Jain NK (1993) Micropropagation of tea using cotyledon culture and encapsulated somatic embryos. *J Plant Crops* 21: 295-300.

Starrantino A (1992) *In vitro* micrografting of citrus. *Petria* 2 (Suppl): 27-35.

Tahardi JS, Shu W (1992) Commercialization of clonal micropropagation of superior tea genotypes using tissue culture technology USAID/ CDR Network meeting on tea crop biotech Costa Rica

Takeo T (1983) Effects of clonal specificity of the monoterpene alcohol composition of tea shoots on black tea aroma profile. *Jap Agric Res Duart.* 17: 120-124.

Thomas J, Ajay D, Kumar RR, Mandal AKA (2010) Influence of beneficial microorganisms during *in vivo* acclimatization of *in vitro* derived tea (*Camellia sinensis*) plants. *Plant Cell Tiss Org Cult* 101(3): 365-370, Doi: 10.1007/s11240-010-9687-7.

Thomas K, Sergio R, David M, Jeanine DL, Rosalind EWC, Peggy GL (2001) Transposon-mediated single-copy gene delivery leads to increased transgene expression stability in barley. *Plant Physiol.* 125(3): 1354-1362.

Tian-Ling (1982) Regeneration of plantlets in cultures of immature cotyledons and young embryos of *Camellia oleifera* Abel. *Acta Biol Exp Sin.* 15: 393-403.

Tosca A, Pondofi R, Vasconi S (1996) Organogenesis in *Camellia x williamsii:* cytokinin requirement and susceptibility to antibiotics. *Plant Cell Rep.*15: 541–544.

Vieitez AM (1995) Somatic embryogenesis in *Camellia* spp. In: Jain SM, Gupta PK and Newton R (eds) Somatic Embryogenesis in Woody Plants, Vol. 2. Angiosperms. Kluwer Academic Publishers, Dordrecht, The Netherlands, pp 235-276.

Vieitez AM Barciela J, Ballester A (1989a). Propagation of *Camellia japonica* cv. Alba Plena by tissue culture. *J. Hort. Sci.* 64: 177-182.

Vieitez AM, San-Jose MC, Ballester A (1989b). Progress towards clonal propagation of *Camellia japonica* cv. Alba Plena by tissue culture techniques. *J. Hort. Sci.* 64: 605-610.

Vieitez AM, Vieitez ML, Ballester A, Vieitez E (1992) Micropropagation of *Camellia* spp. In: Bajaj YPS. (ed) Biotechnology in Agriculture and Forestry, Vol. 19. High Tech and Micropropagation III. Springer-Verlag, Berlin Heidelberg, pp 361-387.

Vyas D (2003) Biochemical and molecular aspects of oxidative stress in tea (*Camellia sinensis* (L.) O. Kuntze). Guru Nanak Dev University, Amritsar. 134 p.

Vyas D, Kumar S (2005) Purification and partial characterization of a low temperature responsive Mn-SOD from tea (*Camellia sinensis* (L.) O. Kuntze). *Biochem Bioph Res Co.* 329: 831-838.

Vyas D, Kumar S (2005) Tea (*Camellia sinensis* (L.) O. Kuntze) clone with lower period of winter dormancy exhibits lesser cellular damage in response to low temperature. *Plant Physiol Bioch (France)* 43: 383-388.

Vyas D, Kumar S, Ahuja PS (2007) Tea (*Camellia sinensis*) clones with shorter periods of winter dormancy exhibit lower accumulation of reactive oxygen species. *Tree Physiol* 27: 1253-1259.

Wachira F, Ogado J (1995b) *In vitro* regeneration of *Camellia sinensis* (L.) O. Kuntze by somatic embryo. *Plant Cell Rep.* 14: 463-466.

Wachira FN, Powell W, Waugh R (1997) An assessment of genetic diversity among *Camellia sinensis* L. (cultivated tea) and its wild relatives based on randomly amplified polymorphic DNA and organelle specific STS. *Heredity* 78 (6): 603-611.

Wachira FN, Waugh R, Hackett CA, Powell W (1995a) Detection of genetic diversity in tea (*Camellia Sinensis*) using RAPD markers. *Genome* 38: 201-210.

Wang G, Mahalingam R, Knap HT (1998) (C-A) and (G-A) anchored simple sequence repeats (ASSRs) generated polymorphism in soybean, *Glycine max* (L.). Merr *Theor Appl Genet.* 96: 1086-1096.

Wickramaratne MR (1981) Variation in some leaf in tea (*Camellia sinensis* L.) and their use in the identification of clones. *Tea Quarterly* 50: 183-189.

Wiso RR (1995) Chilling-enhanced photooxidation the production, action and study of oxygen reactive species produced during chilling in the light, *Photosyn. Res.* 45: 79–97.

Wu CT, Huang TK, Chen GR, Chen SY (1981) A review on the tissue culture of tea plants and on the utilization of callus derived plantlets. In: Rao AN (ed) Tissue Culture of Economically Important plants. Proceedings Costed. Symposium, Singapore, pp 104-106.

Xin Y, Zhi Gang W, FuGen G, AiYing S, WeiPing S, BingYao S, ZhongXing Y (2008) Micropropagation of tea tree of Biluochun. *Jiangsu J Agric Sci.* 24(1): 95-96.

Yamagishi M, Abe H, Nakano M, Nakatsuka A (2002) PCR-based molecular markers in Asiatic hybrid lily. *Sci Hortic.* 96: 225-234.

Yepes LM, Aldwinckle HS (1998) Factors that affect leaf regeneration efficiency in apples and effect of antibiotics in morphogenesis. *Plant Cell Tiss. Org. Cult.* 37: 257-269.

Zehra M, Banerjee S, Mathur AK, Kukreja AK (1996) Induction of hairy roots in tea (*Camellia sinensis* (L.)) using *Agrobacterium rihzogenes*. *Curr Sci.* 70: 84-86.

Zhuang C, Duan J, Zhou J (1988) Somatic embryogenesis and plantlet regeneration of *Camellia sasanqua*. *Acta Botanica Yunn* 10: 241-244.